Flora Europaea check-list and chromosome index

Flora Europaea check-list and chromosome index

D.M. MOORE

CAMBRIDGE UNIVERSITY PRESS
Cambridge
London New York New Rochelle
Melbourne Sydney

Published by the Press Syndicate of the University of Cambridge
The Pitt Building, Trumpington Street, Cambridge CB2 1RP
32 East 57th Street, New York, NY 10022, USA
296 Beaconsfield Parade, Middle Park, Melbourne 3206, Australia
© Cambridge University Press 1982
first published 1982
printed in Great Britain by Redwood Burn Ltd
Library of Congress catalogue card number: 82-1335

British Library cataloguing in publication data
Moore, D.M.
Flora Europaea check-list and chromosome index.
1. Botany - Europe - Catalogs
2. Chromosome numbers - Catalogs
1. Title   II. Flora Europaea
581.87'322   QK281
ISBN 0 521 23759 9

CONTENTS

Introduction

This book has grown out of the requests by many users of Flora Europaea
(Tutin et al., 1964, 1968, 1972, 1976, 1980) for information on the sources
of the chromosome numbers cited in that work. During discussions about the
publication of these data it became apparent that there is also a demand for
a basic checklist of the taxa recognised in the Flora, and that these two
requirements could readily be met in the same volume.

The Checklist

Apart from the fundamental value of Flora Europaea to all modern taxon-
omic studies of European plants, the names and sequence of taxa it presents
are widely used for arranging herbaria and libraries concerned with these plants.
It is intended that the handy format of this checklist will facilitate these
latter activities, as well as being useful in recording and assessing local
floras within Europe.

The sequence of taxa is exactly the same as that used in Flora
Europaea except that the Rubiaceae, which for technical reasons had to be
published in volume 4 rather than in volume 3, have been positioned correct-
ly in the Checklist. All species and subspecies 'accepted' in Flora
Europaea (i.e. those prefixed by a number or a letter) are included here.
Those unnumbered taxa of doubtful status are only included if a chromosome
number is given for them in Flora Europaea, as is also the case for the
agamospecies ('microspecies') listed, but not described, within the
'species-groups' of apomictic or partially apomictic genera. It has not
been possible to include any synonyms in the Checklist but access to these
is provided by the Consolidated Index of Flora Europaea (Halliday and
Beadle, 1982) as well as the volumes of the Flora itself.

The Chromosome Index

As noted in the Introductions to the five volumes of Flora Europaea,
diploid ($2n$) chromosome numbers were given when they were derived from

material of known wild origin. In the case of naturalized or cultivated taxa
chromosome numbers were included when they were known to be based on material
with a status which justified their inclusion in the Flora: aliens had to
be common and to have been naturalized for at least 25 years, while crops had
to be widely cultivated. As can be seen from the Index, between 98% and 99%
of the chromosome numbers included in Flora Europaea fulfil these criteria,
whether they were based on a single observation from one locality or on many
counts from a number of areas. In many instances the Editors were provided with
unpublished chromosome data; some of these were subsequently published, hence
references to them often postdate the publication of the relevant volume of
the Flora, others are apparently still unpublished. The provenance of the
material is indicated by using the abbreviations for the territories
recognised in Flora Europaea (see p. iii), except where the data only permit a
wider area to be cited. In several cases the chromosome numbers were published
under names different from those adopted in the Flora Europaea, but these can
readily be traced by reference to the synonymy given in the Flora.

Ideally, of course, a definitive chromosome index would give due weight
to the number of observations on which each number is based, but the practical-
ities of Flora-writing preclude this. The Index, consequently, cites only one
reference for each chromosome number given, even though in many instances far
more data were available to the editors of Flora Europaea; this has been done
to obviate any impression of completeness in the coverage, which would be
illusory. Less than 2% of the chromosome numbers cited in the Flora cannot be
substantiated or are shown to be derived from material which does not fulfil
the criteria referred to above. The commonest reason for this is that pre-
liminary accounts of genera included chromosome numbers taken uncritically
from the various worldwide compilations and their credentials were mis-
interpreted or inadequately checked during the various stages of the
editorial process; in other cases, far fewer, the sources cannot now be
traced in the records of the Editorial Committee; occasionally, the
numbers included in the Flora are simple misprints, undetected during proof

reading.

The Chromosome Index reveals, therefore, the problems of fully documenting multi-author Floras of substantial areas produced within a realistic, or realisable, time-scale. It is felt that the level of accuracy achieved fully justifies the approach adopted by the editors of Flora Europaea and the publication of this documented list of chromosome numbers. The extra effort required to achieve complete accuracy (if such were possible) would not have been worthwhile.

Abbreviations of geographical territories

(For precise definitions of these territories, see maps in Flora Europaea)

Al	Albania	Hs	Spain
Amer.*	America	Hu	Hungary
Au	Austria	Is	Iceland
Az	Açores (Azores	It	Italy
Be	Belgium and Luxembourg	Ju	Jugoslavia
Bl	Islas Baleares (Balearic Islands)	Lu	Portugal
		No	Norway
Br	Britain	Po	Poland
Bu	Bulgaria	Rm	Romania
Co	Corse (Corsica)	Rs	U.S.S.R. (European part), subdivided thus:
Cr	Kriti (Crete)		
Cz	Czechoslovakia		(N) Northern region
Da	Denmark		(B) Baltic region
Fa	Faeröer (Faeroes)		(C) Central region
Fe	Finland		(W) South-western region
Ga	France		(K) Krym (Crimea)
Ge	Germany		(E) South-eastern region
Gr	Greece	Sa	Sardegna (Sardinia)
Hb	Ireland	Sb	Svalbard (Spitsbergen)
He	Switzerland	Scand.*	Scandinavia
Ho	Netherlands	Si	Sicilia (Sicily)

ix

Su   Sweden   Tu   Turkey (European part)

Acknowledgements

This book is founded on the endeavours of the Editorial Committee of
Flora Europaea (T.G. Tutin, V.H. Heywood, N.A. Burges, D.M. Moore, D.H.
Valentine, S.M. Walters and D.A. Webb), to which I am the most recent
recruit. Professor D.H. Valentine and later Dr. S.M. Walters were primarily
responsible for the chromosome numbers cited in volume 1 of Flora Europaea,
and Dr. Walters for a significant part of those included in the second volume.
They, and my other colleagues on the Editorial Committee, have been a constant
source of support and information. Dr. P.W. Ball, Mr. A.O. Chater, Dr. R.A.
de Fillips and Dr. I.B.K. Richardson, Research Associates during the various
phases of the Flora Europaea project, checked hundreds of chromosome references,
while Dr. Áskell Löve freely gave of his wide cytotaxonomical knowledge. Above
all, however, the Chromosome Index is a testament to the many authors of Flora
Europaea who provided accurate documentation of their chromosomal data; the
cooperation of them and other friends and colleagues who made available unpublished information is greatly appreciated. The onerous task of typing the
camera-ready copy was undertaken by Miss Valerie Norris, whose efforts are grate-
fully acknowledged, whilst I am also indebted to the Flora Europaea Trust Fund
of the Linnean Society of London for a grant towards the costs of preparing
the manuscript.

I LYCOPODIACEAE

1 Huperzia Bernh.
  1 selago (L.) Bernh. ex Schrank & Mart.   90   Fe   Sorsa, 1963
                                         264   Br   Manton, 1950
                                      c.272   Is   Löve & Löve, 1961a
  (a) selago
  (b) arctica (Grossh.) Á. & D. Löve
  (c) dentata (Herter) Valentine

2 Lepidotis Beauv.
  1 cernua (L.) Beauv.
  2 inundata (L.) Börner   156   Br Fe   V. Sorsa, 1958

3 Lycopodium L.
  1 annotinum L.   68   Br He Su Manton, 1950
  2 dubium Zoega   68   Is   Löve & Löve, 1958
  3 clavatum L.   68   Br He   Manton, 1950

4 Diphasium C. Presl.
  1 complanatum (L.) Rothm.   46   Scand. Löve & Löve, 1961a
  2 tristachyum (Pursh) Rothm.   46   Scand. Löve & Löve, 1961a
  3 issleri (Rouy) J. Holub
  4 alpinum (L.) Rothm.   46   Su   Wilce, 1961
                            48-50   Cz   Manton, 1950

II SELAGINELLACEAE

1 Selaginella Beauv.
  1 selaginoides (L.) Link   18   Br   Manton, 1950
  2 helvetica (L.) Spring   18   He   Manton, 1950
  3 denticulata (L.) Link   18   It   Manton, 1950
  4 apoda (L.) Fernald
  5 kraussiana (G. Kunze) A. Braun

III ISOETACEAE

1 Isoetes L.
  1 lacustris L.   c.110   Is   Löve & Löve, 1961d
  2 setacea Lam.   22   Is   Löve & Löve, 1961d
                            100   Hb   Manton, 1950
  3 brochonii Moteley
  4 azorica Durieu ex Milde
  5 boryana Durieu
  6 delilei Rothm.
  7 malinverniana Cesati & De Not.
  8 heldreichii Wettst.
  9 velata A. Braun
  10 tenuissima Boreau
  11 histrix Bory   20   Br   Manton, 1950
  12 durieui Bory

IV EQUISETACEAE

1 Equisetum L.
  1 hyemale L.   216   Is   Löve & Löve, 1961c
  2 ramosissimum Desf.   c.216   It   Manton, 1950
  3 variegatum Schleicher ex Weber &
    Mohr   216   Is   Löve & Löve, 1961c

4 scirpoides Michx.	c.216	No	Manton, 1950
5 fluviatile L.	c.216	Ho	Bir, 1960
6 palustre L.	c.216	Ho	Bir, 1960
7 sylvaticum L.	c.216	Is	Löve & Löve, 1961c
8 pratense Ehrh.	c.216	Is	Löve & Löve, 1961c
9 arvense L.	c.216	Ho	Bir, 1960
10 telmateia Ehrh.	c.216	Br	Manton, 1950

V OPHIOGLOSSACEAE

1 Ophioglossum L.			
1 lusitanicum	250-260	Ga	Manton, 1950
2 azoricum C. Presl.	c.480		
3 vulgatum L.	480	Is	Löve & Löve, 1961c
	c.496	Ho	Verma, 1958
	500-520	Br	Manton, 1950

2 Botrychium Swartz			
1 simplex E. Hitchc.			
2 lunaria (L.) Swartz.	90	Br	Manton, 1950
	96	Da	Hagerup, 1941
3 boreale Milde	90	Is	Löve & Löve, 1961c
4 matricariifolium A. Braun ex Koch			
5 lanceolatum (S.G. Gmelin) Angström.	90	Is	Löve & Löve, 1961c
6 multifidum (S.G. Gmelin) Rupr.			
7 virginianum (L.) Swartz			

VI OSMUNDACEAE

1 Osmunda L.			
1 regalis L.	44	Br	Manton, 1950

VII SINOPTERIDACEAE

1 Cheilanthes Swartz			
1 marantae (L.) Domin	58	It	Fabbri, 1957
2 fragrans (L. fil.) Swartz			
3 hispanica Mett.			
4 catanensis (Cosent.) H.P. Fuchs			
5 persica (Bory) Mett. ex Kuhn			
2 Pellaea Link			
1 calomelanos (Swartz) Link			

VIII ADIANTACEAE

1 Adiantum L.				
1 capillus-veneris L.	60	Hb Hs It	Manton, 1950

IX PTERIDACEAE

1 Pteris L.			
1 serrulata Forskål			
2 cretica L.	58	It	Manton, 1950
3 vittata L.			

X CRYPTOGRAMMACEAE

1 Cryptogramma R. Br.
   1 crispa (L.) R. Br. ex Hooker   120   Br   Manton, 1950
   2 stelleri (S.G. Gmelin) Prantl

XI GYMNOGRAMMACEAE

1 Anogramma Link
   1 leptophylla (L.) Link

XII DICKSONIACEAE

1 Culcita C. Presl
   1 macrocarpa C. Presl   132-136   Az   Manton, 1958

XIII HYPOLEPIDACEAE

1 Pteridium Scop.
   1 aquilinum (L.) Kuhn   104   Br   Manton, 1950

XIV DAVALLIACEAE

1 Davallia Sm.
   1 canariensis (L.) Sm.

XV HYMENOPHYLLACEAE

1 Hymenophyllum Sm.
   1 tunbrigense (L.) Sm.   26   Br   Manton, 1950
   2 wilsonii Hooker   36   Br   Manton, 1950

2 Trichomanes L.
   1 speciosum Willd.   144   Hb   Manton, 1950

XVI THELYPTERIDACEAE

1 Thelypteris Schmidel
   1 limbosperma (All.) H.P. Fuchs   68   Br   Manton, 1950
   2 palustris Schott   70   Fe   Sorsa, 1958
   3 phegopteris (L.) Slosson   90   Fe   Sorsa, 1958
   4 pozoi (Lag.) C.V. Morton

2 Cyclosorus Link
   1 dentatus (Forskål) R.-C. Ching

XVII ASPLENIACEAE

1 Asplenium L.
   1 hemionitis L.
   2 marinum L.   72   Br   Manton, 1950
   3 petrarchae (Guerin) DC.   144   Ge   Manton, 1950
   4 monanthes L.
   5 trichomanes L.

(a) trichomanes	73	Ge	Meyer, 1957
(b) quadrivalens	144	Ge	Meyer, 1957
6 adulterinum Milde	144	Au Ge	Meyer, 1957
7 viride Hudson	72	Br	Manton, 1950
8 jahandiezii (Litard.) Rouy	72	Ge	Meyer, 1960
9 fontanum (L.) Bernh.	72	He	Manton, 1950
majoricum Litard.	144	Bl	Lovis, 1963
10 forisiense Le Grand	144	Ge	Meyer, 1961
11 obovatum Viv.	72	Sa	Meyer, 1962
12 billotii F.W. Schultz	144	Br	Manton, 1950
13 adiantum-nigrum L.	144	Br	Manton, 1950
14 onopteris L.	72	Hb	Manton, 1955
15 cuneifolium Viv.	72	Ge	Meyer, 1967
16 septentrionale (L.) Hoffm.	144	Fe	Sorsa, 1958
17 seelosii Leybold	72	It	Meyer, 1957
18 ruta-muraria L.	144	Br	Manton, 1950
19 lepidum C. Presl	144	Europe	Meyer, 1960
20 fissum Kit. ex Willd.	72	Ge	Meyer, 1958

2 Ceterach DC.

1 officinarum DC.	144	Br	Manton, 1950

3 Pleurosorus Fée

1 hispanicus (Cosson) C.V. Morton	72	Hs	Lovis, 1963

4 Phyllitis Hill

1 scolopendrium (L.) Newman	72	Br	Manton, 1950
2 sagittata (DC.) Guinea & Heywood	72	Ga	Manton, 1950
3 hybrida (Milde) C. Chr.	144	It	Manton, 1950

XVIII ATHYRIACEAE

1 Athyrium Roth

1 filix-femina (L.) Roth	80	Br	Manton, 1950
2 distentifolium Tausch ex Opiz	80	Br	Manton, 1950

2 Diplazium Swartz

1 caudatum (Cav.) Jermy			
2 sibiricum (Turcz. ex G. Kunze) Jermy	82	No	Brøgger, 1960

3 Cystopteris Bernh.

1 fragilis (L.) Bernh.	168	Br He	Manton, 1950
	252	Br	Manton, 1950
2 dickieana R. Sim	168	Br	Manton, 1950
3 montana (Lam.) Desv.	168	He	Manton, 1950
4 sudetica A. Braun & Milde	168	No	Brøgger, 1960

4 Woodsia R. Br.

1 ilvensis (L.) R. Br.	82	Br	Manton, 1950
2 alpina (Bolton) S.F. Gray	164	Is	Löve & Löve, 1961c
3 pulchella Bertol.	78	It	Meyer, 1959
4 glabella R. Br.	c.80	Su	Löve & Löve, 1961b
	164	Fe	Sorsa, 1963

5 Matteuccia Tod.

1 struthiopteris (L.) Tod.	c.80	Fe	Sorsa, 1958

6 Onoclea L.

1 sensibilis L.

4

XIX ASPIDIACEAE

1 Polystichum Roth
  1 lonchitis (L.) Roth   82   Br He   Manton, 1950
  2 aculeatum (L.) Roth   164   Br He   Manton, 1950
  3 setiferum (Forskal) Woynar   82   Br He   Manton, 1950
  4 braunii (Spenner) Fée   164   Ge   Manton &
                                        Reichstein, 1961

2 Dryopteris Adanson
  1 filix-mas (L.) Schott   164   Fe   Sorsa, 1958
  2 borreri Newman   82   Br He   Manton, 1950
                   123   Br He No Manton, 1950
                   130   No   Knaben, 1948
  3 abbreviata (DC.) Newman   82   Br   Manton, 1950
  4 villarii (Bellardi) Woynar ex Schinz
    & Thell.   82   He No Su Manton, 1950
                   164   Br   Manton, 1950
    (a) villarii
    (b) pallida (Boey) Heywood
  5 cristata (L.) A. Gray   164   Br He   Manton, 1950
  6 carthusiana (Vill.) H.P. Fuchs 164   Fe   Sorsa, 1958
  7 dilatata (Hoffm.) A. Gray 164   Br   Manton, 1950
  8 assimilis S. Walker   82   Fe   Sorsa, 1958
  9 aemula (Aiton) O. Kuntze   82   Br   Manton, 1950
 10 fragrans (L.) Schott

3 Gymnocarpium Newman
  1 dryopteris (L.) Newman   160   Fe   Sorsa, 1958
  2 robertianum (Hoffm.) Newman 160-168 Br   Manton, 1950

XX ELAPHOGLOSSACEAE

1 Elaphoglossum Schott ex Sm.
  1 hirtum (Swartz) C. Chr.

XXI BLECHNACEAE

1 Blechnum L.
  1 spicant (L.) Roth   68   Br   Manton, 1950

2 Woodwardia Sm.
  1 radicans (L.) Sm.   68   Br   Manton, 1950

XXII POLYPODIACEAE

1 Polypodium L.
  1 australe Fée   74   Ga He It Manton, 1950
  2 vulgare L.   148   Br He   Manton, 1950
  3 interjectum Shivas   222   Br   Manton, 1950

XXIII MARSILEACEAE

1 Marsilea L.
  1 quadrifolia L.
  2 aegyptiaca Willd.
  3 strigosa Willd.

5

2 Pilularia L.
  1 globulifera L.
  2 minuta Durieu ex A. Braun

XXIV SALVINIACEAE

1 Salvinia Adanson
  1 natans (L.) All.   8   Rs(W)   Arnoldi, 1910
                                       18   It   D'Amato-Avanzi,
                                                     1957
  2 rotundifolia Willd.

XXV AZOLLACEAE

1 Azolla Lam.
  1 filiculoides Lam.
  2 caroliniana Willd.

SPERMATOPHYTA-GYMNOSPERMAE
CONIFEROPSIDA

XXVI PINACEAE

1 Abies Miller
  1 procera Rehder
  2 grandis (D. Don) Lindley
  3 lasiocarpa (Hooker) Nutt.
  4 sibirica Ledeb
  5 nordmanniana (Steven) Spach
  6 alba Miller   24   ?Ge   Seitz, 1951
  7 nebrodensis (Lojac.) Mattei
  8 cephalonica Loudon
  9 pinsapo Boiss.

2 Pseudotsuga Carrière
  1 menziesii (Mirbel) Franco

3 Tsuga (Antoine) Carrière
  1 heterophylla (Rafin.) Sarg.
  2 canadensis (L.) Carrière

4 Picea A. Dietr.
  1 abies (L.) Karsten
    (a) abies   22
                                       24   Su   Lövkvist, 1963
    (b) obovata (Ledeb.) Hultén
  2 orientalis (L.) Link
  3 glauca (Moench) Voss
  4 engelmannii Parry ex Engelm.
  5 pungens Engelm.
  6 sitchensis (Bong.) Carrière
  7 omorika (Pančić) Purkyne

5 Larix Miller
  1 kaempferi (Lamb.) Carrière
  2 decidua Miller
  3 russica (Endl.) Sabine ex Trautv.

6

    4 gmelinii (Rupr.) Kuzeneva

6 Cedrus Trew
    1 deodara (D. Don) G. Don fil.
    2 atlantica (Endl.) Carrière

7 Pinus L.
    1 contorta Douglas ex Loudon
    2 pinaster Aiton   24   Ga   Dangeard, 1941
      (a) pinaster
      (b) atlantica H. del Villar
    3 radiata D. Don
    4 banksiana Lamb.
    5 ponderosa Douglas ex P. & C. Lawson
    6 nigra Arnold
      (a) nigra
      (b) salzmannii (Dunal) Franco
      (c) laricio (Poiret) Maire
      (d) dalmatica (Vis.) Franco
      (e) pallasiana (Lamb.) Holmboe
    7 sylvestris L.   24   Su   Lövkvist, 1963
    8 mugo Turra
    9 uncinata Miller ex Mirbel
   10 leucodermis Antoine
   11 heldreichii Christ
   12 halepensis Miller
   13 brutia Ten.
   14 canariensis Sweet ex Sprengel
   15 pinea L.
   16 cembra L.
   17 sibirica Du Tour
   18 wallichiana A.B. Jackson
   19 peuce Griseb.
   20 strobus L.

XXVII TAXODIACEAE

1 Sequoia Endl.
    1 sempervirens (Lamb.) Endl.

2 Sequoiadendron Buchholz
    1 giganteum (Lindley) Buchholz

3 Taxodium L.C.M. Richard
    1 distichum (L.) L.C.M. Richard

4 Cryptomeria D. Don
    1 japonica (L. fil.) D. Don

XXVIII CUPRESSACEAE

1 Cupressus L.
    1 sempervirens L.
    2 macrocarpa Hartweg
    3 lusitanica Miller
    4 arizonica E.L. Greene

2 Chamaecyparis Spach
    1 lawsoniana (A. Murray) Parl.

3 Thuja L.
  1 plicata D. Don ex Lambert
    orientalis L.   22 Bot. Gard. Dark, 1932

4 Tetraclinis Masters
  1 articulata (Vahl) Masters

5 Juniperus L.
  1 drupacea Labill.
  2 communis L.
    (a) communis   22   Su   Löwkvist, 1963
    (b) hemisphaerica (J. & C. Presl) Nyman
    (c) nana Syme   22   Is   Löve & Löve, 1956
  3 oxycedrus L.
    (a) oxycedrus
    (b) macrocarpa (Sibth. & Sm.) Ball
    (c) transtagana Franco
  4 brevifolia (Seub.) Antoine
  5 phoenicea L.
  6 thurifera L.
  7 foetidissima Willd.
  8 excelsa Bieb.
  9 sabina L.   22 Bot. Gard. Seitz, 1951
 10 virginiana L.

TAXOPSIDA

XXIX TAXACEAE

1 T???? L?
  1 baccata L.   24   Su   Löwkvist, 1963

GNETOPSIDA

XXX EPHEDRACEAE

1 Ephedra L.
  1 fragilis Desf.
    (a) fragilis
    (b) campylopoda (C.A. Meyer) Ascherson & Graebner
  2 distachya L.
    (a) distachya   24   Ga   Berudye &
                                   Jandar, 1907
                     28   Ga   Delay, 1947
    (b) helvetica (C.A. Meyer) Ascherson & Graebner
  3 major Host
    (a) major   14   Ju   Geitler, 1929
    (b) procera (Fischer & C.A. Meyer) Markgraf

SPERMATOPHYTA-ANGIOSPERMAE

XXXI SALICACEAE

1 Salix L.
  1 pentandra L.

2 fragilis L.	79	Br	Wilkinson, 1941
3 alba L.			
(a) alba			
(b) coerulea (Sm.) Rech. fil.			
(c) vitellina (L.) Arcangeli			
(d) micans (N.J. Andersson) Rech. fil.			
4 babylonica L.			
5 triandra L.			
(a) triandra	38	Br	Blackburn & Heslop-Harrison, 1924
(b) discolor (Koch) Arcangeli			
6 reticulata L.	38	Rs(N)	Sokolovskaya & Strelkova, 1960
7 herbacea L.	38	Is	Löve & Löve, 1956
8 polaris Wahlenb.			
9 retusa L.			
10 kitaibeliana Willd.			
11 serpyllifolia Scop.			
12 rotundifolia Trautv.			
13 myrsinites L.	190	No	Wilkinson, 1944
14 breviserrata B. Flod.			
15 alpina Scop.			
16 arctica Pallas			
17 pulchra Cham.			
18 glauca L.			
callicarpaea Trautv.	190	Is	Löve & Löve, 1956
19 stipulifera B. Flod. ex Hayren	152	Su	Löve & Löve, 1956
20 glaucosericea B. Flod			
21 pyrenaica Gouan			
22 reptans Rupr.			
23 lanata L.			
24 glandulifera B. Flod.	38	Scand.	Marklund, 1931
25 phylicifolia L.	114	Is	Löve & Löve, 1956
26 bicolor Willd.			
27 hegetschweileri Heer			
28 hibernica Rech. fil.			
29 nigricans Sm.	114	Su	Håkansson, 1955
30 borealis Fries			
31 mielichhoferi Sauter			
32 glabra Scop.			
33 crataegifolia Bertol.			
34 pedicellata Desf.			
35 silesiaca Willd.			
36 aegyptiaca L.			
37 appendiculata Vill.			
38 laggeri Wimmer			
39 cinerea L.	76	Br	Blackburn & Heslop-
40 atrocinerea Brot.	76	Br	Harrison, 1924
41 aurita L.	76	Br	Blackburn & Heslop-Harrison, 1924
42 caprea L.			
43 coaetanea (Hartman) B. Flod			
44 salvifolia Brot.			
45 cantabrica Rech. fil.			
46 starkeana Willd.			
47 xerophila B. Flod.			
48 myrtilloides L.			
49 repens L.	38	Br	Blackburn & Heslop-Harrison, 1924
50 rosmarinifolia L.	38	Su	Lövkvist, 1963
51 arenaria L.	38	Su	Lövkvist, 1963

```
52 arbuscula L.
53 foetida Schleicher
54 waldsteiniana Willd.
55 hastata L.                      c.110      Rs(N)     Sokolovskaya &
                                                         Strelkova, 1960
   (a) hastata
   (b) vegeta N.J. Andersson
   (c) subintegrifolia (B. Flod.) B. Flod.
56 pyrolifolia Ledeb.
57 helvetica Vill.
58 marrubifolia Tausch ex N.J. Andersson
59 lapponum L.
60 viminalis L.
61 rossica Nasarov
62 elaeagnos Scop.
   (a) elaeagnos                     38        Europe     Håkansson, 1955
   (b) angustifolia (Cariot) Rech. fil.
63 purpurea L.
   (a) purpurea
   (b) lambertiana (Sm.) A. Neumann ex Rech. fil.
64 amplexicaulis Bory
65 caspica Pallas
66 wilhelmsiana Bieb.
67 caesia Vill.
68 tarraconensis Paul
69 daphnoides Vill.                  38        Scand.     Marklund, 1931
70 acutifolia Willd.                 38        Scand.     Marklund, 1931


2 Populus L.
  1 alba L.                          38        Ge         Wettstein
                                                          Wettstein, 1933

  2 canescens (Aiton) Sm.            38        Au         Seitz, 1954
  3 grandidentata Michx
  4 tremula L.
  5 simonii Carrière
  6 gileadensis Rouleau
  7 x berolinensis C. Koch
  8 nigra L.                         38        Br         Blackburn & Heslop-
                                                          Harrison, 1924

  9 x canadensis Moench
 10 deltoides Marshall
 11 euphratica Olivier


XXXII MYRICACEAE

1 Myrica L.
  1 gale L.                          48        Su         Hagerup, 1941
  2 faya Aiton
  3 caroliniensis Miller


XXXIII JUGLANDACEAE

1 Juglans L.
  1 regia L.
  2 nigra L.
  3 cinerea L.

2 Carya Nutt.
```

```
1 cordiformis (Wangenh.) C. Koch
2 glabra (Miller) Sweet
3 alba (L.) Nutt.

3 Pterocarya Kunth
  1 fraxinifolia (Poiret) Spach


XXXIV BETULACEAE

1 Betula L.
  1 pendula Roth                   28    Da    Helms &
                                                Jörgensen, 1925
  2 pubescens Ehrh.
    (a) pubescens                  56    Da    Helms &
                                                Jörgensen, 1925
    (b) carpatica (Willd.)         56    Po    Skalińska, 1959
    (c) tortuosa (Ledeb.) Nyman    56    Is    Löve & Löve, 1956
  3 humilis Schrank                28    Ge    Natho, 1959
  4 nana L.                        28    Sb    Flovik, 1940

2 Alnus Miller
  1 viridis (Chaix) DC.
    (a) viridis
    (b) fruticosa (Rupr.) Nyman
    (c) suaveolens (Req.) P.W. Ball
  2 glutinosa (L.) Gaertner        28    Su    Eklundh-Ehrenburg,
                                                1946
  3 rugosa (Duroi) Sprengel
  4 incana (L.) Moench
    (a) incana                     28    Su    Eklundh-Ehrenburg,
                                                1946
    (b) kolaensis (Orlova) Á. & D. Löve
  5 cordata (Loisel.) Loisel.


XXXV CORYLACEAE

1 Carpinus L.
  1 betulus L.                     64    Su    Johnsson, 1942
  2 orientalis Miller

2 Ostrya Scop.
  1 carpinifolia Scop.

3 Corylus L.
  1 avellana L.                    22    Su    Lövkvist, 1963
  2 colurna L.
  3 maxima Miller


XXXVI FAGACEAE

1 Fagus L.
  1 sylvatica L.                   24    Su    Lövkvist, 1963
  2 orientalis Lipsky

2 Castanea Miller
  1 sativa Miller
  2 crenata Siebold & Zucc.
```

3 Quercus L.
  1 rubra L.
  2 palustris Muenchh.

3 coccifera L.	24	Lu	Natividade, 1937
4 ilex L.	24	Lu	Natividade, 1937
5 rotundifolia Lam.	24	Lu	Natividade, 1937
6 suber L.	24	Lu	Natividade, 1937

  7 trojana Webb
  8 macrolepis Kotschy
  9 cerris L.
  10 mas Thore
  11 polycarpa Schur

12 petraea (Mattuschka) Liebl.	24	Br	Jones, 1959

  13 dalechampii Ten.
  14 hartwissiana Steven
  15 robur L.

(a) robur	24	Br	Jones, 1959

    (b) brutia (Ten.) O. Schwarz
  16 pedunculiflora C. Koch
  17 sicula Borzi
  18 frainetto Ten.

19 pyrenaica Willd.	24	Lu	Natividade, 1937

  20 congesta C. Presl
  21 brachyphylla Kotschy
  22 virgiliana (Ten.) Ten.
  23 pubescens Willd.

(a) pubescens	24	It	Vignoli, 1933

    (b) anatolica O. Schwarz
    (c) palensis (Palassou) O. Schwarz
  24 canariensis Willd

25 faginea Lam.	24	Lu	Natividade, 1937
26 frutescens Brot.	24	Lu	Natividade, 1937

  27 infectoria Oliver

XXXVII ULMACEAE

1 Ulmus L.

1 glabra Hudson	28	Su	Lövkvist, 1963

  2 procera Salisb.

3 minor Miller	28	Br	Maude, 1939

  4 canescens Melville

5 laevis Pallas	28	Su	Löve & Löve, 1942

2 Zelkova Spach
  1 abelicea (Lam.) Boiss.

3 Celtis L.
  1 australis L.
  2 caucasica Willd.
  3 glabrata Steven ex Planchon
  4 tournefortii Lam.

XXXVIII MORACEAE

1 Broussonetia L.'Hér. ex Vent.
  1 papyrifera (L.) Vent.

2 Morus L.
  1 nigra L.

2 alba L.

3 Maclura Nutt.
  1 pomifera (Rafin.) C.K. Schneider

4 Ficus L.
  1 carica L.

XXXIX CANNABACEAE

1 Humulus L.
  1 lupulus L.   20   Hu   Pólya, 1949

2 Cannabis L.
  1 sativa L.   20   Hu   Pólya, 1949

XL URTICACEAE

1 Urtica L.
  1 atrovirens Req. ex Loisel.   26   Co   Contandriopoulos, 1962
  2 dioica L.   48   Is   Löve & Löve, 1956
                                  52   Br   Fothergill, 1936
  3 kioviensis Rogow.   22   Hu   Baksay, 1956
  4 cannabina L.
  5 rupestris Guss
  6 urens L.   24   Is   Löve & Löve, 1956
                                  26   Su   Lövkvist, 1963
                                  52   Su or Is   Löve & Löve, 1942
  7 dubia Forskål
  8 pilulifera L.

2 Pilea Lindley
  1 microphylla (L.) Liebm.

3 Parietaria L.
  1 officinalis L.
  2 diffusa Mert. & Koch
  3 mauritanica Durieu
  4 lusitanica L.
   (a) lusitanica
   (b) serbica (Pančić) P.W. Ball
  5 debilis Forster fil.
  6 cretica L.

4 Soleirolia Gaud.-Beaup.
  1 soleirolii (Req.) Dandy   20   Co   Contandriopoulos, 1962

5 Forsskalea L.
  1 tenacissima L.

XLI PROTEACEAE

1 Hakea Schrader
  1 sericea Schrader
  2 salicifolia (Vent.) B.L. Burtt

13

XLII SANTALACEAE

1 Comandra Nutt.
  1 elegans (Rochel ex Reichenb.) Reichenb. fil

2 Osyris L.
  1 alba L.
  2 quadripartita Salzm. ex Decne

3 Thesium L.
  1 alpinum L.   12   Au   Mattick, 1950
  2 pyrenaicum Pourret
    (a) pyrenaicum
    (b) alpestre O. Schwarz
  3 ebracteatum Hayne
  4 rostratum Mert. & Koch   26   Au Ge   Rutishauser, 1936
  5 kerneranum Simonkai
  6 auriculatum Vandas
  7 arvense Horvatovszky
  8 parnassi A. DC.
  9 italicum A. DC.
 10 dollineri Murb.
    (a) dollineri
    (b) simplex (Velen.) Stoj. & Stefanov.
 11 bavarum Schrank   24   Ge   Schulle, 1933
 12 linophyllon L.   14   Hu   Baksay, 1961
 13 divaricatum Jan ex Mert. & Koch
 14 humifusum DC.
 15 bergeri Zucc.
 16 humile Vahl
 17 procumbens C.A. Meyer
 18 brachyphyllum Boiss

XLIII LORANTHACEAE

1 Loranthus L.
  1 europaeus Jacq.

2 Viscum L.
  1 cruciatum Sieber ex Boiss.
  2 album L.
    (a) album
    (b) abietis (Wiesb.) Abromeit

3 Arceuthobium Bieb.
  1 oxycedri (DC.) Bieb.

XLIV ARISTOLOCHIACEAE

1 Asarum L.
  1 europaeum L.   26   Po   Skalińska et al.,
                                         1959
                   40   Su   Ehrenberg, 1945
2 Aristolochia L.
  1 sempervirens L.
  2 baetica L.
  3 cretica Lam.
  4 pontica Lam.

```
 5 bodamae Dingler
 6 clematitis L.
 7 pistolochia L.
 8 rotunda L.
 9 pallida Willd.
10 sicula Tineo
11 longa L.
12 bianorii Sennen & Pau
13 microstoma Boiss. & Spruner
```

XLV RAFFLESIACEAE

```
1 Cytinus L.
  1 hypocistis (L.) L.
    (a) hypocistis
    (b) orientalis Wettst.
  2 ruber (Fourr.) Komarov
```

XLVI BALANOPHORACEAE

```
1 Cynomorium L.
  1 coccineum L.
```

XLVII POLYGONACEAE

1 Koenigia L.			
1 islandica L.	28	Sb	Löve & Löve, 1961b
2 Polygonum L.			
1 scoparium Req. ex Loisel.	20	Co	Contandriopoulos, 1962
2 equisetiforme Sibth. & Sm.			
3 icaricum Rech. fil.			
4 romanum Jacq.			
5 maritimum L.	20	Ge Lu	Styles, 1962
6 idaeum Hayek			
7 cognatum Meissner			
8 floribundum Schlecht. ex Sprengel			
9 salsugineum Bieb.			
10 aschersonianum H. Gross			
11 patulum Bieb.	20	Rm	Tarnavschi, 1948
12 arenarium Waldst. & Kit.			
(a) arenarium			
(b) pulchellum (Loisel.) D.A. Webb & Chater	20	Lu	Löve, 1954
13 oxyspermum Meyer & Bunge ex Ledeb.			
(a) oxyspermum	40	Da	Styles, 1962
(b) raii (Bab.) D.A. Webb & Chater	40	Br	Styles, 1962
14 graminifolium Wierzb. ex Heuffel			
15 aviculare L.	60	Br	Styles, 1962
16 boreale (Lange) Small	40	Br	Styles, 1962
17 rurivagum Jordan ex Boreau	60	Br	Styles, 1962
18 arenastrum Boreau			
19 acetosum Bieb.			
20 minus Hudson			
21 foliosum H. Lindb.	20	Fe	Löve, ined.
22 mite Schrank	40	Be	Pauwels, 1959

23 hydropiper L.	20	Is Su	Löve & Löve, 1942
24 salicifolium Brouss. ex Willd.			
25 persicaria L.	44	Be	Pauwels, 1959
26 lapathifolium L.	22	Be	Pauwels, 1959
27 amphibium L.	66	Is	Löve & Löve, 1956
28 hydropiperoides Michx			
29 orientale L.			
30 sagittatum L.			
31 tinctorium Aiton			
32 bistorta L.	44	Ge	Jaretsky, 1928b
33 amplexicaule D. Don			
34 viviparum L.	c.100	Sb	Flovik, 1940
	c.132	Po	Skalińska, 1950
35 alpinum All.			
36 polystachyum Wall. ex Meissner			

3 Bilderdykia Dumort.			
1 convolvulus (L.) Dumort.	40	Be	Pauwels, 1959
2 dumetorum (L.) Dumort.	20	Be	Pauwels, 1959
3 aubertii (Louis Henry) Moldenke			

4 Reynoutria Houtt.			
1 japonica Houtt.			
2 sachalinensis (Friedrich Schmidt Petrop.) Nakai			

5 Fagopyrum Miller			
1 esculentum Moench	16	Rs( )	Mansurova, 1948
2 tataricum (L.) Gaertner	16	Rs( )	Mansurova, 1948

6 Oxyria Hill			
1 digyna (L.) Hill	14	Rr(N)	Sokolovskaya & Strelkova, 1960

7 Rheum L.			
1 tataricum L. fil.			
2 rhaponticum L.			

8 Rumex L.			
1 angiocarpus Murb.	14	Lu	Löve, 1941
2 tenuifolius (Wallr.) Á. Löve	28	Su	Löve, 1940
3 acetosella L.	42	Su	Löve, 1940
4 graminifolius Rudolph ex Lamb	56	Rs(N)	Löve, 1940
5 lunaria L.			
6 tingitanus L.			
7 suffruticosus Gay ex Meissner			
8 scutatus L.	20	Au	Mattick, 1950
9 tuberosus L.			
(a) tuberosus			
(b) creticus (Boiss.) Rech. fil.			
10 nivalis Hegetschw.	♀ 14	Alps	Löve, 1954b
	♂ 15	Alps	Löve, 1954b
11 arifolius All.	♀ 14	Is	Löve & Löve, 1956
	♂ 15	Is	Löve & Löve, 1956
12 amplexicaulis Lapeyr.			
13 gussonei Arcangeli			
14 acetosa L.	♀ 14	Is	Löve, 1940
	♂ 15	Is	Löve, 1940
15 thyrsiflorus Fingerh.	14	Su	Löve, 1940
16 rugosus Campd.			
17 papillaris Boiss. & Reuter			
18 intermedius DC.			

16

```
19 thyrsoides Desf.
20 vesicarius L.
21 triangulivalvis (Danser) Rech. fil.
22 frutescens Thouars
23 alpinus L.                                20      Ga        Larsen, 1954
24 aquaticus L.                           c.200      Su        Löve, 1942
25 azoricus Rech. fil.
26 arcticus Trautv.                       c.200      Su        Löve, 1942
27 balcanicus Rech. fil.
28 pseudonatronatus Borbás                  40      Au Su     Löve, 1961b
29 cantabricus Rech. fil.
30 longifolius DC.                          60      Su        Löve, 1942
31 confertus Willd.                         40  C. Europe     Löve, 1961b
32 hydrolapathum Hudson                    200      Su        Löve, 1942
33 cristatus DC.
34 kerneri Borbás
35 patientia L.
   (a) patientia                           60  C. Europe     Löve, 1961b
   (b) orientalis (Bernh.) Danser          60  C. Europe     Löve, 1961b
   (c) recurvatus (Rech.) Rech. fil.       60  C. Europe     Löve, 1961b
36 crispus L.                              60      Su        Löve, 1942
37 stenophyllus Ledeb.                     60      Hu        Pólya, 1950
38 conglomeratus Murray                    20      Su        Löve, 1942
39 sanguineus L.                           20      Su        Löve, 1942
40 rupestris Le Gall
41 nepalensis Sprengel                     60
42 pulcher L.
   (a) pulcher                             20  C. Europe     Löve, 1961b
   (b) divaricatus (L.) Murb.              20  C. Europe     Löve, 1961b
   (c) anodontus (Hausskn,) Rech. fil.
   (d) raulinii (Boiss.) Rech. fil.
43 obtusifolius L.
   (a) obtusifolius                        40      Su        Löve, 1961b
   (b) subalpinus (Schur) Čelak
   (c) transiens (Simonkai) Rech. fil.
   (d) sylvestris (Wallr.) Rech.           40  C. Europe     Löve, 1961b
44 dentatus L.                             40      Su        Löve, 1967
   (a) halacsyi (Rech.) Rech. fil.
   (b) reticulatus (Besser) Rech. fil.
45 palustris Sm.                           40      Su        Löve, 1942
46 maritimus L.                            40      Su        Löve, 1942
47 rossicus Murb.                          40      Rs(C)     Löve & Löve, 1961b
48 marschallianus Reichenb.
49 ucranicus Besser ex Sprengel           40      Rs(W)     Löve & Löve, 1961b
50 bucephalophorus L.
   (a) bucephalophorus                     16      It        Löve, 1967
   (b) hispanicus (Steinh.) Rech. fil.
   (c) graecus (Steinh.) Rech. fil.
   (d) aegaeus Rech. fil.

 9 Emex Campd.
   1 spinosa (L.) Campd.

10 Muehlenbeckia Meissner
   1 sagittifolia (Ortega) Meissner

11 Calligonum L.
   1 aphyllum (Pallas) Gurke

12 Atraphaxis L.
   1 spinosa L.
```

```
  2 replicata Lam.
  3 frutescens (L.) C. Koch
  4 billardieri Jaub. & Spach
```

XLVIII CHENOPODIACEAE

```
1 Polycnemum L.
  1 majus A. Braun
  2 arvense L.
  3 verrucosum A.F. Láng
  4 heuffelii A.F. Láng

2 Beta L.
  1 vulgaris L.
    (a) vulgaris                          18      Ge      Wulff, 1936
    (b) maritima (L.) Arcangeli           18      Rm      Tarnavschi, 1948
  2 macrocarpa Guss.
  3 patellaris Moq.
  4 trigyna Waldst. & Kit.
  5 nana Boiss. & Heldr.

3 Chenopodium L.
  1 botrys L.
  2 schraderanum Schultes                 18      Su      Kjellmark, 1934
  3 aristatum L.
  4 ambrosioides L.                        32      Su      Kjellmark, 1934
  5 multifidum L.
  6 bonus-henricus L.                      36      Su      Lövkvist, 1963
  7 foliosum Ascherson
  8 capitatum (L.) Ascherson              18      Su      Kjellmark, 1934
  9 glaucum L.                             18      Su      Lövkvist, 1963
 10 rubrum L.                              36      Su      Lövkvist, 1963
 11 botryodes Sm.
 12 hybridum L.                            18      Su      Lövkvist, 1963
 13 polyspermum L.                         18      Su      Lövkvist, 1963
 14 vulvaria L.                            18      Su      Lövkvist, 1963
 15 urbicum L.                             36      Br      Cole, 1962
 16 murale L.                              18      Br      Cole, 1962
 17 ficifolium Sm.                         18      Su      Kjellmark, 1934
 18 acerifolium Andrz.
 19 jenissejense Aellen & Iljin
 20 opulifolium Schrader ex Koch & Ziz     54      Br      Cole, 1962
 21 album L.
    (a) album                             18      Su      Lövkvist, 1963
                                          54      Su      Lövkvist, 1963
    (b) striatum (Krasan.) J. Murr
 22 suecicum J. Murr                       18      Su      Lövkvist, 1963
 23 giganteum D. Don

4 Cycloloma Moq.
  1 atriplicifolia (Sprengel) Coulter

5 Spinacia L.
  1 oleracea L.

6 Atriplex L.
  1 halimus L.                            18      Lu      Castro &
                                                          Fontes, 1946
  2 glauca L.
  3 cana Ledeb.
```

```
 4 hortensis L.
 5 nitens Schkuhr
 6 aucheri Moq.
 7 oblongifolia Waldst. & Kit.
 8 heterosperma Bunge
 9 sphaeromorpha Iljin
10 rosea L.
11 laciniata L.              18    Ge         Wulff, 1937a
12 tatarica L.               18    Ge Hu Rm   Wulff, 1937a
13 littoralis L.             18    Rm         Tarnavschi, 1948
14 patens (Litv.) Iljin
15 patula L.                 36    Br         Hulme, ined.
16 calotheca (Rafn) Fries    18    Ge         Wulff, 1937a
17 hastata L.                18    Br         Hulme, ined.
18 glabriuscula Edmondston   18    Br         Hulme, ined.
19 longipes Drejer           18    Br         Hulme, ined.


 7 Halimione Aellen
 1 portulacoides (L.) Aellen 36    Lu Rm      Tarnavschi, 1948
 2 verrucifera (Bieb.) Aellen
 3 pedunculata (L.) Aellen   18    Ge         Wulff, 1937a


 8 Krascheninnikovia Gueldenst.
 1 ceratoides (L.) Gueldenst.


 9 Ceratocarpus L.
 1 arenarius L.


10 Axyris L.
 1 amaranthoides L.


11 Camphorosma L.
 1 monspeliaca L.
 2 lessingii Litv.
 3 annua Pallas
 4 songorica Bunge


12 Bassia All.
 1 hyssopifolia (Pallas) Volk.
 2 hirsuta (L.) Ascherson    18    Rm         Tarnavschi, 1947
 3 sedoides (Pallas) Ascherson


13 Kochia Roth
 1 prostrata (L.) Schrader
 2 saxicola Guss.
 3 laniflora (S.G. Gmelin) Borbás
 4 scoparia (L.) Schrader


14 Corispermum L.
 1 uralense (Iljin) Aellen
 2 marschallii Steven
 3 algidum Iljin
 4 canescens Kit.
 5 aralocaspicum Iljin
 6 nitidum Kit.
 7 filifolium C.A. Meyer
 8 declinatum Stephan ex Steven
 9 hyssopifolium L.
10 intermedium Schweigger
11 leptopterum (Ascherson) Iljin
12 orientale Lam.
```

15 Agriophyllum Bieb. ex C.A. Meyer
 1 squarrosum (L.) Moq.

16 Kalidium Moq.
 1 foliatum (Pallas) Moq.
 2 caspicum (L.) Ung.-Sternb.

17 Halopeplis Bunge ex Ung.-Sternb.
 1 amplexicaulis (Vahl) Ung.-Sternb.
 2 pygmaea (Pallas) Bunge ex Ung.-Sternb.

18 Halostachys C.A. Meyer
 1 belangerana (Moq.) Botsch.

19 Halocnemum Bieb.
 1 strobilaceum (Pallas) Bieb.

20 Arthrocnemum Moq.
 1 perenne (Miller) Moss   18   Br   Maude, 1940
 2 fruticosum (L.) Moq.   54   Ga   Contandriopoulos, 1968
 3 glaucum (Delile) Ung.-Sternb.   36   Lu   Castro & Fontes, 1946

21 Salicornia L.
 1 europaea L.   18   Br   Ball & Tutin, 1959
 2 ramosissima J. Woods   18   Br   Ball & Tutin, 1959
 3 prostrata Pallas   18   Rm   Tarnavschi, 1947
 4 pusilla J. Woods   18   Br   Ball & Tutin, 1959
 5 nitens P.W. Ball & Tutin   36   Br   Ball & Tutin, 1959
 6 fragilis P.W. Ball & Tutin   36   Br   Ball & Tutin, 1959
 7 dolichostachya Moss   36   Br   Ball & Tutin, 1959
   (a) dolichostachya
   (b) strictissima (K. Gram) P W Dall

22 Microcnemum Ung.-Sternb.
 1 coralloides (Loscos & Pardo) Font Quer

23 Suaeda Forskål ex Scop.
 1 vera J.F. Gmelin
 2 pruinosa Lange
 3 physophora Pallas
 4 dendroides (C.A. Meyer) Moq.
 5 confusa Iljin
 6 baccifera Pallas
 7 maritima (L.) Dumort.
   (a) maritima   36   Ge   Wulff, 1937a
   (b) salsa (L.) Soó   36   Hu   Pólya, 1948
   (c) pannonica (G. Beck) Soó ex P.W. Ball
 8 corniculata (C.A. Meyer) Bunge
 9 splendens (Pourret) Gren. & Godron 18   Lu   Castro &
                                                 Fontes, 1946
 10 altissima (L.) Pallas
 11 linifolia Pallas
 12 eltonica Iljin
 13 heterophylla (Kar. & Kit.) Bunge ex Boiss.
 14 kossingskyi Iljin

24 Bienertia Bunge ex Boiss.
 1 cycloptera Bunge ex Boiss.

25 Salsola L.
 1 foliosa (L.) Schrader

20

    2 brachiata Pallas
    3 paulsenii Litv.
    4 pellucida Litv.
    5 soda L.
    6 acutifolia (Bunge) Botsch.
    7 tamariscina Pallas
    8 kali L.
     (a) kali
     (b) tragus (L.) Nyman
     (c) ruthenica (Iljin) Soó
    9 collina Pallas
   10 lanata Pallas
   11 turcomanica Litv.
   12 crassa Bieb.
   13 nitraria Pallas
   14 affinis C.A. Meyer
   15 nodulosa (Moq.) Iljin
   16 carpatha P.H. Davis
   17 arbuscula Pallas
   18 genistoides Juss. ex Poiret
   19 webbii Moq.
   20 verticillata Schousboe
   21 papillosa (Cosson) Willk.
   22 vermiculata L.
   23 aegaea Rech. fil.
   24 dendroides Pallas
   25 laricina Pallas

26 Noaea Moq.
   1 mucronata (Forskål) Ascherson & Schweinf.

27 Ofaiston Rafin.
   1 monandrum (Pallas) Moq.

28 Girgensohnia Bunge
   1 oppositifolia (Pallas) Fenzl in Ledeb.

29 Anabasis L.
   1 salsa (C.A. Meyer) Paulsen
   2 aphylla L.
   3 cretacea Pallas
   4 articulata (Forskål) Moq.

30 Haloxylon Bunge
   1 articulatum (Moq.) Bunge

31 Nanophyton Less.
   1 erinaceum (Pallas) Bunge

32 Petrosimonia Bunge
   1 brachiata (Pallas) Bunge
   2 triandra (Pallas) Simonkai
   3 litwinowii Korsh.
   4 monandra (Pallas) Bunge
   5 oppositifolia (Pallas) Litv.
   6 brachyphylla (Bunge) Iljin
   7 glaucescens (Bunge) Iljin

33 Halimocnemis C.A. Meyer
   1 sclerosperma (Pallas) C.A. Meyer

34 Halogeton C.A. Meyer
  1 sativus (L.) Moq.

XLIX AMARANTHACEAE

1 Celosia L.
  1 argentea L.

2 Amaranthus L.
  1 hybridus L.   32   Lu   Grant, 1959
  2 cruentus L.   34   non-Europe   Grant, 1959
  3 bouchonii Thell.
  4 retroflexus L.
  5 muricatus (Moq.) Gillies ex Hicken
  6 blitoides S. Watson
  7 crispus (Lesp. & Thev.) N. Terracc.
  8 standleyanus Parodi ex Covas
  9 albus L.
  10 graecizans L.
  11 deflexus L.
  12 lividus L.

3 Achyranthes L.
  1 aspera L.
  2 sicula (L.) All.

4 Alternanthera Forskål
  1 pungens Kunch
  2 peploides (Humb. & Bonpl.) Urban
  3 nodiflora R.Br,

L NYCTAGINACEAE

1 Commicarpus Standley
  1 plumbagineus (Cav.) Standley

2 Mirabilis L.
  1 jalapa L.
  2 longiflora L.

3 Oxybaphus L'Hér. ex Willd.
  1 nyctagineus (Michx) Sweet

LI PHYTOLACCACEAE

1 Phytolacca L.
  1 americana L.
  2 dioica L.

LII AIZOACEAE

1 Aizoon L.
  1 hispanicum L.

2 Sesuvium L.
  1 portulacastrum L.

3 Carpobrotus N.E. Br.
  1 acinaciformis (L.) L. Bolus
  2 edulis (L.) N.E. Br.   18   Lu   Rodrigues, 1953
  3 chilensis (Molina) N.E. Br.

4 Mesembryanthemum L.
  1 nodiflorum L.
  2 crystallinum L.

5 Aptenia N.E. Br.
  1 cordifolia (L. fil.) N.E. Br.

6 Disphyma N.E. Br.
  1 crassifolium (L.) L. Bolus

7 Lampranthus N.E. Br.
  1 glaucus (L.) N.E. Br.

LIII MOLLUGINACEAE

1 Mollugo L.
  1 verticillata L.
  2 cerviana (L.) Ser.

2 Glinus L.
  1 lotoides L.

LIV TETRAGONIACEAE

1 Tetragonia L.
  1 tetragonoides (Pallas) O. Kuntze   32   Lu   Rodrigues, 1953

LV PORTULACACEAE

1 Portulaca L.
  1 oleracea L.
    (a) oleracea   54   Africa Hagerup, 1932
    (b) sativa (Haw.) Čelak.   54

2 Montia L.
  1 fontana L.   18   No   Hagerup, 1941
                                            20   Ge   Scheerer, 1940
    (a) fontana   18   Is   Löve & Löve, 1956
                                            20   Su   Lövkvist, 1963
    (b) variabilis Walters
    (c) amporitana Sennen
    (d) chondrosperma (Fenzl) Walters   18   Ge   Scheerer, 1940
                                            20   Su   Lövkvist, 1963
  2 perfoliata (Donn ex Willd.) Howell   36   Br   Rutland, 1941
  3 sibirica (L.) Howell

LVI BASELLACEAE

1 Boussingaultia Humb., Bonpl. & Kunth
  1 cordifolia Ten.

23

1 Arenaria L.
 1 rigida Bieb.
 2 cephalotes Bieb.
 3 gypsophiloides L.
 4 procera Sprengel
  (a) procera
  (b) glabra (F.N. Williams) J. Holub
 5 longifolia Bieb.
 6 purpurascens Ramond ex DC.
 7 lithops Heywood ex McNeill
 8 tetraquetra L.
 9 tomentosa Willk.
 10 armerina Bory
 11 aggregata (L.) Loisel.
  (a) aggregata
  (b) erinacea (Boiss.) Font Quer
 12 pungens Clemente ex Lag.
 13 valentina Boiss.
 14 grandiflora L. — 44 — He — Favarger, 1959
 15 bertolonii Fiori — 30 — Co — Favarger, 1962
 16 huteri Kerner
 17 montana L.
  (a) montana — 28 — Ga — Favarger, 1962
  (b) intricata (Dufour) Pau
 18 balearica L.
 19 biflora L. — 22 — Au — Favarger, 1962
 20 humifusa Wahlenb. — 40 — No — Horn, 1948
 21 norvegica Gunnerus
  (a) norvegica — 80 — Ic — Flba & Love, 1956
  (b) anglica Halliday — 80 — Br — Halliday, 1960
 22 ciliata L.
  (a) ciliata — 80 — He — Favarger, 1962
   — 120 — Au — Favarger, 1965
   — 240 — Au — Favarger, 1965
    ssp. pseudofrigida Ostenf. & O. Dahl — 40 — No — Horn, 1948
  (b) moehringioides (J. Murr) Br.-Bl. — 40 — Ga — Favarger, 1969
 23 gothica Fries — 100 — Su — Halliday, 1960
 24 gracilis Waldst. & Kit.
 25 filicaulis Fenzl
  (a) filicaulis
  (b) graeca (Boiss.) McNeill
 26 deflexa Decne — 44 — Ju — Favarger, 1962
 27 fragillima Rech. fil.
 28 ligericina Lecoq & Lamotte
 29 cretica Sprengel
 30 halacsyi Bald.
 31 cinerea DC.
 32 hispida L. — 40 — Ga — Favarger, 1962
 33 serpyllifolia L. — 40 — Su — Lövkvist, 1963
 34 leptoclados (Reichenb.) Guss — 20 — Su — Lövkvist, 1963
 35 conferta Boiss.
 36 serpentini A.K. Jackson
 37 nevadensis Boiss. & Reuter
 38 muralis (Link) Sieber ex Sprengel
 39 saponarioides Boiss. & Balansa
 40 guicciardii Heldr. ex Boiss.
 41 cerastioides Poiret
 42 emarginata Brot.
 43 algarbiensis Welw. ex Willk.

```
44 conimbricensis Brot.                         22    Ga     Favarger, 1962
45 obtusiflora G. Kunze
   (a) obtusiflora
   (b) ciliaris (Loscos) Font Quer
46 controversa Boiss.
47 conica Boiss.
48 modesta Dufour                               26    Ga     Favarger, 1962
49 retusa Boiss.
50 capillipes Boiss.
51 provincialis Chater & Halliday              40    Ga     Favarger, 1962

2 Moehringia L.
  1 trinervia (L.) Clairv.                      24    Br Lu  Blackburn &
                                                             Morton, 1957
  2 pentandra Gay                               48    Lu     Blackburn &
                                                             Morton, 1957
  3 lateriflora (L.) Fenzl
  4 minutiflora Bornm.
  5 intricata Willk.
  6 fontqueri Pau
  7 tejedensis Huter, Porta & Rigo ex Willk.
  8 diversifolia Dolliner ex Koch               24    Au     Blackburn, 1950
  9 pendula (Waldst. & Kit.) Fenzl
 10 villosa (Wulfen) Fenzl
 11 jankae Griseb. ex Janka
 12 grisebachii Janka
 13 dielsiana Mattf.
 14 papulosa Bertol.
 15 tommasinii Marchesetti
 16 markgrafii Merxm. & Guterm.
 17 bavarica (L.) Gren.
   (a) bavarica                                 24    Au     Mattick, 1950
   (b) insubrica (Degen) Sauer
 18 sedifolia Willd.
 19 glaucovirens Bertol.
 20 muscosa L.                                   24    Au     Mattick, 1950
 21 ciliata (Scop.) Dalla Torre                 24    Au     Mattick, 1950

3 Minuartia L.
  1 thymifolia (Sibth. & Sm.) Bornm.
  2 mesogitana (Boiss.) Hand.-Mazz.
  3 viscosa (Schreber) Schinz & Thell.          46    Ju     Favarger, 1962
  4 bilykiana Klokov
  5 hybrida (Vill.) Schischkin
   (a) hybrida                                  46    Ga     Favarger, 1962
                                                70    Ga     Favarger, 1962
   (b) lydia L. (Boiss.) Rech. fil.
  6 velenovskyi (Rohlena) Hayek
  7 mediterranea (Link) K. Maly                 24    Lu     Favarger, 1962
  8 regeliana (Trautv.) Mattf.
  9 hamata (Hausskn.) Mattf.
 10 dichotoma L.
 11 montana L.
 12 campestris L.
 13 globulosa (Labill.) Schinz & Thell.
 14 glomerata (Bieb.) Degen                      28    Rs(K)  Favarger, 1962
   (a) glomerata
   (b) velutina (Boiss. & Orph.) Mattf.
 15 rubra (Scop.) McNeill                        30    He     Favarger, 1962
 16 funkii (Jordan) Graebner
 17 mutabilis Schinz & Thell. ex Becherer        28    He     Favarger, 1959
```

25

```
18 setacea (Thuill.) Hayek              30    Cz        Favarger, 1962
   (a) setacea
   (b) banatica (Reichenb.) Prodan
19 bosniaca (G. Beck) K. Maly
20 krascheninnikovii Schischkin
21 adenotricha Schischkin
22 anatolica (Boiss.) Graebner
23 trichocalycina (Ten. & Guss.) Grande
24 hirsuta (Bieb.) Hand.-Mazz.
   (a) hirsuta
   (b) falcata (Griseb.) Mattf.
   (c) frutescens (Kit.) Hand.-Mazz.    30    Europe    Favarger, 1961
                                        32    Hu        Baksay, 1958
25 eurytanica (Boiss. & Heldr.) Hand.-Mazz.
26 recurva (All.) Schinz & Thell.
   (a) recurva                          30    He        Favarger, 1962
   (b) juressi (Willd. ex Schlecht.) Mattf.
27 bulgarica (Velen.) Graebner
28 saxifraga (Friv.) Graebner
29 graminifolia (Ard.) Jáv.
   (a) graminifolia
   (b) clandestina (Portenschl.) Mattf.
30 stellata (E.D. Clarke) Maire & Petitmengin
31 cerastiifolia (Lam. & DC.) Graebner
32 rupestris (Scop.) Schinz & Thell.    c.72  Ga        Favarger, 1962
33 lanceolata (All.) Mattf.             36    He        Favarger, 1959
34 grignensis (Reichenb.) Mattf.        36    It        Favarger, 1959
35 cherlerioides (Hoppe) Becherer
   (a) cherlerioides
   (b) rionii (Gremli) Friedrich
36 austriaca (Jacq.) Hayek              26    Au        Favarger, 1959
37 helmii (Ser.) Schischkin             28    Rs(C)     Favarger, 1962
38 villarii (Balbis) Chenevard          26    Ga        Favarger, 1962
39 taurica (Steven) Graebner
40 juniperina (L.) Maire & Petitmengin
41 pichleri (Boiss.) Maire & Petitmengin
42 verna (L.) Hiern
   (a) verna                            24    Br Ga He  Halliday, 1961
   (b) collina (Neilr.) Halliday        48    Hu        Favarger, 1962
   (c) valentina (Pau) Font Quer
   (d) oxypetala (Woloszczak) Halliday
   (e) attica (Boiss. & Spruner) Hayek
   (f) idaea (Halácsy) Hayek
43 rubella (Wahlenb.) Hiern             24    Br        Blackburn &
                                                          Morton, 1957
                                        26    Is        Löve & Löve, 1956
44 stricta (Swartz) Hiern               22    Br        Blackburn &
                                                          Morton, 1957
                                        26    Is        Löve & Löve, 1956
45 rossii (R. Br.) Graebner
46 wettsteinii Mattf.
47 capillacea (All.) Graebner           26    Ga        Favarger, 1959
48 baldaccii (Halácsy) Mattf.
   (a) baldaccii
   (b) doerfleri (Hayek) Hayek
   (c) skutariensis Hayek
49 garckeana (Ascherson & Sint.) Mattf.
50 handelii Mattf.
51 laricifolia (L.) Schinz & Thell.
   (a) laricifolia
   (b) kitaibelii (Nyman) Mattf.        26    Po        Favarger, 1962
52 macrocarpa (Pursh) Ostenf.
```

```
53 arctica (Ser.) Graebner
54 biflora (L.) Schinz & Thell.
55 olonensis (Bonnier) P. Fourn.
56 sedoides (L.) Hiern                26      He      Favarger, 1959
57 geniculata (Poiret) Thell.

4 Honkenya Ehrh.
  1 peploides (L.) Ehrh.             66      Sb      Flovik, 1940
                                     68-70   Rs(N)   Sokolovskaya &
                                                     Strelkova, 1960

5 Bufonia L.
  1 paniculata F. Dubois
  2 tenuifolia L.
  3 tuberculata Loscos
  4 macropetala Willk.
  5 willkommiana Boiss.
  6 perennis Pourret
 ▾7 stricta (Sibth. & Sm.) Gurke

6 Stellaria L.
  1 nemorum L.
    (a) nemorum
    (b) glochidisperma Murb.
  2 bungeana Fenzl
  3 media (L.) Vill.
    (a) media                        40 44   Br      Blackburn &
                                                     Morton, 1957
                                     42      Is      Löve & Löve, 1956
    (b) postii Holmboe
    (c) cupaniana (Jordan & Fourr.) Nyman
  4 neglecta Weihe                   22      Su      Peterson, 1933
  5 pallida (Dumort.) Pire           22      Su      Peterson, 1935
  6 holostea L.                      26      Br Lu   Blackburn &
                                                     Morton, 1957
  7 alsine Grimm                     24      Br Lu   Blackburn &
                                                     Morton, 1957
  8 palustris Retz.
  9 hebecalyx Fenzl
 10 graminea L.                      26      Br      Blackburn &
                                                     Morton, 1957
 11 longipes Goldie
 12 crassipes Hultén                 104     No      Knaben, 1950
 13 ciliatisepala Trautv.
 14 longifolia Muhl. ex Willd.
 15 calycantha (Ledeb.) Bong.        52      Is      Löve & Löve, 1956
 16 crassifolia Ehrh.                26      No      Knaben, 1950
 17 humifusa Rottb.                  26      Sb      Flovik, 1940

7 Pseudostellaria Pax
  1 europaea Schaeftlein             32      Au      Favarger, 1961

8 Holosteum L.
  1 umbellatum L.
    (a) umbellatum                   20      Su      Lövkvist, 1963
    (b) glutinosum (Bieb.) Nyman

9 Cerastium L.
  1 cerastoides (L.) Britton         36      Au      Mattick, 1950
                                     38      He      Söllner, 1954
  2 dubium (Bast.) O. Schwarz        38      Ga      Söllner, 1954
```

27

```
 3 maximum L.
 4 dahuricum Fischer ex Sprengel
 5 nemorale Bieb.
 6 perfoliatum L.                          c.37      Hs     Söllner, 1954
 7 dichotomum L.
 8 grandiflorum Waldst. & Kit.
 9 candidissimum Correns
10 biebersteinii DC.
11 tomentosum L.                           c.108     It     Söllner, 1954
12 moesiacum Friv.
13 decalvans Schlosser & Vuk.
14 boissieri Gren.                          72        Co     Söllner, 1954
15 lineare All.
16 gibraltaricum Boiss.
17 banaticum (Rochel) Heuffel
   (a) banaticum
   (b) alpinum (Boiss.) Buschm.
18 julicum Schellm.                         36        It     Söllner, 1954
19 soleirolii Ser. ex Duby                  72        Co     Söllner, 1954
20 scaranii Ten.
21 vagans Lowe
22 arvense L.
   (a) thomasii (Ten.) Rouy & Fouc.
   (b) strictum (Haenke) Gaudin             36        He     Söllner, 1954
   (c) allasii (Vest) Walters
   (d) suffruticosum (L.) Hegi              36        Hs     Favarger, 1969c
                                            72        Ga     Küpfer, 1969
   (e) ciliatum (Waldst. & Kit.) Reichenb.c.72
   (f) arvense                              72        Br He  Söllner, 1954
   (g) lerchenfeldianum (Schur) Ascherson & Graebner
   (h) glandulosum (Kit.) Soó
23 alsinifolium Tausch
24 jenisejense Hultén
25 regelii Ostenf.                          72        Sb     Flovik, 1940
26 alpinum L.                              c.54       Rs(N)  Sokolovskaya &
                                                             Strelkova, 196
                                            72        Br     Söllner,1954
                                           144        He     Brett, 1955
   (a) squalidum (Lam.) Hultén
   (b) lanatum (Lam.) Ascherson & Graebner
   (c) alpinum
   (d) glabratum (Hartman) Á. & D. Löve
27 arcticum Lange
   (a) arcticum                            108        Br     Brett, 1955
   (b) edmondstonii (H.C. Watson) Á. &
       D. Löve                             108        Br     Söllner, 1954
28 uniflorum Clairv.                        36        Au Ile Söllner, 1954
29 latifolium L.                            36        He     Söllner, 1954
30 pyrenaicum Gay
31 runemarkii Möschl & Rech. fil.
32 pedunculatum Gaudin                      36        Au     Söllner, 1954
33 dinaricum G. Beck & Szysz.
34 carinthiacum Vest
   (a) carinthiacum                         36        Au     Söllner, 1954
   (b) austroalpinum (H. Kunz) H. Kunz      36        Au     Sollner, 1954
35 transsilvanicum Schur
36 subtriflorum (Reichenb.) Pacher          36        Ju     Söllner, 1954
37 sylvaticum Waldst. & Kit.                36        Au     Söllner, 1954
38 fontanum Baumg.
   (a) macrocarpum (Schur) Jalas           144        Au     Söllner, 1954
   (b) scandicum H. Gartner                144        Is     Löve & Löve, 1956
```

28

(c) fontanum	144	Is	Löve & Löve, 1956
(d) hispanicum H. Gartner			
(e) triviale (Link) Jalas	144	Co	Söllner, 1954
	134-152	It	Söllner, 1954

39 pauciflorum Steven ex Ser.
40 illyricum Ard.

(a) comatum (Desv.) P.D. Sell & Whitehead	34 (not 68)	Co Gr	Söllner, 1954

    (b) decrescens (Lonsing) P.D. Sell &
       Whitehead
    (c) prolixum (Lonsing) P.D. Sell &
       Whitehead
    (d) illyricum
    (e) crinitum (Lonsing) P.D. Sell &
       Whitehead
41 pedunculare Bory & Chaub.
42 scaposum Boiss. & Heldr.
43 brachypetalum Pers.
    (a) pindigenum (Lonsing) P.D. Sell &
       Whitehead
    (b) corcyrense (Möschl) P.D. Sell &
       Whitehead
    (c) doerfleri (Halácsy ex Hayek) P.D.
       Sell & Whitehead

(d) brachypetalum	c.90	He	Söllner, 1954
(e) tenoreanum (Ser.) Soó	c.52	Au He	Söllner, 1954

    (f) atheniense (Lonsing) P.D. Sell &
       Whitehead

(g) tauricum (Sprengel) Murb.	c.90	Ga	Söllner, 1954

    (h) roeseri (Boiss. & Heldr.) Nyman

44 glomeratum Thuill.	72	Br Da Ga He	Söllner, 1954

45 rectum Friv.
    (a) rectum
    (b) petricola (Pančić) H. Gartner
46 ligusticum Viv.

(a) ligusticum	34	It	Söllner, 1954

    (b) palustre (Moris) P.D. Sell &
       Whitehead
    (c) granulatum (Huter, Porta & Rigo)
       P.D. Sell & Whitehead
    (d) trichogynum (Möschl) P.D. Sell
       & Whitehead
47 semidecandrum L.

(a) semidecandrum	36	Hs	Söllner, 1954

    (b) macilentum (Aspegren) Möschl

48 pumilum Curtis	90-100	Au Br He	Söllner, 1954

    (a) pumilum
    (b) pallens (F.W. Schultz) Schinz &
       Thell.
    (c) litigiosum (De Lens) P.D. Sell
       & Whitehead
49 diffusum Pers.
    (a) gussonei (Tod. ex Lojac.) P.D.
       Sell & Whitehead

(b) diffusum	36	Ge	Wulff, 1937
	72	Br	Brett, 1955

    (c) subtetrandrum (Lange) P.D. Sell

& Whitehead	72	Da	Hagerup, 1941

50 siculum Guss.

51 gracile Dufour	88-92	Lu	Söllner, 1954

10 Moenchia Ehrh.
   1 erecta (L.) P. Gaertner, B. Meyer   36   Br Lu   Blackburn &
      & Scherb.   Morton, 1957
      (a) erecta
      (b) octandra (Ziz) Coutinho
   2 graeca Boiss. & Heldr.
   3 mantica (L.) Bartl.
      (a) mantica
      (b) caerulea (Boiss.) Clapham

11 Myosoton Moench
   1 aquaticum (L.) Moench   28   Hs Lu   Blackburn &
                                                                    Morton, 1957

12 Sagina L.
   1 nodosa (L.) Fenzl   22-24   Ge   Wulff, 1937b
                                                 44   Is   Löve & Löve, 1956
                                                 56   Br   Blackburn &
                                                                    Morton, 1957
   2 intermedia Fenzl   88   Is   Löve & Löve, 1956
   3 caespitosa (J. Vahl) Lange   88   Is   Löve & Löve, 1956
   4 glabra (Willd.) Fenzl
   5 pilifera (DC.) Fenzl
   6 subulata (Swartz) C. Presl   22   Is   Löve & Löve, 1956
   7 sabuletorum (Gay) Lange
   8 nevadensis Boiss. & Reuter
   9 saginoides (L.) Karsten   22   Is   Löve & Löve, 1956
  10 x normaniana Lagerh.
  11 procumbens L.   22   Is   Löve & Löve, 1956
      (a) procumbens
      (b) muscosa (Jordan) Nyman
         boydii Buchanan-White   22   Br   Blackburn &
                                                                    Morton, 1957
  12 apetala Ard.   12   Br Lu   Blackburn &
                                                                    Morton, 1957

      (a) apetala
      (b) erecta (Hornem.) F. Hermann
  13 maritima G. Don.   22-24   Ge   Wulff, 1937a
                                                 28   Br   Blackburn &
                                                                    Morton, 1957

13 Scleranthus L.
   1 perennis L.
      (a) perennis   22   Su   Lövkvist, 1963
      (b) prostratus P.D. Sell
      (c) burnatii (Briq.) P.D. Sell
      (d) polycnemoides (Willk. & Costa)
          Font Quer
      (e) vulcanicus (Strobl) Béguinot
      (f) marginatus (Guss.) Arcangeli
      (g) dichotomus (Schur) Stoj. &
          Stefanov
   2 annuus L.   44   Br Lu   Blackburn &
                                                                    Morton, 1957

      (a) annuus
      (b) polycarpos (L.) Thell.
      (c) verticillatus (Tausch) Arcangeli
      (d) ruscinonensis (Gillot & Coste)
          P.D. Sell
   3 uncinatus Schur

14 Corrigiola L.
  1 litoralis L.   18   Br Lu   Blackburn &
                                                        Morton, 1957
  2 telephiifolia Pourret   18   Lu   Blackburn &
                                                        Morton, 1957

15 Paronychia Miller
  1 cymosa (L.) DC.
  2 echinulata Chater
  3 rouyana Coincy
  4 suffruticosa (L.) Lam.
  5 argentea Lam.
  6 polygonifolia (Vill.) DC.
  7 kapela (Hacq.) Kerner
    (a) kapela
    (b) serpyllifolia (Chaix) Graebner
    (c) chionaea (Boiss.) Borhidi
  8 taurica Borhidi & Sikura
  9 cephalotes (Bieb.) Besser   36   Hu   Baksay, 1956
10 aretioides DC.
11 capitata (L.) Lam.

16 Herniaria L.
  1 alpina Chaix   18   He   Favarger, 1959a
  2 boissieri Gay
  3 latifolia Lapeyr.
  4 parnassica Heldr. & Sart. ex Boiss.
  5 glabra L.   18   Br   Blackburn &
                                                    Morton, 1957
  6 ciliolata Melderis   72   Br   Blackburn &
                                                Morton, 1957
                               108   ?Br   Blackburn &
                               126   Lu   Adams, 1955
  7 maritima Link
  8 scabrida Boiss.
  9 baetica Boiss. & Reuter
10 incana Lam.
11 hirsuta L.   36   Lu   Blackburn &
                                                    Morton, 1957
12 polygama Gay
13 nigrimontium F. Hermann
14 fontanesii Gay
    (a) fontanesii
    (b) almeriana Brummitt & Heywood
15 fruticosa L.
    (a) fruticosa
    (b) erecta (Willk.) Batt.

17 Illecebrum L.
  1 verticillatum L.   10   Br Lu   Blackburn &
                                                    Morton, 1957

18 Pteranthus Forskål
  1 dichotomus Forskål

19 Polycarpon Loefl. ex L.
  1 tetraphyllum (L.) L.
  2 diphyllum Cav.
  3 alsinifolium (Biv.) DC.
  4 polycarpoides (Biv.) Zodda

20 Ortegia L.
  1 hispanica L.

21 Loeflingia L.
  1 hispanica L.
  2 baetica Lag.
  3 tavaresiana Samp.

22 Spergula L.

1 arvensis L.	18	Br Lu	Blackburn & Morton, 1957
2 morisonii Boreau	18	Ge	Rohweder, 1939
3 pentandra L.			
4 viscosa Lag.			

23 Spergularia (Pers.) J. & C. Presl
  1 azorica (Kindb.) Lebel
  2 fimbriata Boiss.

3 rupicola Lebel ex Le Jolis	36	Br	Blackburn & Morton, 1957
4 macrorhiza (Req.) Heynh.	36	?Co	Monnier, ined.
5 australis Samp.	40	Lu	Monnier, ined.
6 media (L.) C. Presl	18	Br	Ratter, 1964
	36	Br	Blackburn & Morton, 1957
7 marina (L.) Griseb.	36	Is	Löve & Löve, 1956
8 tangerina P. Monnier	18	S. Europe	Ratter, 1976
9 segetalis (L.) G. Don fil.	18	Lu	Blackburn & Morton, 1957
10 diandra (Guss.) Boiss.	18	Lu	Blackburn & Morton, 1957
11 purpurea (Pers.) G. Don fil	18	Lu	Blackburn & Morton, 1957
	36	Hs Lu	Ratter, 1964
12 rubra (L.) J. & C. Presl	36	Br Fe Lu	Ratter, 1964
	54	Ga	Ratter, 1964
13 nicaeensis Sarato ex Burnat	36	Ga Hs	Ratter, 1964
14 capillacea (Kindb. & Lange) Willk.	18	Lu	Blackburn & Morton, 1957
15 bocconii (Scheele) Ascherson & Graebner	36	Lu	Blackburn & Morton, 1957
16 heldreichii Fouc. ex E. Simon secundus & P. Monnier	36	N. Africa	Ratter, 1964
17 echinosperma Celak.			

24 Telephium L.
  1 imperati L.
    (a) imperati
    (b) orientale (Boiss.) Nyman

25 Lychnis L.
  1 chalcedonica L.
  2 coronaria (L.) Desr.
  3 flos-jovis (L.) Desr.
  4 flos-cuculi L.

(a) flos-cuculi	24	He	Favarger, 1946
(b) subintegra Hayek			

  5 sibirica L.
  6 viscaria L.

(a) viscaria	24	Ge Su	Rohweder, 1939
(b) atropurpurea (Griseb.) Chater			
7 alpina L.	24	Is	Löve & Löve, 1956
8 nivalis Kit.			

26 Agrostemma L.
  1 githago L.   48   Su   D. Löve, 1942
  2 linicola Terechov
  3 gracilis Boiss.

27 Petrocoptis A. Braun
  1 pyrenaica (J.P. Bergeret) A. Braun
  2 viscosa Rothm.
  3 glaucifolia (Lag.) Boiss.
  4 grandiflora Rothm.
  5 hispanica (Willk.) Pau
  6 crassifolia Rouy
  7 pardoi Pau

28 Silene L.
  1 italica (L.) Pers.
   (a) italica   24   S. Europe D. Löve, 1942
   (b) nemoralis (Waldst. & Kit.) Nyman   24   Au   Mattick, 1950
  2 patula Desf.   24   Lu   Blackburn &
                                                 Morton, 1957
  3 hifacensis Rouy ex Willk.
  4 mollissima (L.) Pers.
  5 velutina Pourret ex Loisel.
  6 pseudovelutina Rothm.
  7 mellifera Boiss. & Reuter
  8 paradoxa L.
  9 nodulosa Viv.
 10 cythnia (Halácsy) Walters
 11 dictaea Rech. fil.
 12 spinescens Sibth. & Sm.
 13 fruticosa L.
 14 gigantea L.
 15 nutans L.
   (a) nutans   24   He   Favarger, 1946
   (b) dubia (Herbich) Zapal.   24   Br   Blackburn &
 16 brachypoda Rouy   Morton, 1957
 17 viridiflora L.
 18 catholica (L.) Aiton fil.
 19 viscariopsis Bornm.
 20 longipetala Vent.
 21 niederi Heldr. ex Boiss.
 22 guicciardii Boiss. & Heldr.
 23 bupleuroides L.
   (a) bupleuroides   24   Cz   Májovský et al.,
   (b) staticifolia (Sibth. & Sm.)   1974
      Chowdhuri
 24 chlorifolia Sm.
 25 chlorantha (Willd.) Ehrh.
 26 frivaldszkyana Hampe
 27 multiflora (Waldst. & Kit.) Pers.   24   Au   Larsen, 1956a
 28 viscosa (L.) Pers.   24   Su   Löve & Löve, 1942
 29 radicosa Boiss. & Heldr.
 30 reichenbachii Vis.
 31 skorpilii Velen.
 32 tatarica (L.) Pers.
 33 paucifolia Ledeb.
 34 graminifolia Otth
 35 wahlbergella Chowdhuri   24   Su   Nygren, 1949
 36 furcata Rafin.   48   No Su   Nygren, 1951
 37 sibirica (L.) Pers.
 38 roemeri Friv.

```
39 sendtneri Boiss.
40 ventricosa Adamović
41 borysthenica (Gruner) Walters        24     Hu      Baksay, 1958
42 media (Litv.) Kleopow
43 hellmannii Claus
44 otites (L.) Wibel                     24     Rm      Vladesco, 1940
45 densiflora D'Urv.
46 exaltata Friv.
47 wolgensis (Willd.) Besser ex Sprengel 24     Rs( )   F. Wrigley, ined.
48 baschkirorum Janisch.
49 auriculata Sibth. & Sm.
50 zawadzkii Herbich
51 elisabetha Jan
52 requienii Otth
53 cordifolia All.
54 foetida Link ex Sprengel              24     Lu      Blackburn &
                                                        Morton, 1957

55 macrorhiza Gay & Durieu ex Lacaita
56 vulgaris (Moench) Garcke
   (a) vulgaris                          24     Au      Griesinger, 1937
   (b) angustifolia (Miller) Hayek
   (c) commutata (Guss.) Hayek           48     It Si   K. Larsen, ined.
   (d) macrocarpa Turrill
   (e) glareosa (Jordan) Marsden-Jones & 24     Ga      Marsden-Jones &
       Turrill                                          Turrill, 1957
   (f) prostrata (Gaudin) Chater &
       Walters                           24     Au      Griesinger, 1937
   (g) maritima (With.) A. & D. Löve     24     Su      Griesinger, 1937
   (h) thorei (Duf.) Chater & Walters
57 csereii Baumg.
58 fabaria (L.) Sibth. & Sm.
59 caesia Sibth. & Sm
60 variegata (Desf.) Boiss. & Heldr.
61 fabarioides Hausskn.
62 procumbens Murray
63 flavescens Waldst. & Kit.
64 congesta Sibth. & Sm.
65 vallesia L.
   (a) vallesia
   (b) graminea (Vis. ex Reichenb.)
       Ascherson & Graebner
66 boryi Boiss.
67 repens Patrin
68 succulenta Forskål
69 supina Bieb.                          24     Rm      Vladesco, 1940
70 thymifolia Sibth. & Sm.               48     Rm      Vladesco, 1940
71 altaica Pers.
72 cretacea Fischer ex Sprengel
73 suffrutescens Bieb.
74 taliewii Kleopow
75 linifolia Sibth. & Sm.
76 schwarzenbergeri Halácsy
77 campanula Pers.
78 saxifraga L.
79 schmuckeri Wettst.
80 multicaulis Guss.
81 waldsteinii Griseb.
82 pindicola Hausskn.
83 orphanidis Boiss.
84 barbeyana Heldr. ex Boiss.
85 falcata Sibth. & Sm.
```

86 cephallenia Heldr.			
87 ciliata Pourret	24	Hs	Blackburn & Morton, 1957
	36	Hs	Blackburn & Morton, 1957
	48	Hs	Blackburn & Morton, 1957
	72		
	84		
	c.96		
	c.120	Bot. Card.	Favarger, 1946
88 legionensis Lag.			
89 borderi Jordan			
90 acaulis (L.) Jacq.	24	Au Su	Jörgensen et al., 1958
(a) longiscapa (Kerner ex Vierh.) Hayek			
(b) exscapa (All.) J. Braun			
91 dinarica Sprengel			
92 rupestris L.	24	It	Griesinger, 1937
93 lerchenfeldiana Baumg.			
94 pusilla Waldst. & Kit.	24	It	Rohweder, 1939
95 veselskyi (Janka) Béguinot			
96 macrantha (Pančić) Neumayer			
97 tommasinii Vis.			
98 retzdorffiana (K. Maly) Walters			
99 chromodonta Boiss. & Reuter			
100 alpestris Jacq.			
101 armeria L.			
102 compacta Fischer			
103 asterias Griseb.			
104 noctiflora L.	24	Su	Löve & Löve, 1942
105 alba (Miller) E.H.L. Krause			
(a) alba	24	Br	Blackburn, 1923
(b) divaricata (Reichenb.) Walters	24	Lu	Blackburn & Morton, 1957
(c) eriocalycina (Boiss.) Walters			
106 dioica (L.) Clairv.	24	Br	Baker, 1948
107 heuffelii Soó			
108 diclinis (Lag.) M. Laínz			
109 portensis L.			
110 echinosperma Boiss. & Heldr.			
111 pinetorum Boiss. & Heldr.			
112 inaperta L.			
113 fuscata Link			
114 pseudatocion Desf.			
115 rubella L.			
(a) rubella			
(b) turbinata (Guss.) Chater & Walters			
116 bergiana Lindman			
117 insularis W. Barbey			
118 divaricata Clemente			
119 integripetala Bory & Chaub.			
120 sedoides Poiret			
121 pentelica Boiss.			
122 haussknechtii Heldr. ex Hausskn.			
123 laconica Boiss. & Orph.			
124 cretica L.			
125 ungeri Fenzl			
126 graeca Boiss. & Spruner			
127 muscipula L.			

```
128 stricta L.
129 behen L.
130 holzmannii Heldr. ex Boiss.
131 linicola C.C. Gmelin
132 echinata Otth
133 squamigera Boiss.
134 trinervia Sebastiani & Mauri
135 laeta (Aiton) Godron
136 coeli-rosa (L.) Godron
137 pendula L.
138 psammitis Link ex Sprengel
139 littorea Brot.
140 adscendens Lag.
141 boissieri Gay
142 almolae Gay
143 dichotoma Ehrh.                          24      Rm      Vladesco, 1940
144(a) dichotoma
    (b) racemosa (Otth) Graebner
144 remotiflora Vis.
145 nicaeensis All.
146 scabriflora Brot.
147 micropetala Lag.
148 discolor Sibth. & Sm.
149 obtusifolia Willd.
150 sericea All.
151 nocturna L.
    (a) nocturna
    (b) neglecta (Ten.) Arcangeli
152 ramosissima Desf.
153 gallica L.                               24      Br Lu   Blackburn &
                                                             Morton, 1957

154 giraldii Guss.
155 bellidifolia Juss. ex Jacq.
156 cerastoides L.
157 tridentata Desf.
158 disticha Willd.
159 colorata Poiret
160 secundiflora Otth
161 apetala Willd.
162 longicaulis Pourret ex Lag.
163 ammophila Boiss. & Heldr.
    (a) ammophila
    (b) carpathae Chowdhuri
164 conica L.
    (a) conica                              20      Su      Lövkvist, 1963
    (b) subconica (Friv.) Gavioli
    (c) sartorii (Boiss. & Heldr.) Chater
        & Walters
165 lydia Boiss.
166 conoidea L.


29 Cucubalus L.
   1 baccifer L.                            24      Lu      Blackburn &
                                                            Morton, 1957


30 Drypis L.
   1 spinosa L.                             60      Bot. Gard.  Favarger, 1946
    (a) spinosa
    (b) jacquiniana Murb. & Wettst. ex Murb.
```

31 Gypsophila L.
  1 nana Bory & Chaub.
  2 achaia Bornm.
  3 spergulifolia Griseb.
  4 repens L.   34   Po   Skalińska, 1950
  5 altissima L.
  6 litwinowii Kos.-Pol.
  7 fastigiata L.   34   Hu   Baksay, 1956
  8 papillosa Porta
  9 hispanica Willk.
10 struthium Loefl.
11 collina Steven ex Ser.
12 patrinii Ser.
13 uralensis Less.
14 petraea (Baumg.) Reichenb.
15 glomerata Pallas ex Bieb.
16 globulosa Steven ex Boiss.
17 paniculata L.
18 arrostii Guss.
19 belorossica Barkoudah
20 acutifolia Steven ex Sprengel
21 perfoliata L.
22 scorzonerifolia Ser.
23 tomentosa L.
24 linearifolia (Fischer & C.A. Meyer) Boiss.
25 elegans Bieb.
26 muralis L.
27 macedonica Vandas
28 pilosa Hudson

32 Bolanthus (Ser.) Reichenb.
  1 laconicus (Boiss.) Barkoudah
  2 fruticulosus (Bory & Chaub.) Barkoudah
  3 thessalus (Jaub. & Spach) Barkoudah
  4 graecus (Schreber) Barkoudah

33 Saponaria L.
  1 bellidifolia Sm.
  2 lutea L.
  3 caespitosa DC.
  4 pumilio (L.) Fenzl ex A. Braun
  5 glutinosa Bieb.
  6 sicula Rafin.
   (a) sicula
   (b) intermedia (Simmler) Chater
   (c) stranjensis (Jordanov) Chater
  7 calabrica Guss.
  8 ocymoides L.   28   He   Favarger, 1946
  9 officinalis L.   28   He   Favarger, 1946
10 orientalis L.

34 Vaccaria Medicus
  1 pyramidata Medicus   30   He   Favarger, 1946

35 Petrorhagia (Ser. ex DC.) Link
  1 illyrica (L.) P.W. Ball & Heywood
   (a) illyrica
   (b) haynaldiana (Janka) P.W. Ball &
      Heywood
   (c) taygetea (Boiss.) P.W. Ball &
      Heywood
  2 ochroleuca (Sibth. & Sm.) P.W. Ball & Heywood

3 armerioides (Ser.) P.W. Ball & Heywood
4 candica P.W. Ball & Heywood
5 cretica (L.) P.W. Ball & Heywood
6 alpina (Habl.) P.W. Ball & Heywood
7 phthiotica (Boiss. & Heldr.) P.W. Ball
  & Heywood
8 fasciculata (Margot & Reuter) P.W. &
  Heywood
9 saxifraga (L.) Link   60   Au   Larsen, 1954a
10 graminea (Sibth. & Sm.) P.W. Ball &
   Heywood
11 thessala (Boiss.) P.W. Ball & Heywood
12 dianthoides (Sibth. & Sm.) P.W. Ball
   & Heywood
13 prolifera (L.) P.W. Ball & Heywood   30   Ga   Carolin, 1957
14 nanteuilii (Burnat) P.W. Ball &
   Heywood   60   Hs   Böcher et al., 1953
15 velutina (Guss.) P.W. Ball & Heywood 30   Lu   Blackburn &
                                        Morton, 1957
16 glumacea (Chaub. & Bory) P.W. Ball
   & Heywood

36 Dianthus L.
 1 seguieri Vill.
  (a) seguieri   90   Alps   Carolin, 1957
  (b) guatieri (Sennen) Tutin
  (c) glaber Čelak.
  (d) italicus Tutin
 2 collinus Waldst. & Kit.
  (a) collinus
  (b) glabriusculus (Kit.) Soó
 3 fischeri Sprengel
 4 pratensis Bieb.
  (a) pratensis
  (b) racovitzae (Prodan) Tutin
 5 versicolor Fischer ex Link
 6 pseudoversicolor Klokov
 7 furcatus Balbis
  (a) furcatus
  (b) tener (Balbis) Tutin
  (c) geminiflorus (Loisel.) Tutin   30   Hs   Carolin, 1957
  (d) gyspergerae (Rouy) Burnat ex Briq.
 8 viridescens Vis.
 9 trifasciculatus Kit.
  (a) trifasciculatus
  (b) deserti (Prodan) Tutin
  (c) euponticus (Zapal.) Kleopow
10 urumoffii Stoj. & Acht.
11 eugeniae Kleopow
12 tesquicola Klokov
13 guttatus Bieb.
14 knappii (Pant.) Ascherson & Kanitz
   ex Borbás
15 membranaceus Borbás
16 dobrogensis Prodan
17 barbatus L.
  (a) barbatus
  (b) compactus (Kit.) Heuffel
18 monspessulanus L.   60   It   Carolin, 1957
  (a) monspessulanus

(b) marsicus (Ten.) Novák
(c) sternbergii Hegi
19 repens Willd.
20 alpinus L.   30   Alps   Carolin, 1957
21 nitidus Waldst. & Kit.
22 scardicus Wettst.
23 callizonus Schott & Kotschy
24 glacialis Haenke
   (a) glacialis   30   Alps   Carolin, 1957
   (b) gelidus (Schott, Nyman & Kotschy
       Tutin
25 freynii Vandas
26 microlepis Boiss.
27 pavonius Tausch   30   Alps   Carolin, 1957
28 gratianopolitanus Vill.   60   Ga   Carolin, 1957
29 xylorrhizus Boiss. & Heldr.
30 sylvestris Wulfen   30   Alps   Carolin, 1957
                                             c.60   Ga   K. Jones, ined.
   (a) sylvestris
   (b) siculus (C. Presl) Tutin
   (c) nodosus (Tausch) Hayek
   (d) bertisceus Rech. fil.
   (e) tergestinus (Reichenb.) Hayek
31 caryophyllus L.
32 subacaulis Vill. var rusconinensis   60   Hs   Carolin, 1957
   (a) subacaulis
   (b) brachyanthus (Boiss.) P. Fourn.   30   Hs   Carolin, 1957
33 minutiflorus (Borbás) Halácsy
34 pungens L.
35 hispanicus Asso
36 costae Willk.
37 planellae Willk.
38 langeanus Willk.
39 laricifolius Boiss. & Reuter
40 serratifolius Sibth. & Sm.
41 pyrenaicus Pourret   30   Hs   Carolin, 1957
                                               60   Ga   K. Jones, 1964
   (a) pyrenaicus
   (b) catalaunicus (Willk. & Costa) Tutin
42 lusitanus Brot.   30   Hs   Carolin, 1957
43 malacitanus Haenseler ex Boiss.   60   Hs   Carolin, 1957
44 scaber Chaix
   (a) scaber
   (b) cutandae (Pau) Tutin   30   Hs   K. Jones,ined.
   (c) toletanus (Boiss. & Reuter) Tutin
45 crassipes R. de Roemer
46 graniticus Jordan   30   Ga   K. Jones,ined.
47 cintranus Boiss. & Reuter
   (a) cintranus   60   Lu   K. Jones,ined.
   (b) multiceps (Costa ex Willk.) Tutin
   (c) charidemii (Pau) Tutin
48 anticarius Boiss. & Reuter
49 petraeus Waldst. & Kit.
   (a) petraeus
   (b) integer (Vis.) Tutin
   (c) simonkaianus (Péterfi) Tutin
   (d) noeanus (Boiss.) Tutin
50 spiculifolius Schur
51 stefanoffii Eig
52 plumarius L.
53 serotinus Waldst. & Kit.

54 acicularis Fischer ex Ledeb.
55 rigidus Bieb.
56 uralensis Korsh.
57 squarrosus Bieb.
58 arenarius L.
  (a) arenarius   60   Su   Lövkvist, 1963
  (b) borussicus (Vierh.) Kleopow
  (c) pseudoserotinus (Blocki) Tutin
  (d) pseudosquarrosus (Novák) Kleopow
59 krylovianus Juz.
60 volgicus Juz.
61 gallicus Pers.
62 superbus L.
  (a) superbus
  (b) speciosus (Reichenb.) Pawl.   30   Ju   Carolin, 1957
  (c) stenocalyx (Trautv.) Kleopow
63 deltoides L.
64 degenii Bald.
65 myrtinervius Griseb.
66 sphacioticus Boiss. & Heldr.
67 pallidiflorus Ser.
68 campestris Bieb.
  (a) campestris
  (b) laevigatus (Gruner) Klokov
  (c) steppaceus Širj.
69 hypanicus Andrz.
70 carbonatus Klokov
71 aridus Griseb. ex Janka
72 marschallii Schischkin
73 cinnamomeus Sibth. & Sm.
74 pallens Sibth. & Sm.
75 lanceolatus Steven ex Reichenb,
76 leptopetalus Willd.
77 roseoluteus Velen.
78 mercurii Heldr.
79 gracilis Sibth. & Sm.   60   Ju   Carolin, 1957
  (a) gracilis
  (b) xanthianus (Davidov) Tutin
  (c) armerioides (Griseb.) Tutin
  (d) friwaldskyanus (Boiss.) Tutin
80 drenowskianus Rech. fil.
81 haematocalyx Boiss. & Heldr.
  (a) haematocalyx
  (b) pruinosus (Boiss. & Orph.) Hayek
  (c) sibthorpii (Vierh.) Hayek
  (d) pindicola (Vierh.) Hayek
82 biflorus Sibth. & Sm.
83 strictus Banks & Solander   30   Balkans Carolin, 1957
84 tripunctatus Sibth. & Sm.
85 corymbosus Sibth. & Sm.   30   Bu   Carolin, 1957
86 viscidus Bory & Chaub.
87 diffusus Sibth. & Sm.
88 formanekii Borbás ex Form.
89 armeria L.
  (a) armeria   30   Br   Carolin, 1957
  (b) armeriastrum (Wolfner) Velen. 30   Ju   Carolin, 1957
90 pseudarmeria Bieb.
91 humilis Willd. ex Ledeb.
92 ciliatus Guss.
  (a) ciliatus
  (b) dalmaticus (Čelak.) Hayek

```
93 arpadianus Ade & Bornm.
94 nardiformis Janka                     30      Bu      Carolin, 1957
95 ingoldbyi Turrill
96 juniperinus Sm.                       30      Cr      Carolin, 1957
97 rupicola Biv.
98 arboreus L.                           30      Gr      Carolin, 1957
99 fruticosus L.
100 ferrugineus Miller
    (a) ferrugineus                      30 Bot. Gard.  Carolin, 1957
    (b) liburnicus (Bartl.) Tutin
    (c) vulturius (Guss. & Ten.) Tutin
101 capitatus Balbis ex DC.
    (a) capitatus
    (b) andrzejowskianus Zapal.
102 pinifolius Sibth. & Sm.
    (a) pinifolius                       30      Bu      Carolin, 1957
    (b) serbicus Wettst.
    (c) lilacinus (Boiss. & Heldr.) Wettst.
103 giganteus D'Urv.
    (a) giganteus
    (b) croaticus (Borbás) Tutin
    (c) banaticus (Heuffel) Tutin
    (d) haynaldianus (Borbás) Tutin
    (e) italicus Tutin
    (f) leucophoeniceus (Dörfler & Hayek) Tutin
104 pontederae Kerner
    (a) pontederae                       30      Au      Carolin, 1957
    (b) giganteiformis (Borbás) Soó
    (c) kladovanus (Degen) Stoj. & Acht.
105 diutinus Kit.
106 platyodon Klokov
107 bessarabicus (Kleopow) Klokov
108 carthusianorum L.                    30      Alps    Carolin, 1957
109 puberulus (Simonkai) Kerner
110 tenuifolius Schur
111 borbasii Vandas
    (a) borbasii
    (b) capitellatus (Klokov) Tutin
112 henteri Heuffel ex Griseb
113 cruentus Griseb.                     30      Bu      Carolin, 1957
    (a) cruentus
    (b) turcicus (Velen.) Stoj. & Acht.
114 quadrangulus Velen.                  30      Bu      Carolin, 1957
115 tristis Velen.                       30 Bot. Gard.  Rohweder, 1934
116 stribrnyi Velen.
117 brachyzonus Borbás & Form.           30      Al      Carolin, 1957
118 pelviformis Heuffel
119 stenopetalus Griseb.
120 burgasensis Tutin
121 moesiacus Vis. & Pančić


37 Velezia L.
   1 rigida L.                           28      Lu      Blackburn &
                                                         Morton, 1957

   2 quadridentata Sibth. & Sm.


LVIII NYMPHAEACEAE

1 Nymphaea L.
  1 alba L.                              84      Br      Y. Heslop-
                                                         Harrison, 1955
```

41

		c.105	Su	Ehrenberg, 1945
		112	Scand.	Langlet & Söderberg, 1927
2 candida C. Presl		c.160	Scand.	Lohammon, 1942
3 tetragona Georgi				
4 lotus L.				

2 Nuphar Sm.
  1 lutea (L.) Sibth. & Sm.   34   Br   Y. Heslop-Harrison,
  2 pumila (Timm) DC.   34   Br   1953b

LIX NELUMBONACEAE

1 Nelumbo Adanson
  1 nucifera Gaertner

LX CERATOPHYLLACEAE

1 Ceratophyllum L.
  1 demersum L.   34   Scand.   Langlet & Söderberg, 1927

   (a) demersum
   (b) platyacanthum (Cham.) Nyman
  2 submersum L.   24   Ge   Strasburger, 1902
                    40   Ge   Wulff, 1938

  3 tanaiticum Sapjegin
  4 muricatum Cham.

LXI RANUNCULACEAE

1 Helleborus L.
  1 foetidus L.
  2 lividus Aiton
   (a) lividus
   (b) corsicus (Willd.)
  3 cyclophyllus Boiss.
  4 orientalis Lam.
  5 viridis L.
   (a) viridis
   (b) occidentalis (Reuter) Schiffner
  6 odorus Waldst. & Kit.
   (a) odorus
   (b) laxus (Host) Merxm. & Podl.
  7 multifidus Vis.
   (a) multifidus
   (b) serbicus (Adamović) Merxm. & Podl.
   (c) istriacus (Schiffner) Merxm. & Podl.
  8 bocconei Ten.
   (a) bocconei
   (b) siculus (Schiffner) Merxm. & Podl.
  9 dumetorum Waldst. & Kit.
   (a) dumetorum
   (b) atrorubens (Waldst. & Kit.) Merxm. & Podl.
 10 purpurascens Waldst. & Kit.
 11 niger L.
   (a) niger
   (b) macranthus (Freyn) Schiffner

2 Eranthis Salisb.
  1 hyemalis (L.) Salisb.

3 Callianthemum C.A. Meyer
  1 anemonoides (J. Zahlbr.) Endl. ex
    Heynh   32   Au   Mattick, 1950
  2 kerneranum Freyn ex Kerner
  3 coriandrifolium Reichenb.   16   Po   Skalińska et al.,
                                                         1959

4 Nigella L.
  1 arvensis L.
    (a) arvensis
    (b) aristata (Sibth. & Sm.) Nyman
    (c) rechingeri (Tutin) Tutin
  2 hispanica L.
    (a) hispanica
    (b) atlantica Murb.
  3 gallica Jordan
  4 segetalis Bieb.
  5 degenii Vierh.
  6 cretica Miller
  7 fumariifolia Kotschy
  8 doerfleri Vierh.
  9 sativa L.   12   Bot. Gard. Pereira, 1942
 10 elata Boiss.
 11 damascena L.
 12 orientalis L.

5 Garidella I
  1 nigellastrum L.

6 Trollius L.
  1 europaeus L.
    (a) europaeus   16   Rs(N)   Sokolovskaya, 1958
    (b) transsilvanicus (Schur) Jáv.   16   Po   Skalińska et al.,
                                                         1959
  2 asiaticus L.

7 Isopyrum L.
  1 thalictroides L.   14   Po   Skalińska et al.,
                                                         1959
8 Actaea L.
  1 spicata L.   16   Po   Skalińska et al.,
                                                         1959
  2 erythrocarpa Fischer

9 Cimicifuga L.
  1 europaea Schipcz.   16   Hu   Baksay, 1957b

10 Caltha L.
  1 palustris L.   32   Po   Skalińska et al.,
                                                         1959
                                         53-62   Ge   Reese, 1954

11 Aconitum L.
  1 septentrionale Koelle   16   Su   Schafer & La
                                                          Cour, 1934
  2 lasiostomum Reichenb.
  3 moldavicum Hacq. ex Reichenb.
  4 vulparia Reichenb.   16   Alps   Schafer & La
                                                          Cour, 1934

                            43

   5 lamarckii Reichenb.
   6 anthora L.
   7 variegatum L.   16   Au   Schafer & La
                                                    Cour, 1934

   8 toxicum Reichenb.
   9 paniculatum Lam.   16   Au   Mattick, 1950
  10 firmum Reichenb.
  11 napellus L.   32   Br   Schafer & La
                                                    Cour, 1934

  12 tauricum Wulfen
  13 compactum Reichenb.
  14 nevadense Uechtr. ex Gáyer

12 Delphinium L.
   1 montanum DC.
   2 dubium (Rouy & Fouc.) Pawl.
   3 oxysepalum Borbas & Pax   32   Po   Skalińska et al.,
                                                1959

   4 elatum L.
     (a) elatum   32   Rs(N) Sokolovskaya &
                                                Strelkova, 1960

     (b) helveticum Pawl.
     (c) austriacum Pawl.
   5 simonkaianum Pawl.
   6 cuneatum Steven ex DC.
   7 rossicum Litv.
   8 dictyocarpum DC.
     (a) dictyocarpum
     (b) uralense (Nevski) Pawl.
   9 pentagynum Lam.
  10 nevadense G. Kunze
  11 emarginatum C. Presl
  12 fissum Waldst. & Kit.
  13 albiflorum DC.
  14 puniceum Pallas
  15 schmalhausenii Albov
  16 peregrinum L.
  17 hirschfeldianum Heldr. & Holzm.
  18 obcordatum DC.
  19 halteratum Sibth. & Sm.
  20 balcanicum Pawl.
  21 verdunense Balbis
  22 gracile DC
  23 hellenicum Pawl.
  24 staphisagria L.
  25 pictum Willd.
  26 requienii DC.

15 Consolida (DC.) S.F. Gray
   1 aconiti (L.) Lindley
   2 thirkeana (Boiss.) Bornm.
   3 hellespontica (Boiss.) Chater
   4 orientalis (Gay) Schrödinger
     (a) orientalis
     (b) phrygia (Boiss.) Chater
   5 ambigua (L.) P.W. Ball & Heywood
   6 brevicornis (Vis.) Soó
   7 uechtritziana (Pančić) Soó
   8 tuntasiana (Halácsy) Soó
   9 regalis S.F. Gray   16   Löve & Löve, 1942
     (a) regalis

    (b) paniculata (Host) Soó
10 tenuissima (Sibth. & Sm.) Soó
11 pubescens (DC.) Soó
12 rigida (DC.) Hayek

14 Anemone L.
   1 nemorosa L.   28-32   Su   Bernström, 1946
                                                 37   Su   Bernström, 1946
                                                 42   Su   Lövkvist, 1963
                                                 45   Su   Lövkvist, 1963
                                                 46   Su   Lövkvist, 1963
   2 altaica Fischer ex C.A. Meyer
   3 trifolia L.   32   Au   Mattick, 1950
     (a) trifolia
     (b) albida (Mariz) Tutin
   4 ranunculoides L.
     (a) ranunculoides   32   Su   Bernström, 1946
     (b) wockeana (Ascherson & Graebner) Hegi
   5 uralensis Fischer ex DC.
   6 apennina L.
   7 blanda Schott & Kotschy
   8 reflexa Stephan
   9 narcissiflora L.   14   He   Larsen, 1954a
  10 dichotoma L.
  11 sylvestris L.   16   Po   Gajewski, 1947
  12 baldensis L.
  13 pavoniana Boiss.
  14 palmata L.
  15 coronaria L.
  16 hortensis L.
  17 pavonina Lam.

15 Hepatica Miller
   1 nobilis Miller   14   Po   Skalińska et al.,
                                                                1959
   2 transsilvanica Fuss   28   He   Baumberger, 1970

16 Pulsatilla Miller
   1 alpina (L.) Delarbre   16   Au   Rosenthal, 1936
     (a) alpina
     (b) apiifolia (Scop.) Nyman
   2 alba Reichenb.
   3 vernalis (L.) Miller   16   Europe Bücher, 1932
   4 pratensis (L.) Miller   16   Ge   Rosenthal, 1936
                                                 32   Da   Bücher, 1954
   5 montana (Hoppe) Reichenb.   16   It   Rosenthal, 1936
   6 rubra (Lam.) Delarbre
   7 vulgaris Miller
     (a) vulgaris   32   Da   Bücher, 1954
     (b) grandis (Wenderoth) Zamels
   8 halleri (All.) Willd.   32   Rs(K)   Rosenthal, 1936
     (a) rhodopaea K. Krause
     (b) taurica (Juz.) K. Krause
     (c) slavica (G. Reuss) Zamels
     (d) halleri
     (e) styriaca (G.A. Pritzel) Zamels
   9 patens (L.) Miller
     (a) patens   16   Rm   Tarnavschi, 1948
     (b) teklae (Zamels) Zamels
     (c) multifida (Pritzel) Zamels
     (d) flavescens (Zucc.) Zamels

17 Clematis L.
  1 flammula L.
  2 vitalba L.
  3 orientalis L.
  4 recta L.
  5 pseudoflammula Schmalh. ex Lipsky
  6 cirrhosa L.
  7 viticella L.
  8 campaniflora Brot.
  9 integrifolia L.
 10 alpina (L.) Miller

(a) alpina	16	Po	Skalińska et al., 1961
(b) sibirica (L.) O. Kuntze			

18 Adonis L.

1 vernalis L.	16	Po	Skalińska et al., 1961

  2 volgensis Steven
  3 sibirica Patrin ex Ledeb.
  4 pyrenaica DC.
  5 distorta Ten.
  6 cyllenea Boiss., Heldr. & Orph.
  7 annua L.
    (a) annua
    (b) carinata Vierh.
  8 flammea Jacq.
  9 aestivalis L.
 10 microcarpa DC.

19 Ranunculus L.
  1 velutinus Ten.
  2 constantinopolitanus (DC.) D'Urv.
  3 polyanthemos L.

(a) polyanthemos	16	Hu	Baksay, 1956
(b) thomasii (Ten.) Tutin			
(c) polyanthemoides (Boreau) Ahlfvengren			
4 nemorosus DC.	16	He	Langlet, 1932
(a) nemorosus			
(b) polyanthemophyllus (W. Koch & H. Hess) Tutin			
(c) serpens (Schrank) Tutin			
5 caucasicus Bieb.			
6 repens L.	32	Fe	Vaarama, 1948
7 lanuginosus L.	28	Hu	Baksay, 1956
8 acris L.			
(a) acris	14	Fa	Bücher, 1938a
(b) borealis (Trautv.) Nyman	14	Rs(N)	Sokolovskaya, 1958
(c) friesianus (Jordan) Rouy & Fouc.			
(d) granatensis (Boiss.) Nyman			
(e) strigulosus (Schur) Hyl.			
9 serbicus Vis.			
10 brutius Ten.			
11 gouanii Willd.	16	Ga/Hs	Landolt, 1954
12 ruscinonensis Landolt	16	Hs	Landolt, 1956
13 sartorianus Boiss. & Heldr.			
14 carinthiacus Hoppe	16	Alps	Landolt, 1954
15 clethraphilus Litard.			
16 montanus Willd.	32	He	Landolt, 1954
17 pseudomontanus Schur			
18 croaticus Schott			

19 venetus Huter ex Landolt	32	It	Landolt, 1954
20 grenieranus Jordan	16	He	Landolt, 1954
21 oreophilus Bieb.	16	It	Landolt, 1954
22 aduncus Gren.	16	Alps	Landolt, 1954
23 carpaticus Herbich	28	Hu	Baksay, 1956
24 macrophyllus Desf.			
25 demissus DC.			
26 hayekii Dörfler			
27 marschlinsii Steudel			
28 dissectus Bieb.			
29 neapolitanus Ten.	16	Gr	Gregson, ined.
30 pratensis C. Presl			
31 bulbosus L.			
(a) bulbosus	16	Da	Bücher, 1938a
(b) castellanus (Boiss. & Reuter ex Freyn) P.W. Ball & Heywood			
(c) bulbifer (Jordan) Neves	16	Lu	Neves, 1950
(d) aleae (Willk.) Rouy & Fouc.			
(e) adscendens (Brot.) Neves	16	Lu	Neves, 1950
(f) gallecicus (Freyn ex Willk.) P.W. Ball & Heywood			
32 sardous Crantz	16	Ge	Reese, 1961
33 cordiger Viv.			
34 marginatus D'Urv.			
35 trilobus Desf.			
36 muricatus L.	48	Lu	Neves, 1950
	64	Lu	Neves, 1950
37 cornutus DC.			
38 arvensis L.	32	Lu	Neves, 1950
39 parviflorus L.			
40 chius DC.			
41 gracilis E.D. Clarke			
42 oxyspermus Willd.			
43 psilostachys Griseb.	32	?Br	Larter, 1932
44 rumelicus Griseb.			
45 monspeliacus L.			
46 miliarakesii Halácsy			
47 isthmicus Boiss.			
48 illyricus L.			
49 pedatus Waldst. & Kit.			
50 platyspermus Fischer ex DC.			
51 paludosus Poiret	32	Ga	Tutin, 1961
52 cupreus Boiss. & Heldr.			
53 millefoliatus Vahl			
54 garganicus Ten.			
55 millii Boiss. & Heldr.			
56 nigrescens Freyn			
57 cortusifolius Willd.			
58 rupestris Guss.			
59 creticus L.			
60 thasius Halácsy			
61 subhomophyllus (Halácsy) Vierh.			
62 spruneranus Boiss.			
63 gregarius Brot.			
64 asiaticus L.			
65 cassubicus L.	24	Po	Skalińska et al., 1961
	32	Po	Skalińska et al., 1961
	40	Fe	Rousi, 1956
	44	Po	Skalińska et al., 1961

	64	Po	Skalińska et al., 1961
66 fallax (Wimmer & Grab.) Kerner	32	Fe	Rousi, 1956
67 monophyllus Ovcz.	32	Fe	Rousi, 1956
68 auricomus L.	16	He	Häfliger, 1943
	32	Fe	Rousi, 1956
	40	Fe	Rousi, 1956
	48	Fe	Rousi, 1956
69 affinis R. Br.			
70 degenii Kümmerie & Jáv.			
71 flabellifolius Heuffel ex Reichenb.			
72 polyrhizos Stephan ex Willd.			
73 pedatifidus Sm.			
74 pygmaeus Wahlenb.	16	Rs(N)	Sokolovskaya, 1958
	32		
75 nivalis L.	40	Scand.	Nygren, 1948
	48	Scand.	Nygren, 1948
	56	Scand.	Nygren, 1948
76 sulphureus C.J. Phipps	96	Rs(N)	Sokolovskaya & Strelkova, 1960
77 hyperboreus Rottb.			
(a) hyperboreus	32	Rs(N)	Sokolovskaya, 1958
(b) samojedorum (Rupr.) Hultén			
78 sceleratus L.			
(a) sceleratus	32	Po	Skalińska et al., 1961
(b) reptabundus (Rupr.) Hultén			
79 gmelinii DC.			
80 lapponicus L	16	Su	Langlet, 1936
81 pallasii Schlecht.			
82 cymbalaria Pursh			
83 ficaria L.			
(a) calthifolius (Reichenb.) Arcangeli			
(b) bulbifer Lawalrée	32	Br	Heywood & Walker, 1961
(c) ficaria	16	Br	Heywood & Walker, 1961
(d) ficariiformis Rouy & Fouc.			
84 ficarioides Bory & Chaub.			
85 bullatus L.			
86 thora L.			
87 scutatus Waldst. & Kit.			
88 hybridus Biria			
89 brevifolius Ten.			
(a) brevifolius			
(b) pindicus (Hausskn.) E. Mayer			
90 alpestris L.	16	He	Larsen, 1954
91 traunfellneri Hoppe			
92 crenatus Waldst. & Kit.			
93 bilobus Bertol.			
94 aconitifolius L.	14	He	Larsen, 1954
	16	Au	Mattick, 1950
95 platanifolius L.	16	Po	Skalińska, 1950
96 seguieri Vill.			
(a) seguieri			
(b) montenegrinus (Halácsy) Tutin			
97 acetosellifolius Boiss.			
98 glacialis L.	16	Ga He	Larsen, 1954
99 cymbalarifolius Balbis ex Moris			
100 weyleri Marés			
101 flammula L.			

(a) flammula	32	Po	Skalińska et al., 1959
(b) minimus (Ar. Benn.) Padmore	32	Br	Padmore, 1957
(c) scoticus (E.S. Marshall) Clapham			
102 reptans L.	32	Rs(N)	Sokolovskaya, 1958
103 batrachioides Pomel			
104 lingua L.	128	Su	Lövkvist, 1963
105 ophioglossifolius Vill.	16	Lu	Neves, 1944
106 fontanus C. Presl			
107 longipes Lange ex Cutanda			
108 revelieri Boreau			
(a) revelieri			
(b) rodiei (Litard.) Tutin			
109 lateriflorus DC.			
110 nodiflorus L.			
111 polyphyllus Waldst. & Kit. ex Willd.			
112 pyrenaeus L.	32	He	Langlet, 1932
(a) pyrenaeus			
(b) alismoides (Bory) O. Bolós & Font Quer			
113 bupleuroides Brot.	16	Lu	Neves, 1950
114 gramineus L.			
115 abnormis Cutanda & Willk.			
116 parnassifolius L.			
117 wettsteinii Dörfler			
118 amplexicaulis L.			
119 hederaceus L.	16	Br Ge Lu	Cook, 1962
120 omiophyllus Ten.	16	Si	Cook, 1962
	32	Br	Cook, 1962
121 tripartitus DC.	48	Br	Cook, 1962
122 ololeucos Lloyd	16	He	Cook, 1962
123 baudotii Godron	32	Au Br	Cook, 1962
124 peltatus Schrank	32	Da Ge	Cook, 1962
125 pseudofluitans (Syme) Newbould ex Baker & Foggitt	24	Ge	Cook, 1962
	32	Ge	Cook, 1962
	40	Br	Cook, 1962
	48	Br Ge	Cook, 1962
126 sphaerospermus Boiss. & Blanche			
127 aquatilis L.	48	Br Ge	Cook, 1962
128 trichophyllus Chaix			
(a) trichophyllus	32	Br Ge	Cook, 1962
(b) lutulentus (Perr. & Song.) Vierh.	32	Ge	Cook, 1962
129 rionii Lagger	16	Au	Cook, 1962
130 circinatus Sibth.	16	Br Ge	Cook, 1962
131 fluitans Lam.	16	Ge	Cook, 1962
	24	Ge	Cook, 1962
	32	Br	Cook, 1962

20 Ceratocephalus
  1 falcatus (L.) Pers.
  2 testiculatus (Crantz) Roth

21 Myosurus L.
  1 minimus L.   16   Rm   Tarnavschi, 1948
                28   Su   Ehrenberg, 1945
  2 heldreichi Léveillé

22 Aquilegia L.
  1 litardierei Briq.
  2 paui Font Quer

```
 3 transsilvanica Schur
 4 vulgaris L.                              14    Po    Skalińska, 1950
 5 viscosa Gouan
 6 hirsutissima (Lapeyr.) Timb.-Lagr.
   ex Gariod
 7 amaliae Heldr. ex Boiss.
 8 ottonis Orph. ex Boiss.
 9 dinarica G. Beck
10 dichroa Freyn
11 pancicii Degen
12 atrata Koch                             14    Au    Mattick, 1950
13 nigricans Baumg.
   (a) nigricans
   (b) subscaposa (Borbás) Soó
14 aurea Janka
15 kitaibelii Schott
16 einseleana F.W. Schultz
17 thalictrifolia Schott & Kotschy
18 bertolonii Schott
19 pyrenaica DC.
20 aragonensis Willk.
21 discolor Levier & Leresche
22 guarensis Losa
23 cazorlensis Heywood
24 bernardii Gren. & Godron
25 nevadensis Boiss. & Reuter
26 grata F. Maly ex Zimmeter
27 alpina L.


23 Thalictrum L.
 1 aquilegifolium L,                       11    FU    Skalińska et al,
                                                       1959
 2 calabricum Sprengel                     42    It/Si Langlet, 1927
 3 uncinatum Rehman
 4 macrocarpum Gren.
 5 tuberosum L.
 6 orientale Boiss.
 7 alpinum L.                              14    Br    Hedberg &
                                                       Hedberg, 1961
 8 foetidum L.                            14    Au    Mattick, 1950
 9 minus L.
   (a) minus                               42    Hu    Baksay, 1956
   (b) majus (Crantz) Rouy & Fouc.
   (c) saxatile Schinz & Keller
   (d) pubescens (Schleicher ex DC.) Rouy
       & Fouc.
   (e) pseudominus (Borbás) Soó
   (f) kemense (Fries) Tutin
10 simplex L.
   (a) simplex                             56    Su    Lövkvist, 1963
   (b) boreale (F. Nyl.) Tutin
   (c) gallicum (Rouy & Fouc.) Tutin
   (d) bauhinii (Crantz) Tutin
   (e) galioides (Nestler) Borza
11 lucidum L.
12 morisonii C.C. Gmelin
   (a) morisonii
   (b) mediterraneum (Jordan) P.W. Ball
13 flavum L.
   (a) flavum                              84    Ge    Kuhn, 1928
   (b) glaucum (Desf.) Batt.
```

LXII PAEONIACEAE

1 Paeonia L.
  1 tenuifolia L.
  2 anomala L.
  3 peregrina Miller
  4 officinalis L.
    (a) officinalis
    (b) banatica (Rochel) Soó
    (c) humilis (Retz.) Cullen & Heywood
    (d) villosa (Huth) Cullen & Heywood
  5 clusii F.C. Stern   10   Cr   Stern, 1946
  6 broteroi Boiss. & Reuter   10   Lu   Stern, 1946
  7 mascula (L.) Miller
    (a) mascula
    (b) russii (Biv.) Cullen & Heywood
    (c) arietina (G. Anderson) Cullen &
        Heywood
  8 daurica Andrews
  9 coriacea Boiss.
 10 cambessedesii (Willk.) Willk.   10   Bl   Stern, 1946

LXIII BERBERIDACEAE

1 Leontice L.
  1 leontopetalum L.

2 Gymnospermium Spach
  1 altaicum (Pallas) Spach

3 Bongardia C.A. Meyer
  1 chrysogonum (L.) Griseb.

4 Epimedium L.
  1 alpinum L.
  2 pubigerum (DC.) Morren & Decne

5 Berberis L.
  1 vulgaris L.   28   Au   Mattick, 1950
  2 aetnensis C. Presl   28   Co   Contandriopoulos, 1962
  3 hispanica Boiss. & Reuter
  4 cretica L.

6 Mahonia Nutt.
  1 aquifolium (Pursh) Nutt.

LXIV MAGNOLIACEAE

1 Liriodendron L.
  1 tulipifera L.

LXV LAURACEAE

1 Laurus L.
  1 nobilis L.   42   It   Bambacioni-
                                                         Mezzetti, 1936

  2 azorica (Seub.) J. Franco

2 Persea Miller
  1 indica (L.) Sprengel

LXVI PAPAVERACEAE

1 Papaver L.
  1 somniferum L.
    (a) somniferum
    (b) setigerum (DC.) Corb.
    (c) songaricum Basil.
  2 rhoeas L.   14   Su   Lövkvist, 1963
  3 dubium L.   42   Su   Lövkvist, 1963
  4 lecoqii Lamotte   28   Br   McNaughton, 1960
  5 laevigatum Bieb.
  6 arenarium Bieb.
  7 pinnatifidum Moris
  8 stipitatum Fedde
  9 maeoticum Klokov
 10 argemone L.   42   Su   Lövkvist, 1963
 11 nigrotinctum Fedde
 12 apulum Ten.
 13 hybridum L.   14   Br   McNaughton, 1960
 14 macrostomum Boiss. & Huet
 15 rupifragum Boiss. & Reuter
 16 orientale L.
 17 suaveolens Lapeyr.
    (a) suaveolens
    (b) endressii Ascherson
 18 rhaeticum Leresche   14   Ga   Fabergé, 1943
 19 sendtneri Kerner ex Hayek   14   Au   Fabergé, 1943
 20 burseri Crantz
 21 kerneri Hayek   14   Au   Fabergé, 1943
 22 corona-sancti-stephani Zapal.
 23 lapponicum (Tolm.) Nordh.   56   Scand.   Knaben, 1958
 24 dahlianum Nordh.   70   Su   Knaben, 1958
 25 radicatum Rottb.   70   No   Knaben, 1958
 26 laestadianum (Nordh.) Nordh.   56   Scand.   Knaben, 1959

2 Meconopsis Vig.
  1 cambrica (L.) Vig.   22   Br   Maude, 1940

3 Argemone L.
  1 mexicana L.

4 Roemeria Medicus
  1 hybrida (L.) DC.

5 Glaucium Miller
  1 flavum Crantz   12   Ga It   Larsen, 1954
  2 leiocarpum Boiss.
  3 corniculatum (L.) J.H. Rudolph

6 Chelidonium L.
  1 majus L.   12   Su   Lövkvist, 1963

7 Eschscholzia Cham.
  1 californica Cham.

8 Hypecoum L.
  1 procumbens L.

        2 imberbe Sibth. & Sm.
        3 pendulum L.
        4 ponticum Velen.

    9 Dicentra Bernh.
        1 spectabilis (L.) Lemaire

   10 Corydalis Vent.
        1 claviculata (L.) DC.
         (a) claviculata   32   Ge   Reese, 1952
         (b) picta (Samp.) P. Silva & J. Franco
        2 capnoides (L.) Pers.
        3 lutea (L.) DC.
        4 acaulis (Wulfen) Pers.
        5 ochroleuca Koch
         (a) ochroleuca
         (b) leiosperma (Conr.) Hayek
        6 nobilis (L.) Pers.
        7 sempervirens (L.) Pers.
        8 rutifolia (Sibth. & Sm.) DC.
        9 bulbosa (L.) DC.
         (a) bulbosa
         (b) blanda (Schott) Chater
         (c) marschalliana (Pallas) Chater
       10 paczoskii N. Busch
       11 angustifolia (Bieb.) DC.
       12 intermedia (L.) Mérat   20   Au   Mattick, 1950
       13 pumila (Host) Reichenb.
       14 solida (L.) Swartz
         (a) solida
         (b) laxa (Fries) Nordst.
         (c) slivenensis (Velen.) Hayek
         (d) wettsteinii (Adamović) Hayek

   11 Sarcocapnos DC.
        1 enneaphylla (L.) DC.   c.32   Hs   Ryberg, 1960
        2 crassifolia (Desf.) DC.   c.32   Hs   Ryberg, 1960
        3 integrifolia (Boiss.) Cuatrec.
        4 baetica (Boiss. & Reuter) Nyman

   12 Fumaria L.
        1 agraria Lag.
        2 occidentalis Pugsley
        3 rupestris Boiss. & Reuter
        4 bella P.D. Sell
        5 gaillardotii Boiss.
        6 judaica Boiss.
        7 amarysia Boiss. & Heldr.
        8 macrocarpa Parl.
        9 capreolata L.
         (a) capreolata   56   Ga   Ryberg, 1960
         (b) babingtonii (Pugsley) P.D. Sell
       10 flabellata Gaspar.
       11 purpurea Pugsley
       12 macrosepala Boiss.
       13 bicolor Sommier ex Nicotra
       14 bastardii Boreau   c.48   Lu   Ryberg, 1960
       15 martinii Clavaud
       16 sepium Boiss. & Reuter
       17 muralis Sonder ex Koch
         (a) muralis

      (b) boraei (Jordan) Pugsley
      (c) neglecta Pugsley
18 reuteri Boiss.
19 petteri Reichenb.
      (a) petteri
      (b) thuretii (Boiss.) Pugsley
20 transiens P.D. Sell
21 kralikii Jordan
22 mirabilis Pugsley
23 densiflora DC.
24 rostellata Knaf
25 officinalis L.
      (a) officinalis   28   Ge   Wulff, 1937b
                                             32 Da Fe Ge Su Vaarama, 1949
      (b) wirtgenii (Koch) Arcangeli   48   Br   Daker, 1965
26 ragusina (Pugsley) Pugsley
27 jankae Hausskn.
28 caroliana Pugsley
29 schleicheri Soyer-Willimet
30 microcarpa (Hausskn.) Pugsley
31 vaillantii Loisel. in Desv.   32   Su   Ryberg, 1960
32 schrammii (Ascherson) Velen.
33 parviflora Lam.

13 Platycapnos (DC.) Bernh.
   1 spicata (L.) Bernh.
      (a) spicata
      (b) echeandiae (Pau) Heywood
   2 saxicola Willk.

14 Rupicapnos Pomel
   1 africana (Lam.) Pomel

15 Ceratocapnos Durieu
   1 heterocarpa Durieu

LXVII CAPPARIDACEAE

1 Capparis L.
   1 spinosa L.
   2 ovata Desf.

2 Cleome L.
   1 ornithopodioides L.
   2 violacea L.

LXVIII CRUCIFERAE

1 Sisymbrium L.
   1 supinum L.   42   Hs   Baez-Mayor, 1934
   2 strictissimum L.   28   Su   Löve & Löve, 1961d
   3 irio L.   42   Hs   Khoshoo, 1959
   4 loeselii L.
   5 assoanum Loscos & Pardo
   6 polymorphum (Murray) Roth
   7 volgense Bieb. ex E. Fourn.
   8 austriacum Jacq.
      (a) austriacum
      (b) chrysanthum (Jordan) Rouy & Fouc.

(c) contortum (Cav.) Rouy & Fouc.
(d) hispanicum (Jacq.) P.W. Ball &
    Heywood
9 crassifolium Cav.
10 laxiflorum Boiss.
11 arundanum Boiss.
12 altissimum L.   14   Su   Lövkvist, 1963
13 orientale L.   14   Hs   Baez-Mayor, 1934
14 polyceratium L.
15 confertum Steven ex Turcz.
16 runcinatum Lag. ex DC.   56   Bot. Gard. Manton, 1932
17 erysimoides Desf.
18 officinale (L.) Scop.   14   Su   Lövkvist, 1963
19 matritense P.W. Ball & Heywood   14   Hs   Baez-Mayor, 1934

2 Lycocarpus O.E. Schulz
  1 fugax (Lag.) O.E. Schulz

3 Murbeckiella Rothm.
  1 pinnatifida (Lam.) Rothm.   16   He   Favarger, 1949a
  2 boryi (Boiss.) Rothm.
  3 zanonii (Ball) Rothm.
  4 sousae Rothm.

4 Descurainia Webb & Berth.
  1 sophia (L.) Webb ex Prantl   28   Su   Lövkvist, 1963

5 Hugueninia Reichenb.
  1 tanacetifolia (L.) Reichenb.
    (a) tanacetifolia
    (b) suffruticosa (Coste & Soulié) P.W. Ball

6 Alliaria Scop.
  1 petiolata (Bieb.) Cavara & Grande   36   Su   Lövkvist, 1963
                                                    42   Hu   Pólya, 1949

7 Eutrema R.Br.
  1 edwardsii R. Br.   42   Rs(N)   Sokolovskaya &
                                                                     Strelkova, 1960

8 Arabidopsis (DC.) Heynh
  1 thaliana (L.) Heynh   10   Su   Böcher & Larsen,
                                                                     1958
  2 suecica (Fries) Norrlin   26   Su   Hylander, 1957
  3 parvula (Schrenk) O.E. Schulz
  4 toxophylla (Bieb.) N. Busch   24   Rs(E)   Titova, 1935
  5 pumila (Stephan) N. Busch   32   Rs(E)   Titova, 1935

9 Thellungiella O.E. Schulz
  1 salsuginea (Pallas) O.E. Schulz

10 Braya Sternb. & Hoppe
   1 alpina Sternb. & Hoppe
   2 linearis Rouy   42   No   Jørgensen et al.,
                                                                     1958
   3 purpurascens (R. Br.) Bunge   56   Is   Löve & Löve, 1956

11 Chrysochamela (Fenzl) Boiss.
   1 draboides Woronow

12 Sobolewskia Bieb.
  1 sibirica (Willd.) P.W. Ball

13 Myagrum L.
  1 perfoliatum L.

14 Isatis L.
  1 allionii P.W. Ball
  2 platyloba Link ex Steudel
  3 sabulosa Steven ex Ledeb.
  4 laevigata Trautv.
  5 costata C.A. Meyer
  6 praecox Kit. ex Tratt.
  7 tinctoria L.   28   Su   Lövkvist, 1963
  8 athoa Boiss.
  9 arenaria Aznav.
  10 lusitanica L.

15 Tauscheria Fischer ex DC.
  1 lasiocarpa Fischer ex DC.

16 Bunias L.
  1 erucago L.
  2 orientalis L.   14   Su   Lövkvist, 1963
  3 cochlearioides Murray

17 Goldbachia DC.
  1 laevigata DC.

18 Erysimum L.
  1 cylindricum (Boiss.) Boiss.
  2 armatum Puntic
  3 helveticum (Jacq.) DC.   56   He   Favarger, 1965c
  4 suffruticosum Sprengel
  5 grandiflorum Desf.   14   Ga   Favarger, 1965c
  6 raulinii Boiss.
  7 myriophyllum Lange
  8 linariifolium Tausch
  9 decumbens (Schleicher ex Willd.) Dennst.
  10 linifolium (Pers.) Gay
    (a) linifolium
    (b) baeticum Heywood
    (c) cazorlense Heywood
  11 pusillum Bory & Chaub.
    (a) pusillum
    (b) parnassii (Boiss. & Heldr.) Hayek
    (c) hayekii Jav. & Rech. fil.
  12 rechingeri Jav.
  13 mutabile Boiss. & Heldr.
  14 diffusum Ehrh.   c.32   Hu   Baksay, 1961
  15 olympicum Boiss.
  16 calycinum Griseb.
  17 graecum Boiss. & Heldr.
  18 smyrnaeum Boiss. & Balansa
  19 crepidifolium Reichenb.
  20 leptostylum DC.
  21 ucranicum Gay
  22 odoratum Ehrh.   24   Hu   Baksay, 1961
  23 witmannii Zawadzki
    (a) witmannii
    (b) transsilvanicum (Schur) P.W. Ball

```
      (c) pallidiflorum (Jáv.) Jáv.
   24 carniolicum Dolliner
   25 pectinatum Bory & Chaub.
   26 pulchellum (Willd.) Gay
   27 aureum Bieb.
   28 leucanthemum (Stephan) B. Fedtsch.
   29 hieracifolium L.                        32      Is      Löve & Löve, 1956
   30 durum J. & C. Presl
   31 virgatum Roth
   32 wahlenbergii (Ascherson & Engler)
      Borbás
   33 hungaricum Zapal.
   34 pieninicum (Zapal.) Pawl.
   35 cuspidatum (Bieb.) DC.
   36 repandum L.                             16      Hu      Baksay, 1961
   37 incanum G. Kunze
   38 cheiranthoides L.
      (a) cheiranthoides                      16      Su      Löve & Löve, 1956
      (b) altum Ahti                          16      Fe      Ahti, 1962

19 Syrenia Andrz.
    1 cana (Piller & Mitterp.) Neilr.
    2 talijevii Klokov
    3 siliculosa (Bieb.) Andrz. ex C.A. Meyer
    4 montana (Pallas) Klokov

20 Hesperis L.
    1 tristis L.
    2 laciniata All.
      (a) laciniata
      (b) secundiflora (Boiss. & Spruner)
          Breistr.
    3 matronalis L.                           24      Su      Löve & Löve, 1956
      (a) matronalis
      (b) cladotricha (Borbás) Hayek
      (c) candida (Kit.) Hegi & E. Schmid
      (d) voronovii (N. Busch) P.W. Ball
    4 elata Hornem.
    5 sylvestris Crantz
      (a) sylvestris
      (b) velenovskyi (Fritsch) Borza
    6 theophrasti Borbás
    7 oblongifolia Schur
    8 steveniana DC.
    9 macedonica Adamović
   10 pycnotricha Borbás & Degen
   11 dinarica G. Beck
   12 vrabelyiana (Schur) Borbás              24      Hu      Baksay, 1957b
   13 nivea Baumg.
   14 inodora L.

21 Malcolmia R. Br.
    1 littorea (L.) R. Br.                    20      Lu      Rodrigues, 1953
    2 lacera (L.) DC.
    3 ramosissima (Desf.) Thell.              14      Lu      Rodrigues, 1953
    4 africana (L.) R. Br.
    5 taraxacifolia Balbis
    6 maritima (L.) R. Br.
    7 flexuosa (Sibth. & Sm.) Sibth. & Sm.    16      Lu      Rodrigues, 1953
```

```
 8 graeca Boiss. & Spruner
 9 illyrica Hayek
10 orsiniana (Ten.) Ten.
11 angulifolia Boiss. & Orph.
12 macrocalyx (Halácsy) Rech. fil.
   (a) macrocalyx
   (b) scyria (Rech. fil.) P.W. Ball
13 chia (L.) DC.
14 hydraea (Halácsy) Heldr. & Halácsy
15 bicolor Boiss. & Heldr.

22 Torularia (Cosson) O.E. Schulz
   1 torulosa (Desf.) O.E. Schulz
   2 contortuplicata (Stephan ex Willd.)
       O.E. Schulz

23 Maresia Pomel
   1 nana (DC.) Batt.

24 Leptaleum DC.
   1 filifolium (Willd.) DC.

25 Sterigmostemum Bieb.
   1 tomentosum (Willd.) Bieb.

26 Cheiranthus L.
   1 cheiri L.                          14      It      Manton, 1932

27 Clausia Trotzky
   1 aprica (Stephan ex Willd.) Trotzky

28 Parrya R. Br
   1 nudicaulis (L.) Boiss.

29  Matthiola R. Br.
    1 incana (L.) R. Br.
    (a) incana                          14   Bot. Gard.  Manton, 1932
    (b) rupestris (Rafin.) Nyman
    2 sinuata (L.) R. Br.
    3 odoratissima (Bieb.) R. Br.
    4 tatarica (Pallas) DC.
    5 fragrans (Fischer) Bunge
    6 fruticulosa (L.) Maire
    (a) fruticulosa
    (b) valesiaca (Gay ex Gaudin) P.W. Ball
    (c) perennis (P. Conti) P.W. Ball
    7 longipetala (Vent.) DC.
    (a) longipetala
    (b) bicornis (Sibth. & Sm.) P.W. Ball
    (c) pumilio (Sibth. & Sm.) P.W. Ball
    8 lunata DC.
    9 parviflora (Schousboe) R. Br.
   10 tricuspidata (L.) R. Br.

30 Notoceras R. Br.
   1 bicorne (Aiton) Amo

31 Tetracme Bunge
   1 quadricornis (Willd.) Bunge

32 Diptychocarpus Trautv.
```

1 strictus (Fischer) Trautv.

33 Chorispora R. Br. ex DC.
1 tenella (Pallas) DC.

34 Euclidium R. Br.
1 syriacum (L.) R. Br.

35 Litwinowia Woronow
1 tenuissima (Pallas) Woronow ex Pavlov

36 Barbarea R. Br.
1 vulgaris R. Br.   16   Au   Mattick, 1950
2 stricta Andrz.
3 verna (Miller) Ascherson
4 intermedia Boreau
5 sicula C. Presl
6 bosniaca Murb.
7 bracteosa Guss.
8 rupicola Moris   16   Co   Contandriopoulos,
9 conferta Boiss. & Heldr.   1957
10 balcana Pančić
11 longirostris Velen.

37 Sisymbrella Spach
1 dentata (L.) O.E. Schulz
2 aspera (L.) Spach
  (a) aspera
  (b) praeterita Heywood
  (c) boissieri (Cosson) Heywood
  (d) pseudoboissieri (Degen) Heywood

38 Rorippa Scop.
1 austriaca (Crantz) Besser   16   Br   Howard, 1947
2 amphibia (L.) Besser   16   Br   Howard, 1947
                                          32   Ge   Wulff, 1939a
3 prostrata (J.P. Bergeret) Schinz & Thell.
4 sylvestris (L.) Besser
  (a) sylvestris   48   Br   Howard, 1946
  (b) kerneri (Menyh.) Soó
5 islandica (Oeder) Borbás   16   Is   Löve & Löve, 1956
                                          32   Br   Howard, 1947
6 prolifera (Heuffel) Neilr.
7 brachycarpa (C.A. Meyer) Hayek
8 pyrenaica (Lam.) Reichenb.   16   Ga   Manton, 1932
9 lippizensis (Wulfen) Reichenb.   32   Hu   Pólya, 1948
10 thracica (Griseb.) Fritsch

39 Armoracia Gilib.
1 rusticana P. Gaertner, B. Meyer &
  Scherb.   32   Su   Lövkvist, 1963
2 macrocarpa (Waldst. & Kit.) Kit. ex
  Baumg.

40 Nasturtium R. Br.
1 officinale R. Br.   32   Br   Howard, 1947
2 microphyllum (Boenn.) Reichenb..   64   Br   Howard, 1947
                                          48

41 Cardamine L.
1 bulbifera (L.) Crantz   96   Su   Lövkvist, 1963

59

Species			
2 heptaphylla (Vill.) O.E. Schulz	48	Ge	Schwarzenbach, 1922
3 pentaphyllos (L.) Crantz	48	Ge	Schwarzenbach, 1922
4 quinquefolia (Bieb.) Schmalh.			
5 kitaibelii Becherer	48	He	Schwarzenbach, 1922
6 enneaphyllos (L.) Crantz	52-54	Au	Mattick, 1950
	80	Po	Banach-Pogan, 1954
7 glanduligera O. Schwarz	48	Po	Banach-Pogan, 1954
8 waldsteinii Dyer			
9 macrophylla Willd.			
10 trifida (Lam. ex Poiret) B.M.G. Jones			
11 trifolia L.	16	Po	Banach-Pogan, 1955
12 asarifolia L.			
13 amara L.	16	Au	Mattick, 1950
	32	Au	Mattick, 1950
14 barbaraeoides Halácsy			
15 opizii J. & C. Presl	16	Po	Banach-Pogan, 1954
16 raphanifolia Pourret			
(a) raphanifolia	44		Lövkvist, 1956
	46	?Br	Ellis & Jones, 1970
(b) acris (Griseb.) O.E. Schulz			
17 crassifolia Pourret			
18 nymanii Gand.	60	Su	Lövkvist, 1957
	62		
	64	Su	Lövkvist, 1957
	68	Su	Lövkvist, 1956
	72 74	Su	Lövkvist, 1956
	80	No	Lövkvist, 1956
	90	Su	Lövkvist, 1957
19 matthiolii Moretti	16	Ga	Lövkvist, 1956
20 pratensis L.	28-34	Su	Lövkvist, 1956
	38-44	Scand.	Lövkvist, 1956
	48	Da	Lövkvist, 1956
21 rivularis Schur	16	Alps	Lövkvist, 1956
22 palustris (Wimmer & Grab.) Peterm.	56	Scand.	Lövkvist, 1957
	64	Scand.	Lövkvist, 1957
	72	Scand.	Lövkvist, 1957
	76	Scand.	Lövkvist, 1957
	80	Scand.	Lövkvist, 1957
	84	Da	Lövkvist, 1956
	c.96	Scand.	Lövkvist, 1957
23 tenera J.G. Gmelin ex C.A. Meyer			
24 plumieri Vill.			
25 glauca Sprengel			
26 resedifolia L.	16	Au	Mattick, 1950
27 bellidifolia L.			
(a) bellidifolia	16	Rs(N)	Sokolovskaya & Strelkova, 1960
(b) alpina (Willd.) B.M.G. Jones	16	Au	Mattick, 1950
28 maritima Portenschl. ex DC.			
29 carnosa Waldst. & Kit.			
30 graeca L.	16	?Ju	Ball, ined.
31 chelidonia L.			
32 parviflora L.			
33 impatiens L.	16	Po	Banach-Pogan, 1955
34 caldeirarum Guthnick			
35 flexuosa With.	32	Ho	Gadella & Kliphuis 1955
	c.50		
	16	Su	Lövkvist, 1963

42 Cardaminopsis (C.A. Meyer) Hayek

```
 1 arenosa (L.) Hayek                            16      Cz        Jones, 1963
                                                 28      Au        Mattick, 1950
                                                 32      Su        Hylander, 1957
 2 petraea (L.) Hiitonen                         16      Br        Jones, 1963
 3 neglecta (Schultes) Hayek
 4 croatica (Schott, Nyman & Kotschy) Jáv.
 5 halleri (L.) Hayek                            16      Po        Jones, 1963
   (a) halleri
   (b) ovirensis (Wulfen) Hegi & E. Schmid

43 Arabis L.
 1 glabra (L.) Bernh.                            12      Su        Lövkvist, 1963
                                                 16      Au        Mattick, 1950
                                                 32      Ge        Jaretsky, 1928a
 2 pseudoturritis Boiss. & Heldr.
 3 laxa Sibth. & Sm.
 4 pauciflora (Grimm) Garcke
 5 planisiliqua (Pers.) Reichenb.               16      Cz        Novotná, 1962
 6 sagittata (Bertol.) DC.                       16      Au        Titz, 1964
 7 borealis Andrz. ex Ledeb.
 8 hirsuta (L.) Scop.                            32      Br No Su  Jones, 1963
   brownii Jordan                                32      Hb        Jones, 1963
 9 allionii DC.
10 corymbiflora Vest.                            16      Au        Mattick, 1950
11 serpillifolia Vill.
   (a) serpillifolia
   (b) nivalis (Guss.) B.M.G. Jones
   (c) cretica (Boiss. & Heldr.) B.M.G.
       Jones
12 muralis Bertol.
13 collina Ten.
14 turrita L.                                    16      Hu        Baksay, 1957b
15 mollis Steven
16 pendula L.
17 recta Vill.
18 nova Vill.
19 parvula Dufour in DC.
20 verna (L.) R. Br.
21 procurrens Waldst. & Kit.
22 vochinensis Sprengel
23 ferdinandi-coburgii J. Kellerer & Sund.
24 scopoliana Boiss.
25 bryoides Boiss.
26 longistyla Rech. fil.
27 stricta Hudson                               16      Br        Jones, 1963
28 subflava B.M.G. Jones
29 caerulea (All.) Haenke                       16      He        Favarger, 1953
30 pumila Jacq.
31 soyeri Reuter & Huet
   (a) soyeri
   (b) jacquinii (G. Beck) B.M.G. Jones         16      Au        Mattick, 1950
32 alpina L.                                     16      Au Br Ga  Jones, 1963
                                                 32      Au        Mattick, 1950
33 caucasica Schlecht.                           16      Gr Ju     Bordet, 1967
34 cebennensis DC.
35 pedemontana Boiss.

44 Aubrieta Adanson
 1 deltoidea (L.) DC.
 2 columnae Guss.
   (a) columnae
```

(b) italica (Boiss.) Mattf.
(c) croatica (Schott, Nyman & Kotschy) Mattf.
3 scyria Halácsy
4 erubescens Griseb.
5 intermedia Heldr. & Orph. ex Boiss.
6 gracilis Spruner ex Boiss.

45 Ricotia L.
1 cretica Boiss. & Heldr.
2 isatoides (W. Barbey) B.L. Burtt

46 Lunaria L.
1 rediviva L.   30   Au   Mattick, 1950
2 telekiana Jav.
3 annua L.
  (a) annua
  (b) pachyrhiza (Borbás) Hayek

47 Peltaria Jacq.
1 alliacea Jacq.
2 emarginata (Boiss.) Hausskn.

48 Alyssoides Miller
1 utriculata (L.) Medicus
2 cretica (L.) Medicus
3 sinuata (L.) Medicus

49 Degenia Hayek
1 velebetica (Degen) Hayek   16   Ju   Lovka et al.,
                                                               1971

50 Alyssum L.
1 corymbosum (Griseb.) Boiss.
2 leucadeum Guss.
3 petraeum Ard.
4 saxatile L.
  (a) saxatile   48
  (b) orientale (Ard.) Rech. fil.
  (c) megalocarpum (Hausskn.) Rech. fil.
5 linifolium Stephan ex Willd.
6 dasycarpum Stephan ex Willd.
7 alyssoides (L.) L.   32   Da   Böcher & Larsen,
                                                               1958
8 granatense Boiss. & Reuter
9 desertorum Stapf
10 foliosum Bory & Chaub.
11 smyrnaeum C.A. Meyer
12 minutum Schlecht. ex DC.
13 umbellatum Desv.   16   Gr   P.W. Ball, ined.
14 minus (L.) Rothm.   16   Ju   Böcher & Larsen,
                                                               1958
15 strigosum Banks & Solander
16 hirsutum Bieb.
17 rostratum Steven
18 repens Baumg.
  (a) repens   16   Bot. Gard.   Manton, 1932
  (b) trichostachyum (Rupr.) Hayek
19 lenense Adams
20 fischeranum DC.
21 scardicum Wettst.
22 wulfenianum Bernh.
23 ovirense Kerner

24 wierzbickii Heuffel
25 calycocarpum Rupr.
26 pulvinare Velen.
27 montanum L.
   (a) montanum   16   Hu   Bücher &
                                                                 Larsen, 1958
   (b) gmelinii (Jordan) Hegi & E. Schmid
28 fastigiatum Heywood
29 diffusum Ten.
30 cuneifolium Ten.
31 arenarium Loisel.
32 atlanticum Desf.
33 moellendorfianum Ascherson ex G. Beck
34 stribrnyi Velen.
35 sphacioticum Boiss. & Heldr.
36 lassiticum Halácsy
37 idaeum Boiss. & Heldr.
38 handelii Hayek
39 doerfleri Degen
40 taygeteum Heldr.
41 densistellatum T.R. Dudley
42 murale Waldst. & Kit.
43 tenium Halácsy
44 argenteum All.
45 fallacinum Hausskn.
46 corsicum Duby   16   Co   Contandriopoulos, 1957
47 heldreichii Hausskn.
48 robertianum Bernard ex Gren. & Godron 16   Co   Contandriopoulos, 1957
49 smolikanum E.I. Nyárády
50 bertolonii Desv.
   (a) bertolonii
   (b) scutarinum E.I. Nyárády
51 markgrafii O.E. Schulz
52 corymbosoides Form.
53 euboeum Halácsy
54 sibiricum Willd.
55 caliacrae E.I. Nyárády
56 borzaeanum E.I. Nyárády
57 obtusifolium Steven ex DC.
   (a) obtusifolium
   (b) helioscopioides E.I. Nyárády
58 obovatum (C.A. Meyer) Turcz.
59 tortuosum Willd.
60 longistylum (Sommier & Levier) Grossh.
61 alpestre L.
62 serpyllifolium Desf.
63 nebrodense Tineo
64 fragillimum (Bald.) Rech. fil.

51 Fibigia Medicus
   1 clypeata (L.) Medicus
   2 lunarioides (Willd.) Sibth. & Sm.
   3 triquetra (DC.) Boiss. ex Prantl

52 Berteroa DC.
   1 obliqua (Sibth. & Sm.) DC.
   2 mutabilis (Vent.) DC.
   3 orbiculata DC.
   4 incana (L.) DC.   16   Ge   Wulff, 1939b
   5 gintlii Rohlena

53 Lepidotrichum Velen. & Bornm.
  1 uechtritzianum (Bornm.) Velen.

54 Ptilotrichum C.A. Meyer
  1 pyrenaicum (Lapeyr.) Boiss.
  2 reverchonii Degen & Hervier
  3 lapeyrousianum (Jordan) Jordan
  4 halimifolium Boiss.
  5 cyclocarpum Boiss.
  6 longicaule (Boiss.) Boiss.
  7 macrocarpum (DC.) Boiss.
  8 spinosum (L.) Boiss.
  9 purpureum (Lag. & Rodr.) Boiss.

55 Bornmuellera Hausskn.
  1 dieckii Degen
  2 baldaccii (Degen) Heywood
  3 tymphaea (Hausskn.) Hausskn.

56 Lobularia Desv.
  1 maritima (L.) Desv.   24   It   Larsen, 1955b
  2 libyca (Viv.) Webb & Berth.

57 Clypeola L.
  1 jonthlaspi L.
  2 eriocarpa Cav.

58 Schivereckia Andrz.
  1 podolica (Besser) Andrz.
  2 doerfleri (Wettst.) Bornm.

59 Draba L.
  1 aizoides L.   16   Ga He   Duckert &
                                        Favarger, 1960

  2 hoppeana Reichenb.
  3 aspera Bertol.
  4 scardica (Griseb.) Degen & Dörfler
  5 cuspidata Bieb.
  6 athoa (Griseb.) Boiss.   16   Gr   Duckert &
                                        Favarger, 1960
  7 lasiocarpa Rochel   16   Hu   Baksay, 1951

  8 compacta Schott, Nyman & Kotschy
  9 lacaitae Boiss.
 10 haynaldii Stur
 11 hispanica Boiss.
 12 parnassica Boiss. & Heldr.
 13 loiseleurii Boiss.   16   Co   Contandriopoulos, 1957
 14 cretica Boiss. & Heldr.
 15 dedeana Boiss. & Reuter
 16 sauteri Hoppe
 17 heterocoma Fenzl
 18 alpina L.   62-64   No   Heilborn, 1927
                      80   Sb Su   Flovik, 1940
 19 gredinii Elis. Ekman
 20 bellii Holm
 21 kjellmanii Lid ex Elis. Ekman
 22 oblongata R. Br. ex DC.
 23 glacialis Adams
 24 ladina Br.-Bl.
 25 sibirica (Pallas) Thell.

```
26 stellata Jacq.
27 nivalis Liljeblad                              16      No        Heilborn, 1927
28 norvegica Gunnerus                             48      No Su     Heilborn, 1927
29 subcapitata Simmons
30 cacuminum Elis. Ekman                          64      No        Knaben &
                                                                    Engelskjøn, 1967

31 carinthiaca Hoppe
32 dubia Suter
33 kotschyi Stur
34 tomentosa Clairv.
35 fladnizensis Wulfen                            16      No        Heilborn, 1927
36 daurica DC.                                    64      Su        Heilborn, 1927
   norica                                         64      Au        Heilborn, 1941
37 cinerea Adams                                  48      Rs(N)     Heilborn, 1927
38 incana L.                                      32      No Su     Heilborn, 1927
39 muralis L.                                     32      Ge        Reese, 1952
40 nemorosa L.
41 lutescens Cosson
42 crassifolia R.C. Graham


60 Erophila DC.
   1 verna (L.) Chevall.                        14-64     Da        Winge, 1940
     (a) verna
         simplex                                  14      Da        Winge, 1940
     (b) macrocarpa (Boiss. & Heldr.) Walters
     (c) praecox (Steven) Walters
     (d) spathulata (A.F. Lang) Walters
         duplex                                  30-40    Da        Winge, 1940
         quadruplex                             52-64     Da        Winge, 1940
   2 minima C.A. Meyer

61 Petrocallis R. Br.
   1 pyrenaica (L.) R. Br.                         14      He        Favarger, 1953

62 Andrzeiowskia Reichenb.
   1 cardamine Reichenb.

63 Cochlearia L.
   1 danica L.                                     42      Su        Løvkvist, 1963
   2 officinalis L.                                24      Su        Saunte, 1955
   3 pyrenaica DC.                                 12      Ga        Rohner, 1954
                                                   28
     alpina                                        12      Br        Gill, 1965
                                                   28      ?Br       Crane &
                                                                     Gairdner, 1923
   4 polonica Frohlich                             36      Po        Bajer, 1950
   5 aestuaria (Lloyd) Heywood                     12      Hs        Gill, 1965
   6 tatrae Borbis                                 42      Po        Bajer, 1950
   7 anglica L.                                    48      Da        Saunte, 1955
                                                   54
   8 aragonensis Coste & Soulie
   9 fenestrata R. Br.
  10 groenlandica L.                               14      Is        Løve & Løve, 1956
  11 scotica Druce                                 14      Br        Maude, 1940
  12 glastifolia L.

64 Kernera Medicus
   1 saxatilis (L.) Reichenb.                      16      Au        Mattick, 1950
                                                   32      Au        Mattick, 1950
```

65

65 Rhizobotrya Tausch
   1 alpina Tausch   14   It   Chiarugi, 1933

66 Camelina Crantz
   1 sativa (L.) Crantz   40   Is   Löve & Löve, 1956
   2 microcarpa Andrz. ex DC.   40   Is   Löve & Löve, 1956
   3 rumelica Velen.   12   Hu   Baksay, 1957b
   4 alyssum (Miller) Thell.   40   Hu   Baksay, 1961
   5 macrocarpa Wierzb. ex Reichenb.

67 Neslia Desv.
   1 paniculata (L.) Desv.
     (a) paniculata   14   Su   Lövkvist, 1963
     (b) thracica (Velen.) Bornm.

68 Capsella Medicus
   1 bursa-pastoris (L.) Medicus   32   Is   Löve & Löve, 1956
   2 rubella Reuter
   3 thracica Velen.
   4 grandiflora (Fauche & Chaub.) Boiss.
   5 orientalis Klokov

69 Hutchinsia R. Br.
   1 alpina (L.) R. Br.
     (a) alpina   12   He   Larsen, 1954a
     (b) brevicaulis (Hoppe) Arcangeli   12   Au   Mattick, 1950
     (c) auerswaldii (Willk.) Laínz

70 Hymenolobus Nutt. ex Torrey & A. Gray
   1 procumbens (L.) Nutt. ex Torrey &
     Gray   12   Si   Manton, 1932
                                               24   Si   Manton, 1932

   ? pauciflorus (Hook.) Schinz & Thell.

71 Hornungia Reichenb.
   1 petraea (L.) Reichenb.   12   Su   Lövkvist, 1963
   2 aragonensis (Loscos & Pardo) Heywood

72 Ionopsidium Reichenb.
   1 acaule (Desf.) Reichenb.   24   Bot. Gard. Chiarugi, 1928
   2 albiflorum Durieu
   3 prolongoi (Boiss.) Batt.
   4 abulense (Pau) Rothm.
   5 savianum (Caruel) Ball ex Arcangeli   32   It   Chiarugi, 1928

73 Teesdalia R. Br.
   1 nudicaulis (L.) R. Br.   36   Su   Lövkvist, 1963
   2 coronopifolia (J.P. Bergeret) Thell.

74 Thlaspi L.
   1 arvense L.   14   Su   Lövkvist, 1963
   2 alliaceum L.
   3 perfoliatum L.   70   Su   Lövkvist, 1963
   4 brachypetalum Jordan
   5 alpestre L.
     (a) alpestre   14   Su   Lövkvist, 1963
     (b) virens (Jordan) Hooker fil.
   6 stenopterum Boiss. & Reuter
   7 dacicum Heuffel
     (a) dacicum
     (b) banaticum (Uechtr.) Jáv.

```
  8 brevistylum (DC.) Jordan
  9 rivale J. & C. Presl
 10 graecum Jordan
 11 microphyllum Boiss. & Orph.
 12 stylosum (Ten.) Mutel
 13 bulbosum Spruner ex Boiss.
 14 praecox Wulfen
    (a) praecox
    (b) cuneifolium (Griseb.) Clapham
 15 jankae Kerner
 16 goesingense Halácsy
 17 epirotum Halácsy
 18 ochroleucum Boiss. & Heldr.
 19 montanum L.                          28      Hu       Baksay, 1961
 20 alpinum Crantz                      c.54     Au       Mattick, 1950
    (a) alpinum
    (b) sylvium (Gaudin) Clapham
 21 kerneri Huter
 22 avalanum Pančić                      14      Balkans  Manton, 1932
 23 macranthum N. Busch
 24 rotundiflolium (L.) Gaudin
    (a) rotundifolium                    14      Au       Mattick, 1950
    (b) cepaeifolium (Wulfen) Rouy & Fouc.
 25 bellidifolium Griseb.
 26 nevadense Boiss. & Reuter

75 Aethionema R. Br.
  1 orbiculatum (Boiss.) Hayek
  2 polygaloides DC.
  3 cordatum (Desf.) Boiss.
  4 iberideum (Boiss.) Boiss.
  5 arabicum (L.) Andrz. ex O.E. Schulz
  6 saxatile (L.) R. Br.                 48      It       Larsen, 1955b

76 Teesdaliopsis (Willk.) Gand.
  1 conferta (Lag.) Rothm.

77 Iberis L.
  1 semperflorens L.                     22      Si       Manton, 1932
                                         44      Si       Manton, 1932
  2 sempervirens L.
  3 saxatilis L.
   (a) saxatilis
   (b) cinerea (Poiret) P.W. Ball &
       Heywood
  4 linifolia Loefl.
   (a) linifolia
   (b) welwitschii (Boiss.) Franco &
       P. Silva
  5 pruitii Tineo                        22  Bot. Gard.   Manton, 1932
  6 spathulata J.P. Bergeret
   (a) spathulata
   (b) nana (All.) Heywood
  7 aurosica Chaix
   (a) aurosica
   (b) cantabrica Franco & P. Silva
  8 procumbens Lange
   (a) procumbens
   (b) microcarpa Franco & P. Silva
  9 gibraltarica L.
 10 amara L.
```

    (a) amara
    (b) forestieri (Jordan) Heywood
11 intermedia Guersent
    (a) intermedia
    (b) timeroyi (Jordan) Rouy & Fouc.
    (c) prostii (Soyer-Willemet ex Godron)
        Rouy & Fouc.
12 stricta Jordan
    (a) stricta
    (b) leptophylla (Jordan) Franco &
        P. Silva
13 umbellata L.
14 ciliata All.
15 fontqueri Pau
16 pinnata L.
17 crenata Lam..
18 odorata L.
19 sampaiana Franco & P. Silva

78 Biscutella L.
  1 laevigata L.
    (a) laevigata   36   Au   Manton, 1937
    (b) austriaca (Jordan) Mach.-Laur.   18   Au   Manton, 1937
    (c) kerneri Mach.-Laur.   18   Ce   Bresinskya &
    (d) gracilis Mach.-Laur.   Grau, 1970
    (e) tenuifolia (Bluff & Fingerh.) Mach.-
        Laur.   18   Ce   Manton, 1937
    (f) guestphalica Mach.-Laur.   18   Ce   Manton, 1937
    (g) varia (Dumort.) Rouy & Fouc.   18   ?error
    (h) subaphylla Mach.-Laur.
    (i) tirolensis (Mach.-Laur.) Heywood
    (j) angustifolia (Mach.-Laur.) Heywood
    (k) illyrica Mach.-Laur.
    (l) montenegrina Rohlena
    (m) lucida (DC.) Mach.-Laur.   18   It   Schönfelder, 1968
  2 scaposa Sennen ex Mach.-Laur.
  3 flexuosa Jordan
  4 variegata Boiss. & Reuter
  5 megacarpaea Boiss. & Reuter
  6 foliosa Mach.-Laur.
  7 gredensis Guinea
  8 neustriaca Bonnet
  9 frutescens Cosson
 10 vincentina (Samp.) Rothm. ex Guinea
 11 sempervirens L.   18   Hs   Schönfelder, 1968
 12 glacialis (Boiss. & Reuter) Jordan   18   Cult.   Manton, 1937
 13 brevifolia Rouy & Fouc.
 14 cuneata (Font Quer) Font Quer ex
        Mach.-Laur.
 15 rotgesii Fouc.   18   Co   Contandriopoulos, 1962
 16 lamottii Jordan   18   Ga   Manton, 1937
 17 sclerocarpa Revel
 18 arvernensis Jordan   18   Ga   Manton, 1937
 19 divionensis Jordan   18   Ga   Manton, 1937
 20 controversa Boreau   18   Ga   Manton, 1937
 21 brevicaulis Jordan
 22 coronopifolia L.   18   Ga   Manton, 1937
 23 intricata Jordan
 24 apricorum Jordan
 25 granitica Boreau ex Perard
 26 pinnatifida Jordan

```
27 polyclada Jordan
28 lusitanica Jordan              54    Hs      Manton, 1937
29 mediterranea Jordan           18    Ga      Manton, 1937
30 nicaeensis Jordan
31 intermedia Gouan
32 guillonii Jordan
33 valentina (L.) Heywood
34 microcarpa DC.
35 baetica Boiss. & Reuter
36 eriocarpa DC.
37 lyrata L.                      16    Si      Manton, 1932
38 didyma L.                      16    N. Afr. Reese, 1957
39 radicata Cosson
40 auriculata L.                  16    Hs      Schönfelder, 1968
41 cichoriifolia Loisel.
```

79 Megacarpaea DC.
  1 megalocarpa (Fischer ex DC.) Schischkin
     ex B. Fedstch.

```
80 Lepidium L.
   1 campestre (L.) R. Br.        16    Is      Löve & Löve, 1956
   2 villarsii Gren. & Godron
    (a) villarsii
    (b) reverchonii (Debeaux) Breistr.
   3 heterophyllum Bentham        16    Br      Manton, 1932
   4 hirtum (L.) Sm.
    (a) hirtum
    (b) nebrodense (Rafin.) Thell.
    (c) petrophilum (Cosson) Thell.
    (d) stylatum (Lag. & Rodr.) Thell.
    (e) oxyotum (DC.) Thell.
    (f) calycotrichum (Kunze) Thell.
   5 spinosum Ard.
   6 sativum L.
   7 subulatum L.
   8 cardamines L.
   9 virginicum L.
  10 bonariense L.
  11 densiflorum Schrader
  12 neglectum Thell.
  13 ruderale L.                   32    Su      Lövkvist, 1963
  14 divaricatum Solander
  15 pinnatifidum Ledeb.
  16 schinzii Thell.
  17 perfoliatum L.                16    Is      Löve & Löve, 1956
  18 cartilagineum (J. Mayer) Thell.
    (a) cartilagineum
    (b) crassifolium (Waldst. & Kit.) Thell.16  Rm  Tarnavschi, 1938
  19 latifolium L.                 24    Is      Löve & Löve, 1956
  20 lyratum L.
    (a) lacerum (C.A.Meyer) Thell.
    (b) coronopifolium (Fischer) Thell.
  21 graminifolium L.
    (a) graminifolium
    (b) suffruticosum (L.) P. Monts.
```

```
81 Cardaria Desv.
   1 draba (L.) Desv.             64    Su      Lövkvist, 1963
```

82 Coronopus Haller

```
     1 navasii Pau
     2 squamatus (Forskål) Ascherson
     3 didymus (L.) Sm.                          32      Br      Manton, 1932

83 Subularia L.
     1 aquatica L.                            c.36      Is      Löve & Löve , 1956

84 Conringia Adanson
     1 orientalis (L.) Dumort.                  14      Is      Löve & Löve, 1956
     2 austriaca (Jacq.) Sweet

85 Moricandia DC.
     1 arvensis (L.) DC.
     2 moricandioides (Boiss.) Heywood
     3 foetida Bourgeau ex Cosson

86 Euzomodendron Cosson
     1 bourgaeanum Cosson

87 Diplotaxis DC.
     1 catholica (L.) DC.
     2 siifolia G. Kunze
     3 crassifolia (Rafin.) DC.
     4 tenuifolia (L.) DC.                      22      Su      Lövkvist, 1963
     5 cretacea Kotov
     6 erucoides (L.) DC.                       14      Si      Manton, 1932
     7 virgata (Cav.) DC.
     8 viminea (L.) DC.
     9 muralis (L.) DC.                         44      Su      Lövkvist, 1963

88 Brassica L.
     1 elongata Ehrh.
      (a) elongata
      (b) integrifolia (Boiss.) Breistr.
     2 balearica Pers.
     3 macrocarpa Guss.
     4 oleracea L.
      (a) oleracea                             18      Ge      Netroufal, 1927
      (b) robertiana (Gay) Rouy & Fouc.        18      It      Netroufal, 1927
     5 rupestris Rafin.                         18      Si      Manton, 1932
     6 villosa Biv.
     7 incana Ten.
     8 insularis Moris                          18      Co      Contandriopoulos,
                                                                 1962
     9 cretica Lam.
    10 napus L.                                 38      Po      Bijok, 1959
       subsp. rapifera                          38
    11 rapa L.
      (a) sylvestris (L.) Janchen              20      Su      Lövkvist, 1963
      (b) rapa
    12 juncea (L.) Czern.
    13 fruticulosa Cyr.
      (a) fruticulosa
      (b) cossoniana (Boiss. & Reuter) Maire
    14 barrelieri (L.) Janka
      (a) barrelieri
      (b) oxyrrhina (Cosson) P.W. Ball &
          Heywood
    15 tournefortii Gouan
    16 cadmea Heldr. ex O.E. Schulz
    17 procumbens (Poiret) O.E. Schulz
```

```
 18 souliei (Batt.) Batt.
 19 nigra (L.) Koch                     16      Su      Löve & Löve, 1942
 20 gravinae Ten.
 21 repanda (Willd.) DC.
    (a) repanda
    (b) saxatilis (DC.) Heywood
    (c) galissieri (Giraud.) Heywood
    (d) maritima (Rouy) Heywood
    (e) confusa (Emberger & Maire) Heywood
    (f) cantabrica (Font Quer) Heywood
    (g) cadevallii (Font Quer) Heywood
    (h) blancoana (Boiss.) Heywood
    (i) latisiliqua (Boiss. & Reuter)
         Heywood
    (j) nudicaulis (Lag.) Heywood

89 Sinapis L.
    1 arvensis L.                        18      Su      Lövkvist, 1963
    2 pubescens L.
    3 alba L.
    (a) alba                             24      Is      Löve & Löve, 1956
    (b) dissecta (Lag.) Bonnier
    4 flexuosa Poiret

90 Eruca Miller
    1 vesicaria (L.) Cav.
    (a) vesicaria
    (b) sativa (Miller) Thell.           22      Hu      Baksay, 1961

91 Erucastrum C. Presl
    1 laevigatum (L.) O.E. Schulz
    2 virgatum (J. & C. Presl) C. Presl
    3 nasturtiifolium Poiret) O.E. Schulz
    4 palustre (Pirona) Vis.
    5 gallicum (Willd.) O.E. Schulz       30      Au      Manton, 1932

92 Rhynchosinapis Hayek
    1 richeri (Vill.) Heywood
    2 nivalis (Boiss. & Heldr.) Heywood
    3 cheiranthos (Vill.) Dandy           48      Br      Wright, 1936
    (a) cheiranthos
    (b) nevadensis (Willk.) Heywood
    4 wrightii (O.E. Schulz) Dandy ex
        Clapham                          24      Br      Wright, 1936
    5 monensis (L.) Dandy ex Clapham      24      Br      Sikka, 1940
    6 hispida (Cav.) Heywood
    (a) hispida
    (b) transtagana (Coutinho) Heywood
    7 granatensis (O.E. Schulz) Heywood
    8 longirostra (Boiss.) Heywood
    9 pseuderucastrum (Brot.) Franco
    (a) pseuderucastrum
    (b) setigera (Gay ex Lange) Heywood
    (c) cintrana (Coutinho) Franco & P.
         Silva
   10 johnstonii (Samp.) Heywood

93 Hirschfeldia Moench
    1 incana (L.) Lagreze-Fossat

94 Hutera Porta
```

1 rupestris Porta
2 leptocarpa González-Albo

95 Carrichtera DC.
1 annua (L.) DC.

96 Vella L.
1 pseudocytisus L.
2 spinosa Boiss.

97 Succowia Medicus
1 balearica (L.) Medicus

98 Boleum Desv.
1 asperum (Pers.) Desv.

99 Erucaria Gaertner
1 hispanica (L.) Druce

100 Cakile Miller
1 maritima Scop.
  (a) maritima   18   Lu   Rodrigues, 1953
  (b) baltica (Jordan ex Rouy & Fouc.)
      Hyl. ex P.W. Ball
  (c) aegyptiaca (Willd.) Nyman
  (d) euxina (Pobed.) E.I. Nyárády
2 edentula (Bigelow) Hooker
  (a) edentula
  (b) islandica (Gand.) Á. & D. Löve   18   Is   Löve & Löve, 1956

101 Rapistrum Crantz
1 pinnatum (L.) All.
2 rugosum (L.) All.
  (a) rugosum   16   ?Su   Manton, 1932
  (b) orientale (L.) Arcangeli
  (c) linnaeanum Rouy & Fouc.

102 Didesmus Desv.
1 aegyptius (L.) Desv.

103 Crambe L.
1 maritima L.   30   Rm   Tarnavschi, 1947
                                60   Su   Lövkvist, 1963

2 tataria Sebeók
3 steveniana Rupr.
4 grandiflora DC.
5 aspera Bieb.
6 koktebelica (Junge) N. Busch
7 hispanica L.
8 filiformis Jacq.

104 Calepina Adanson
1 irregularis (Asso) Thell.

105 Morisia Gay
1 monanthos (Viv.) Ascherson   14   Sa   Martinoli, 1955

106 Guiraoa Cosson
1 arvensis Cosson

107 Enarthrocarpus Labill.

```
  1 arcuatus Labill.
  2 lyratus (Forskål) DC.

108 Raphanus L.
    1 raphanistrum L.
      (a) raphanistrum                    18      Su      Lövkvist, 1963
      (b) microcarpus (Lange) Thell.
      (c) rostratus (DC.) Thell.
      (d) maritimus (Sm.) Thell.          18      Br      Manton, 1932
      (e) landra (Moretti ex DC.) Bonnier &
          Layens                          18      It      Manton, 1932


LXIX RESEDACEAE


1 Reseda L.
  1 luteola L.                            28      Su      Lövkvist, 1963
  2 glauca L.
  3 gredensis (Cutanda & Willk.) Müller Arg.
  4 complicata Bory
  5 virgata Boiss. & Reuter
  6 alba L.
  7 undata L.
  8 decursiva Forskål
  9 suffruticosa Loefl.
 10 phyteuma L.
 11 media Lag.
 12 inodora Reichenb.
 13 tymphaea Hausskn.
 14 odorata L.
 15 orientalis (Müller Arg.) Boiss.
 16 stricta Pers.
 17 lutea L.
 18 lanceolata Lag.
 19 jacquinii Reichenb.
 20 arabica Boiss.

2 Sesamoides Ortega
  1 pygmaea (Scheele) O. Kuntze
    (a) pygmaea
    (b) minor (Lange) Heywood
  2 canescens (L.) O. Kuntze
    (a) canescens
    (b) suffruticosa (Lange) Heywood


LXX SARRACENIACEAE

1 Sarracenia L.
  1 purpurea L.


LXXI DROSERACEAE

1 Aldrovanda L.
  1 vesiculosa L.

2 Drosera L.
  1 rotundifolia L.                       20      Lu      Fernandes, 1950
  2 anglica Hudson                        40      Ge      Tischler, 1934
  3 intermedia Hayne                      20      Ge      Behre, 1929
```

73

3 Drosophyllum Link
  1 lusitanicum (L.) Link

LXXII CRASSULACEAE

1 Crassula L.
  1 tillaea Lester-Garland
  2 aquatica (L.) Schönl.   42   Is   Löve & Löve, 1956
  3 vaillantii (Willd.) Roth

2 Bryophyllum Salisb.
  1 pinnatum (Lam.) Oken

3 Umbilicus DC.
  1 parviflorus (Desf.) DC.
  2 chloranthus Heldr. & Sart. ex Boiss.
  3 erectus DC.
  4 rupestris (Salisb.) Dandy
  5 horizontalis (Guss.) DC.   24   It   Uhl, 1948
  6 heylandianus Webb & Berth.

4 Pistorinia DC.
  1 hispanica (L.) DC.
  2 breviflora Boiss.

5 Mucizonia (DC.) A. Berger
  1 sedoides (DC.) D.A. Webb
  2 hispida (Lam.) A. Berger

6 Sempervivum L.

1 wulfenii Hoppe ex Mert. & Koch	30	He	Zésiger, 1961
2 grandiflorum Haw.	80	Europe	Zésiger, 1961
3 ciliosum Craib	34	Bu	Zésiger, 1961
4 thompsonianum Wale			
5 octopodes Turrill	34	Ju	Zésiger, ined.
6 zeleborii Schott	64	Europe	Zésiger, 1961
7 kindingeri Adamović			
8 leucanthum Pančić	64	Europe	Zésiger, 1961
9 pittonii Schott	64	Europe	Zésiger, 1961
10 arachnoideum L.			
(a) arachnoideum	32	Ga Hs	Zésiger, 1961
(b) tomentosum (C.B. Lehm. &			
Schnittspahn) Schinz & Thell.	32		Zésiger, ined.
	64	Hs	Zésiger, 1961
11 montanum L.			
(a) montanum	42	Ga	Zésiger, 1961
(b) burnatii Wettst. ex Hayek	42	Ga	Zésiger, 1961
(c) stiriacum Wettst. ex Hayek	64	Au	Zésiger, 1961
12 dolomiticum Facch.	72	It	Zésiger, ined.
13 cantabricum J.A. Huber			
14 kosaninii Praeger			
15 macedonicum Praeger	34	Bot. Gard.	Uhl, 1961
16 erythraeum Velen.	34	Bu	Zésiger, ined.
17 giuseppii Wale	72	Hs	Zésiger, ined.
18 ballsii Wale	34	Gr	Zésiger, ined.
19 marmoreum Griseb.	34		Zésiger, ined.
20 nevadense Wale	108	Hs	Zésiger, ined.
21 andreanum Wale			
22 tectorum L.	36	Alps	Zésiger, 1961
	72	Alps	Zésiger, 1961

23 calcareum Jordan	38	Ga	Favarger & Zésiger, ined.

7 Jovibarba Opiz

1 allionii (Jordan & Fourr.) D.A. Webb	38	Europe	Zésiger, 1961
2 arenaria (Koch) Opiz	38	Europe	Zésiger, 1961
3 hirta (L.) Opiz			
(a) hirta	38	Europe	Zésiger, 1961
(b) glabrescens (Sabr.) Soó & Jáv.	38	Europe	Zésiger, 1961
4 sobolifera (J. Sims) Opiz	38	Europe	Zésiger, 1961
5 heuffelii (Schott) Á. & D. Löve	38	Europe	Zésiger, 1961

8 Aeonium Webb & Berth.
  1 arboreum (L.) Webb & Berth.

9 Aichryson Webb & Berth.
  1 villosum (Aiton) Webb & Berth.
  2 dichotomum (DC.) Webb & Berth.

10 Sedum L.
  1 praealtum A.DC.
  2 aizoon L.
  3 hybridum L.

4 telephium L.	24	Fe	Jalas & Ronkko, 1960
	36	Ga	Delay, 1947
	48	Po	Banach-Pogan, 1958

  (a) telephium
  (b) fabaria (Koch) Kirschleger
  (c) maximum (L.) Krocker
  (d) ruprechtii Jalas
  5 ewersii Ledeb.
  6 anacampseros L.
  7 spurium Bieb.

8 sediforme (Jacq.) Pau	32	Lu	Rodrigues, 1953

  9 ochroleucum Chaix
    (a) ochroleucum
    (b) montanum (Song. & Perr.) D.A. Webb

10 reflexum L.	108	Su	Uhl, ined.

  11 forsteranum Sm.
  12 tenuifolium (Sibth. & Sm.) Strobl
  13 pruinatum Link ex Brot.

14 acre L.	40	Au	Uhl, 1961
	48	Su	Löve & Löve, 1944
	80	Cz	Uhl, ined.
15 sartorianum Boiss.	64	Gr	Uhl, ined.

    (a) sartorianum
    (b) stribrnyi (Velen.) D.A. Webb
    (c) hillebrandtii (Fenzl) D.A. Webb
    (d) ponticum (Velen.) D.A. Webb

16 sexangulare L.	74	Su	Uhl, 1961

  17 borissovae Balk.
  18 laconicum Boiss.

19 alpestre Vill.	16	He	Favarger, 1953

  20 flexuosum Wettst.
  21 kostovii Stefanov
  22 idaeum D.A. Webb
  23 tuberiferum Stoj. & Stefanov
  24 zollikoferi F. Hermann & Stefanov

25 album L.	68	He	Uhl, 1961
	136	Lu	Uhl, 1961

  26 serpentini Janchen

27 gypsicola Boiss. & Reuter
28 subulatum (C.A. Meyer) Boiss.
29 anglicum Hudson

(a) anglicum	120	Br	Uhl, 1961
	c.144	Da	Turesson, 1963
(b) pyrenaicum Lange	24-36	Lu	Uhl, 1961

(c) melanantherum (DC.) Maire
30 stefco Stefanov

31 dasyphyllum L.	28	N. Africa	Uhl, 1961
	42	Su	Uhl, 1961
	56	It	Larsen, 1955b

32 brevifolium DC.
33 monregalense Balbis
34 magellense Ten.
35 alsinefolium All.
36 hierapetrae Rech. fil.
37 tristriatum Boiss.
38 hirsutum All.
  (a) hirsutum
  (b) baeticum Rouy

39 villosum L.	30	No	Knaben, 1950

40 creticum Boiss.
41 cepaea L.
42 lagascae Pau
43 pedicellatum Boiss. & Reuter
44 arenarium Brot.

45 atratum L.	16	Ge	Mattick, 1950

  (a) atratum
  (b) carinthiacum (Hoppe ex Pacher)
    D.A. Webb
46 confertiflorum Boiss.
47 stellatum L.
48 rubrum L.   22   Is   Löve & Löve, 1956
49 litoreum Guss.
50 aetnense Tineo
51 rubens L.
52 caespitosum (Cav.) DC.
53 andegavense (DC.) Desv.
54 nevadense Cosson

55 hispanicum L.	40	Tu	Uhl, pers. comm.

56 pallidum Bieb.
57 caeruleum L.

11 Rhodiola L.

1 rosea L.	22	Is	Löve & Löve, 1956

  2 quadrifida (Pallas) Fischer & C.A. Meyer

12 Orostachys (DC.) Fischer ex Sweet
  1 spinosa (L.) Sweet
  2 thyrsiflora (DC.) Fischer ex Sweet

13 Rosularia (DC.) Stapf
  1 serrata (L.) A. Berger

LXXIII SAXIFRAGACEAE

1 Saxifraga L.

1 hieracifolia Waldst. & Kit.	112	Sb	Flovik, 1940
	120	Ga	Hamel, 1953
2 nivalis L.	60	Is	Löve & Löve, 1951

3 tenuis (Wahlenb.) H. Smith ex Lindman	20	Is	Löve & Löve, 1951
4 clusii Gouan			
(a) clusii			
(b) lepismigena (Planellas) D.A. Webb			
5 stellaris L.	28	Fa	Böcher, 1938b
(a) stellaris			
(b) alpigena Temesy			
6 foliolosa R. Br.	56	Is	Löve & Löve, 1951
7 cuneifolia L.	28	Ga	Hamel, 1953
8 spathularis Brot.	28	Hb	Packer, 1961
9 umbrosa L.	28	Ga	Hamel, 1953
10 hirsuta L.	28	Hb	Packer, 1961
(a) hirsuta			
(b) paucicrenata (Gillot) D.A. Webb			
11 x polita (Haw.) Link	28	Hb	Packer, 1961
12 x geum L.	28		
13 x urbium D.A. Webb	28	Br	Packer, 1961
14 nelsoniana D. Don			
15 rotundifolia L.			
16 chrysosplenifolia Boiss.	22	Europe	K. Jones, ined.
17 taygetea Boiss. & Heldr.			
18 aspera L.	c.26	Ga	Hamel, 1953
19 bryoides L.	26	He	Favarger, 1949a
20 bronchialis L.			
21 flagellaris Sternb. & Willd.	32	Sb	Flovik, 1940
22 hirculus L.	32	Su	Löve & Löve, 1951
23 hederacea L.			
24 cymbalaria L.			
subsp. huetiana (Boiss.) Engler & Irmsch.	18	Ga	Hamel, 1953
25 sibthorpii Boiss.			
26 latepetiolata Willk.	c.66	Hs	K. Jones, ined.
27 tridactylites L.	22	Da	Skovsted, 1934
28 osloensis Knaben	44	No	Knaben, 1954
29 adscendens L.			
(a) adscendens	22	Ga	Hamel, 1953
(b) parnassica (Boiss. & Heldr.) Hayek			
(c) blavii (Engler) Hayek			
30 petraea L.			
31 berica (Béguinot) D.A. Webb			
32 irrigua Bieb.	44	Rs( )	K. Jones, ined.
33 arachnoidea Sternb.	56	It	Damboldt & Podlech, 1963
34 paradoxa Sternb.			
35 aizoides L.	26	No	Skovsted, 1934
36 tenella Wulfen			
37 glabella Bertol.			
38 praetermissa D.A. Webb	44	Ga	Hamel, 1953
39 aquatica Lapeyr.	28	Ga	Hamel, 1953
	66	Ga	K. Jones, ined.
40 x capitata Lapeyr.			
41 cuneata Willd.	28	Hs	Hamel, 1953
42 pentadactylis Lapeyr.			
43 corbariensis Timb.-Lagr.			
(a) corbariensis	60-66	Ga	K. Jones, ined.
(b) valentina (Willk.) D.A. Webb	64	Hs	K. Jones, ined.
44 camposii Boiss. & Reuter	64	Hs	K. Jones, ined.
(a) camposii			
(b) leptophylla (Willk.) D.A. Webb			
45 trifurcata Schrader	28	Hs	Hamel, 1954
46 canaliculata Boiss. & Reuter ex Engler	52	Hs	K. Jones, ined.
47 geranioides L.	c.52	Ga	K. Jones, ined.
48 moncayensis	60	Hs	K. Jones, ined.

49 vayredana Luizet	c.64	Hs	K. Jones, ined.
50 nervosa Lapeyr.	34	Ga	K. Jones, ined.
51 pedemontana All			
(a) pedemontana			
(b) cymosa Engler			
(c) cervicornis (Viv.) Engler	26	Co	Hamel, 1954
	44	Co	K. Jones, ined.
(d) prostii (Sternb.) D.A. Webb	32	Ga	Hamel, 1954
52 wahlenbergii Ball			
53 androsacea L.	16	Ga	Hamel, 1954
	c.128	He	Favarger, 1957a
54 seguieri Sprengel			
55 italica D.A. Webb			
56 depressa Sternb.			
57 sedoides L.			
(a) sedoides			
(b) prenja			
58 aphylla Sternb.			
59 presolanensis Engler	16	It	Damboldt & Podlech, 1963
60 muscoides All.			
61 facchinii Koch			
62 moschata Wulfen	26	Au/He	Damboldt & Podlech, 1963
	52	Au	Damboldt, 1968b
63 hariotii Luizet & Soulie	34	Ga	K. Jones, ined.
64 exarata Vill.	20-22	Ga	K. Jones, ined.
65 cebennensis Rouy & Camus	26	Ga	Hamel, 1954
	32	Ga	K. Jones, ined.
66 pubescens Pourret			
(a) pubescens			
(b) iratiana (F.W. Schultz) Engler & Irmscher	26	Ga	Hamel, 1954
67 nevadensis Boiss.			
68 cespitosa L.	80	Sb	Flovik, 1940
69 hartii D.A. Webb	50	Hb	Packer, 1961
70 rosacea Moench			
(a) rosacea	56	Ga	K. Jones, ined.
	64	Is	Löve & Löve, 1951
(b) sponhemica (C.C. Gmelin) D.A. Webb	50	Ga	K. Jones, ined.
	52	Ga	Hamel, 1954
71 hypnoides L.	c.30	Hb	Packer, 1961
	48	Is	Löve & Löve, 1951
72 continentalis (Engler & Irmscher) D.A. Webb	52	Ga	Hamel, 1953
73 conifera Cosson & Durieu			
74 rigoi Porta			
75 globulifera Desf.			
76 reuterana Boiss.			
77 erioblasta Boiss. & Reuter	34	Hs	K. Jones, ined.
78 boissieri Engler	64	Hs	K. Jones, ined.
79 biternata Boiss.	66	Hs	K. Jones, ined.
80 gemmulosa Boiss.	c.64	Hs	K. Jones, ined.
81 haenseleri Boiss. & Reuter			
82 dichotoma Sternb.	32 + 2B	Hs	K. Jones, ined.
(a) dichotoma			
(b) albarracinensis (Pau) D.A. Webb			
83 carpetana Boiss. & Reuter	20	Hs	K. Jones, ined.
84 graeca Boiss.			
85 bulbifera L.	28	Cz	K. Jones, ined.
	32	Br	Whyte, 1930

86 granulata L.
  (a) granulata   46-60   Da   Skovsted, 1934
  (b) graniticola D.A. Webb
87 corsica (Duby) Gren. & Godron
  (a) corsica   52   Co   Contandriopoulos,
                                                                     1962
                                              62-66   Co   K. Jones, ined.
  (b) cossoniana (Boiss.) D.A. Webb   64-66   Hs   K. Jones, ined.
88 cintrana Kuzinsky ex Willk.
89 sibirica L.
90 carpathica Reichenb.
91 rivularis L.   52   Is   Löve & Löve, 1951
   hyperborea R. Br.   26   Sb   Flovik, 1940
92 cernua L.   50   Alps   Chiarugi, 1935
                                              64   Is   Löve & Löve, 1951
93 oppositifolia L.   26   Ga   Hamel, 1953
94 retusa Gouan   26   Ga   Hamel, 1953
  (a) retusa
  (b) augustana (Vacc.) D.A. Webb
95 biflora All.   26   Ga   Hamel, 1953
96 marginata Sternb.
97 scardica Griseb.
98 spruneri Boiss.
99 diapensioides Bellardi
100 tombeanensis Boiss. ex Engler
101 vandellii Sternb.   26   It   Damboldt & Podlech, 1963
102 burserana L.   26   It   Damboldt & Podlech, 1963
103 juniperifolia Adams
  (a) juniperifolia
  (b) sancta (Griseb.) D.A. Webb
104 ferdinandi-coburgi J. Kellerer & Sund.
105 aretioides Lapeyr.   26   Ga   Hamel, 1953
106 caesia L.   26   Ga   Hamel, 1953
107 squarrosa Sieber
108 luteoviridis Schott & Kotschy
109 sempervivum C. Koch
110 grisebachii Degen & Dörfler
111 porophylla Bertol.
112 stribrnyi (Velen.) Podp.
113 media Gouan
114 valdensis DC.   28   Ga   Hamel, 1953
115 callosa Sm.
  (a) callosa   28   Ga   Hamel, 1953
  (b) catalaunica (Boiss.) D.A. Webb
116 cochlearis Reichenb.   28   Ga   Hamel, 1953
117 crustata Vest
118 cotyledon L.   28   Ga   Hamel, 1953
119 paniculata Miller   28   Ga   Hamel, 1953
120 hostii Tausch
  (a) hostii
  (b) rhaetica (Kerner) Br.-Bl.
121 mutata L.
  (a) mutata   28   Ga   Hamel, 1953
  (b) demissa (Schott & Kotschy) D.A. Webb
122 longifolia Lapeyr.   28   Ga   Hamel, 1953
123 florulenta Moretti   28   Ga   Hamel, 1953

2 Bergenia Moench
  1 crassifolia (L.) Fritsch

3 Chrysosplenium L.

```
 1 alternifolium L.                          48      Ga        Hamel, 1953
 2 tetrandrum (N. Lund) Th. Fries            24      Europe    Löve, 1954a
 3 oppositifolium L.                         42      Ga        Hamel, 1953
 4 alpinum Schur
 5 dubium Gay ex Ser.
```

LXXIV PARNASSIACEAE

```
 1 Parnassia L.
   1 palustris L.
     (a) palustris                           18      Ga        Hamel, 1953
     (b) obtusiflora  (Rupr.) D.A. Webb      36      Rs(N)     Sokolovskaya, 1958
```

LXXV HYDRANGEACEAE

```
 1 Philadelphus L.
   1 coronarius L.

 2 Deutzia Thunb.
   1 scabra Thunb.
```

LXXVI ESCALLONIACEAE

```
 1 Escallonia Mutis ex L. fil.
   1 macrantha Hooker & Arnott
   2 rubra (Ruiz & Pavón) Pers.
```

LXXVII CROCCULANIAQDAD

```
 1 Ribes L.
   1 multiflorum Kit. ex Roemer & Schultes
   2 rubrum L.                               16      Is        Löve & Löve, 1956
   3 spicatum Robson                         16      Su        Löve & Löve, 1956
   4 petraeum Wulfen
   5 nigrum L.                               16      Su        Lövkvist, 1963
   6 uva-crispa L.                           16      Su        Lövkvist, 1963
   7 sardoum U. Martelli
   8 alpinum L.                              16      Su        Lövkvist, 1963
   9 orientale Desf.
```

LXXVIII PITTOSPORACEAE

```
 1 Pittosporum Gaertner
   1 crassifolium Putterl.
   2 undulatum Vent.
```

LXXIX. PLATANACEAE

```
 1 Platanus L.
   1 orientalis L.
   2 hybrida Brot.                           42      Europe    Winge, 1917
```

LXXX ROSACEAE

1 Sorbaria (Ser.) A. Braun
  1 sorbifolia (L.) A. Braun
  2 tomentosa (Lindley) Rehder

2 Physocarpus (Camb.) Maxim.
  1 opulifolius (L.) Maxim.

3 Spiraea L.
  1 salicifolia L.
  2 alba Duroi
  3 douglasii Hooker
  4 tomentosa L.
  5 decumbens Koch
   (a) decumbens
   (b) tomentosa (Poech) Dostál
  6 japonica L. fil.
  7 corymbosa Rafin.
  8 chamaedryfolia L.
  9 media Franz Schmidt   10   Hu   Baksay, 1957b
   (a) media
   (b) polonica (Blocki) Pawl.
 10 cana Waldst. & Kit.
 11 crenata L.
 12 x vanhouttei (Briot) Zabel
 13 hypericifolia L.
   (a) hypericifolia
   (b) obovata (Waldst. & Kit. ex Willd.)
     Dostál

4 Sibiraea Maxim.
  1 altaiensis (Laxm.) C.K. Schneider

5 Aruncus L.
  1 dioicus (Walter) Fernald

6 Filipendula Miller
  1 vulgaris Moench   14   Hu   Pólya, 1950
                           16   Su   Lövkvist, 1963
  2 ulmaria (L.) Maxim.   14   Ho   Gadella &
                                      Kliphuis, 1963
                           16   Ho   Gadella &
                                      Kliphuis, 1963
                           24   Su   Lovkvist, 1963
   (a) ulmaria
   (b) picbaueri (Podp.) Smejkal
   (c) denudata (J. & C. Presl) Hayek

7 Rhodotypos Siebold & Zucc.
  1 scandens (Thunb.) Makino

8 Kerria DC.
  1 japonica (L.) DC.

9 Rubus L.
  1 chamaemorus L.   56   Br   Heslop-Harrison,
                                        1953
  2 humulifolius C.A. Meyer
  3 arcticus L.   14   Rs(N) Sokolovskaya &
                                        Strelkova, 1960
  4 saxatilis L.   28   Fe   Sorsa, 1963
  5 odoratus L.

6 spectabilis Pursh			
7 idaeus L.	14	Br	Heslop-Harrison, 1953
loganobaccus L.H. Bailey	42	Br	Thomas, 1940
8 sachalinensis Léveillé			
9 phoenicolasius Maxim.			
10 nessensis W. Hall	28	Su	Heslop-Harrison, 1953
11 scissus W.C.R. Watson	28	Br	Heslop-Harrison, 1953
12 plicatus Weihe & Nees	28	Br	Heslop-Harrison, 1953
bertramii G. Braun ex Focke	28	Br	Maude, 1939
opacus Focke	28	Br	Datta, 1932
13 sulcatus Vest ex Tratt.	28	Scand.	Gustafsson, 1943
14 divaricatus P.J. Mueller	21	Scand.	Gustafsson, 1943
15 affinis Weihe & Nees	28	Br	Heslop-Harrison, 1953
16 lentiginosus Lees	28	Europe	Maude, 1939
17 vulgaris Weihe & Nees	21	Br	Heslop-Harrison, 1953
selmeri Lindeb. ex F. Aresch.	28	Scand.	Gustafsson, 1939
incurvatus Bab.	28	Br	Heslop-Harrison, 1953
18 pedatifolius Genev.	28	Scand.	Gustafsson, 1939
19 gratus Focke	28	Br	Heslop-Harrison, 1953
sciocharis (Sudre) W.C.R. Watson	28	Br	Heslop-Harrison, 1953
20 chaerophyllus Sagorski & W. Schultze			
21 hypomalacus Focke			
22 arrhenii (Lange) Lange	28	Br	Heslop-Harrison, 1953
23 sprengelii Weihe	28	Br	Heslop-Harrison, 1953
24 myricae Focke			
25 chlorothyrsos Focke	28	Br	Heslop-Harrison, 1953
axillaris Lej.	28	Europe	Maude, 1939
26 questieri P.J. Mueller & Lefèvre	28	Br	Maude, 1939
mercicus Bagnall	28	Br	Heslop-Harrison, 1953
scheutzii Lindeb. ex F. Aresch.	28	Scand.	Gustafsson, 1943
27 laciniatus Willd.	28	Br	Heslop-Harrison, 1953
28 rhombifolius Weihe ex Boenn.			
alterniflorus P.J. Mueller & Lefèvre	28	Br	Heslop-Harrison, 1953
silesiacus Weihe	28	Br	Heslop-Harrison, 1953
29 pyramidalis Kaltenb.	28	Br	Heslop-Harrison, 1953
dumnoniensis Bab.	28	Br	Heslop-Harrison, 1953
30 macrophyllus Weihe & Nees	28	Br	Heslop-Harrison, 1953
schlechtendalii Weihe ex Link	28	Europe	Maude, 1939
31 silvaticus Weihe & Nees			
32 egregius Focke	28	Br	Heslop-Harrison, 1953
lindleianus Lees	28	Br	Datta, 1932

33 villicaulis Koehler ex Weihe & Nees	28	Br	Datta, 1932
34 polyanthemus Lindeb.	28	Br	Heslop-Harrison, 1953
35 rhamnifolius Weihe & Nees			
cardiophyllus P.J. Mueller & Lefèvre	28	Br	Heslop-Harrison, 1953
36 hochstetterorum Seub.			
37 lindebergii P.J. Mueller	28	Br	Heslop-Harrison, 1953
38 ulmifolius Schott	14	Br	Heslop-Harrison, 1953
39 godronii Lecoq & Lamotte			
winteri P.J. Mueller ex Focke	28	Europe	Maude, 1939
40 bifrons Vest ex Tratt.			
cuspidifer P.J. Mueller & Lefèvre	28	Br	Heslop-Harrison, 1953
41 discolor Weihe & Nees			
42 chloocladus W.C.R. Watson			
43 candicans Weihe ex Reichenb.	21	Scand.	Gustafsson, 1943
44 canescens DC.			
45 vestitus Weihe & Nees			
podophyllos P.J. Mueller	28	Br	Heslop-Harrison, 1953
46 boraeanus Genev.	28	Br	Heslop-Harrison, 1953
47 adscitus Genev.			
leucostachys Schleicher ex Sm.	28	Br	Heslop-Harrison, 1953
macrothyrsus Lange	28	Br	Heslop-Harrison, 1953
48 mucronulatus Boreau			
49 gremlii Focke			
schmidelyanus Sudre	35	Br	Heslop-Harrison, 1953
50 radula Weihe ex Boenn.	28	Br	Heslop-Harrison, 1953
51 genevieri Boreau			
echinatus Lindley	28	Europe	Maude, 1939
52 apiculatus Weihe & Nees	28	Br	Heslop-Harrison, 1953
53 fuscus Weihe & Nees			
granulatus P.J. Mueller & Lefèvre	28	Br	Heslop-Harrison, 1953
54 foliosus Weihe & Nees			
insericatus P.J. Mueller ex Wirtgen	28	Br	Heslop-Harrison, 1953
55 infestus Weihe ex Boenn.	28	Br	Heslop-Harrison, 1953
56 thyrsiflorus Weihe & Nees	28	Br	Heslop-Harrison, 1953
57 pallidus Weihe & Nees	28	Br	Heslop-Harrison, 1953
bloxamii Lees	28	Br	Datta, 1932
58 obscurus Kaltenb.	28	Br	Heslop-Harrison, 1953
59 menkei Weihe & Nees			
60 rudis Weihe & Nees			
61 melanoxylon P.J. Mueller & Wirtgen ex Genev.			
62 vallisparsus Sudre	28	Br	Heslop-Harrison, 1953

63 fuscater Weihe & Nees	28	Europe	Maude, 1939
adornatus P.J. Mueller ex Wirtgen	28	Br	Heslop-Harrison, 1953
hartmanii Gand. ex Sudre	35	Br	Heslop-Harrison, 1953
64 pilocarpus Gremli			
65 lejeunei Weihe & Nees	35	Br	Heslop-Harrison, 1953
66 rosaceus Weihe & Nees	28	Br	Heslop-Harrison, 1953
67 histrix Weihe & Nees	28	Br	Heslop-Harrison, 1953
rufescens P.J. Mueller & Lefevre	28	Br	Heslop-Harrison, 1953
68 koehleri Weihe & Nees	28	Br	Maude, 1939
apricus Wimmer	28	Br	Heslop-Harrison, 1953
dasyphyllus (Rogers) Rogers	28	Scand.	Gustafsson, 1943
69 incanescens Bertol.			
70 scaber Weihe & Nees	28	Br	Heslop-Harrison, 1953
71 schleicheri Weihe ex Tratt.	28	Br	Maude, 1939
72 glandulosus Bellardi	28	Br	Heslop-Harrison, 1953
73 serpens Weihe ex Lej. & Court.	28	Scand.	Gustafsson, 1943
angustifrons Sudre	28	Br	Heslop-Harrison, 1953
74 hirtus Waldst. & Kit.	28	Br	Heslop-Harrison, 1953
75 caesius L.	28	Br	Heslop-Harrison, 1953

10 Rosa L.
1 sempervirens L.			
2 arvensis Hudson			
3 phoenicia Boiss.			
4 moschata J. Herrmann			
5 pimpinellifolia L.	28	Br	Blackburn & H.-Harrison, 1921
6 foetida J. Herrmann			
7 acicularis Lindley			
8 majalis J. Herrmann	14	Su	Täckholm, 1922
9 glauca Pourret			
10 rugosa Thunb.	14	Br	Blackburn & H.-Harrison, 1921
11 blanda Aiton			
12 pendulina L.			
13 virginiana J. Herrmann			
14 gallica L.			
15 stylosa Desv.			
16 jundzillii Besser			
17 montana Chaix			
18 canina L.	35	Br	Blackburn & H.-Harrison, 1921
19 squarrosa (Rau) Boreau			
20 andegavensis Bast.			
21 nitidula Besser			
22 pouzinii Tratt.			
23 vosagiaca Desportes			
24 subcanina (Christ) Dalla Torre & Sarnth.			
25 caesia Sm.	35		

26 subcolina (Christ) Dalla Torre & Sarnth.
27 rhaetica Gremli
28 obtusifolia Desv.
29 corymbifera Borkh.   35   Br   Blackburn & H.-
30 deseglisei Boreau   Harrison, 1921
31 abietina Gren. ex Christ
32 tomentosa Sm.
33 scabriuscula Sm.
34 sherardii Davies   28   Br   Blackburn & H.-
                                                             Harrison, 1921
35 villosa L.
36 mollis Sm.   28   Br   Blackburn & H.-
                                                             Harrison, 1921
37 heckeliana Tratt.
38 orientalis Dupont ex Ser.
39 rubiginosa L.   35   Br   Blackburn & H.-
                                                             Harrison, 1921
40 elliptica Tausch
41 agrestis Savi
42 micrantha Borrer ex Sm.
43 caryophyllaceae Besser
44 sicula Tratt.
45 glutinosa Sibth. & Sm.
46 turcica Rouy
47 serafinii Viv.

11 Agrimonia L.
  1 pilosa Ledeb.
  2 eupatoria L.
    (a) eupatoria   28   Hu   Pólya, 1950
    (b) grandis (Andrz. ex Ascherson &
        Graebner) Bornm.
    (c) asiatica (Juz.) Skalický
  3 procera Wallr.
  4 repens L.

12 Aremonia Nestler
  1 agrimonoides (L.) DC.
    (a) agrimonoides
    (b) pouzarii Skalický

13 Sanguisorba L.
  1 officinalis L.   28   Su   Nordborg, 1958
                                             56   Su   Nordborg, 1958
  2 albanica Andrasovszky & Jáv.
  3 dodecandra Moretti   56   It   Nordborg, 1966
  4 hybrida (L.) Nordborg   56   Hs Lu   Nordborg, 1966
  5 ancistroides (Desf.) Cesati   28   Hs   Nordborg, 1967
  6 minor Scop.
    (a) minor   28   Su   Nordborg, 1958
    (b) lasiocarpa (Boiss. & Hausskn.)
        Nordborg   56   Tu   Nordborg, 1967
    (c) lateriflora (Cosson) M.C.F. Proctor
    (d) magnolii (Spach) Briq.   28 Ga Hs It Lu Nordborg, 1967
    (e) muricata Briq.   28   Su   Erdtman &
                                                             Nordborg, 1961
                                             56   Su   Erdtman &
                                                             Nordborg, 1961
    (f) rupicola (Boiss. & Reuter)
        Nordborg   28   Hs   Nordborg, 1967
                                             56   Hs   Nordborg, 1967

7 cretica Hayek	28	Cr	Nordborg, 1967

14 Sarcopoterium Spach
1 spinosum (L.) Spach	28	Si	Larsen & Laeggard, 1971

15 Acaena Mutis ex L.
1 anserinifolia (J.R. & G. Forster) Druce

16 Dryas L.
1 octopetala L.	18	Po	Skalińska et al., 1959
	36	He	Bücher & Larsen, 1955

17 Geum L.
1 reptans L.	42	Po	Gajewski, 1957
2 montanum L.	28	Po	Gajewski, 1957
3 bulgaricum Pančić	56	Bu	Gajewski, 1957
4 heterocarpum Boiss.			
5 rivale L.	42	Is Po	Gajewski, 1957
6 sylvaticum Pourret	42	Lu	Gajewski, 1957
7 coccineum Sibth. & Sm.	42	Bu	Gajewski, 1957
8 pyrenaicum Miller			
9 urbanum L.			
10 aleppicum Jacq.	42	Europe	Gajewski, 1957
11 molle Vis. & Pančić	42	Bu	Gajewski, 1957
12 macrophyllum Willd.			
13 hispidum Fries	42	Hs Su	Gajewski, 1957

18 Waldsteinia Willd.
1 geoides Willd.	14	Hu	Pólya 1949

2 ternata (Stephan) Fritsch

19 Potentilla L.
1 fruticosa L.	14	Ga	Elkington & Woodell, 1963
	28	Br	Elkington & Woodell, 1963
2 bifurca L.			
3 palustris (L.) Scop.	28	Su	Lövkvist, 1963
	35	Su	Lövkvist, 1963
	42	Su	Lövkvist, 1963
	62-64	Rs(N)	Sokolovskaya & Strelkova, 1960
4 anserina L.			
(a) anserina	28	Su	Lövkvist, 1963
	35	Su	Lövkvist, 1963
	42	Su	Lövkvist, 1963
(b) egedii (Wormsk.) Hiitonen	28	No	Rousi, 1965a
5 rupestris L.	14	Ga	Gagnieu et al., 1959
6 geoides Bieb.			
7 pulchella R. Br.	28	Sb	Flovik, 1940
8 rubricaulis Lehm.			
9 multifida L.			
10 eversmanniana Fischer ex Ledeb.			
11 pensylvanica L.			
12 longifolia Willd.			
13 pimpinelloides L.			
14 visianii Pančić			
15 nivea L.	56	No	Knaben & Engelskjön, 1967

16 chamissonis Hultén	77	Su	Hultén, 1945
17 hookerana Lehm.			
18 argentea L.	14	Su	Muntzing, 1931b
19 neglecta Baumg.	42	Su	A. & G. Muntzing, 1941
20 calabra Ten.			
21 inclinata Vill.	42	Ga	Gagnieu et al., 1959
22 collina Wibel	42	Su	Muntzing, 1931b
23 supina L.	28	Hu	Pólya, 1949
24 norvegica L.			
25 intermedia L.			
26 astracanica Jacq.			
27 detommasii Ten			
28 recta L.	42	Ga	Gagnieu et al., 1959
29 hirta L.			
30 nevadensis Boiss.			
31 grandiflora L.			
32 montenegrina Pant.			
33 umbrosa Steven ex Bieb.			
34 delphinensis Gren. & Godron			
35 pyrenaica Ramond ex DC.			
36 stipularis L.			
37 longipes Ledeb.			
38 chrysantha Trev.			
39 thuringiaca Bernh. ex Link			
40 brauniana Hoppe	14	Au	Mattick, 1950
41 frigida Vill.	28	He	Favarger, 1965b
42 hyparctica Malte			
43 crantzii (Crantz) G. Beck ex Fritsch	28	Cz	Czapik, 1961
	42	Fe	Sorsa, 1963
	48	Su	Muntzing, 1958
	64	Br	Smith, 1963
44 aurea L.			
(a) aurea	14	Au Cz	Walters, 1967
(b) chrysocraspeda (Lehm.) Nyman			
45 heptaphylla L.	14	Po	Skalińska, 1950
	28	Cz	Walters, 1967
	30	Cz	Walters, 1967
	35	Cz	Walters, 1967
46 humifusa Willd.			
47 patula Waldst. & Kit.	42	Ju	Walters, 1967
48 australis Krašan	42	Cz	Walters, 1967
49 tabernaemontani Ascherson	42	Ho	Gadella & Kliphuis, 1968
	49	Su	Muntzing, 1958
	50	Su	Muntzing, 1958
	56	Su	Muntzing, 1958
	c.60	Su	Muntzing, 1958
	63	Su	Muntzing, 1958
	70	Br	Smith, 1963
	c.84	Su	Muntzing, 1931b
50 pusilla Host			
51 cinerea Chaix ex Vill.	14	Ga	Guinochet, 1964
	28	Hs	Küpfer, 1968
52 erecta (L.) Rauschel	28	Po	Czapik, 1961
53 anglica Laicharding	56	Su	Lövkvist, 1963
54 reptans L.	28	Ho	Gadella & Kliphuis, 1963
55 caulescens L.			

```
56 petrophila Boiss.
57 clusiana Jacq.                        42     It      Contandriopoulos, 1962
58 crassinervia Viv.                     14     Co      Contandriopoulos, 1962
59 nitida L.                             42     Ga      Favarger, 1962
60 alchemilloides Lapeyr.
61 valderia L.                           14     Ga      Contandriopoulos, 1962
62 haynaldiana Janka                     14     Bu      Tombal, 1969
63 doerfleri Wettst.
64 nivalis Lapeyr.
65 grammopetala Moretti
66 speciosa Willd.
67 apennina Ten.
   (a) apennina
   (b) stoianovii Urum. & Jáv.
68 kionaea Halácsy
69 deorum Boiss. & Heldr.
70 alba L.                               28     Po      Czapik, 1961
71 montana Brot.
72 sterilis (L.) Garcke                  28     Ge      Wulff, 1938
73 micrantha Ramond ex DC.               12     Ga      Gagnieu et al.,
                                                        1959
                                         14     Ga      Gagnieu et al.,
                                                        1959
74 carniolica A. Kerner
75 saxifraga Ardoino ex De Not.          14     Ga      Contandriopoulos, 1962

20 Sibbaldia L.
   1 procumbens L.                       14     Fe      Sorsa, 1963
   2 parviflora Willd.

21 Fragaria L.
   1 vesca L.                            14     Ho      Gadella &
                                                        Kliphuis, 1967a
   2 moschata Duchesne
   3 viridis Duchesne
     (a) viridis                         14     Po      Skalińska et al.,
                                                        1959
     (b) campestris (Steven) Pawl.
   4 virginiana Duchesne
   5 x ananassa Duchesne                 56     cult.   Staudt, 1962

22 Duchesnea Sm.
   1 indica (Andrews) Focke

23 Alchemilla L.
   1 pentaphyllea L.                     64     Ga      Strasburger, 1904
   2 saxatilis Buser
   3 transiens (Buser) Buser
   4 alpina L.                         c.120    Au      Turesson, 1958
                                       c.128    Is      Löve & Löve, 1956
                                       c.140    Su      Lövkvist, 1963
                                       c.152    Fa      Bozman & Walters
   5 basaltica Buser
   6 subsericea Reuter
   7 plicatula Gand.
   8 pallens Buser
   9 anisiaca Wettst.
  10 hoppeana (Reichenb.) Dalla Torre
  11 conjuncta Bab.
  12 grossidens Buser                    64     Ga      Strasburger, 1904
```

```
13 faeroensis (Lange) Buser          c.220      Is      Walters &
                                                        Bozman, 1967
                                     c.224      Is      Löve & Löve, 1956
14 paicheana (Buser) Rothm.
15 splendens Christ ex Favrat        c.163      He      Walters &
                                                        Bozman, 1967
16 infravalesiaca (Buser) Rothm.
17 schmidelyana Buser
18 jaquetiana Buser
19 fulgens Buser                     c.140      Ga Hs   Walters &
                                                        Bozman, 1967
20 kerneri Rothm.
21 glaucescens Wallr.                103-110    Su      Turesson, 1957
22 hirsuticaulis H. Lindb.
23 flabellata Buser
24 cinerea Buser
25 erythropoda Juz.
26 lithophila Juz.
27 bulgarica Rothm.
28 lapeyrousii Buser
29 plicata Buser
30 colorata Buser
31 illyrica Rothm.
32 exigua Buser ex Paulin
33 helvetica Brugger
34 vetteri Buser
35 hebescens Juz.
36 gibberulosa H. Lindb.
37 bungei Juz.
38 hirsutissima Juz.
39 propinqua H. Lindb.
40 conglobata H. Lindb.
41 monticola Opiz                    c.101      Su      Ehrenberg, 1945
                                     103-109 He No Su   Turesson, 1957
42 schistophylla Juz.
43 crinita Buser
44 strigosula Buser
45 subglobosa C.G. Westerlund        c.102      Su      Turesson, 1957
                                     108        Su      Turesson, 1957
                                     105-106    Su      Turesson, 1957
46 sarmatica Juz.
47 tytthantha Juz.
48 subcrenata Buser                  96         Is      Löve & Löve, 1956
                                     104-110    He Su   Turesson, 1957
49 cymatophylla Juz.                 106-107    Su      Turesson, 1957
50 heptagona Juz.
51 acutiloba Opiz                    c.100      Su      Ehrenberg, 1945
                                     105-109    Su      Lövkvist, 1963
52 nemoralis Alechin
53 gracilis Opiz                     c.93       Su      Ehrenberg, 1945
                                     104-110    Su      Turesson, 1957
54 xanthochlora Rothm.              c.105      Su      Turesson, 1957
55 heterophylla Rothm.
56 curtiloba Buser
57 filicaulis Buser
   (a) filicaulis Buser              c.96       Is      Löve & Löve, 1956
                                     103-110    Su      Lövkvist, 1963
                                     c.150      Br      Bradshaw, 1963
   (b) vestita (Buser) M.E. Bradshaw c.96       Is      Löve & Löve, 1956
                                     104-110    Br      Bradshaw, 1963
58 minima Walters                    103-108    Br      Bradshaw, 1964
59 compta Buser
```

60 rhododendrophila Buser			
61 polonica Pawl.			
62 decumbens Buser			
63 undulata Buser			
64 rubristipula Buser			
65 tirolensis Buser			
66 tatricola Pawl.			
67 heteropoda Buser			
68 tenuis Buser			
69 glomerulans Buser	101-109	No Su	Turesson, 1957
	c.144	Is	Löve & Löve, ined.
borealis Sam. ex Juz.	130-152	No Su	Turesson, 1957
70 camptopoda Juz.			
71 crebridens Juz.			
72 buschii Juz.			
73 controversa Buser			
74 wallischii Pawl.			
75 connivens Buser			
76 baltica Sam. ex Juz.			
77 wichurae (Buser) Stéfansson	103-106	No Su	Turesson, 1957
78 murbeckiana Buser	102-109	Su	Turesson, 1957
79 acutidens Buser			
80 reniformis Buser			
81 lineata Buser			
82 obtusa Buser	c.103	Su	Turesson, 1957
83 glabra Neygenf.	96	Is	Löve & Löve, 1956
	c.100	Su	Ehrenberg, 1945
	102-110 Fa	No Su	Turesson, 1957
84 glabricaulis H. Lindb.			
85 versipila Buser			
86 coriacea Buser			
87 inconcinna Buser			
88 stanislaae Pawl.			
89 trunciloba Buser			
90 sinuata Buser			
91 straminea Buser			
92 aequidens Pawl.			
93 demissa Buser			
94 pseudincisa Pawl.			
95 longana Buser			
96 achtarowii Pawl.			
97 junrukczalica Pawl.			
98 mollis (Buser) Rothm.			
99 catachnoa Rothm.			
100 viridiflora Rothm.			
101 albanica Rothm.			
102 phegophila Juz.			
103 indivisa (Buser) Rothm.			
104 heterotricha Rothm.			
105 peristerica Pawl.			
106 aroanica (Buser) Rothm.			
107 zapalowiczii Pawl.			
108 haraldii Juz.			
109 fallax Buser			
110 sericoneura Buser			
111 flexicaulis Buser			
112 gorcensis Pawl.			
113 cuspidens Buser			
114 othmarii Buser			
115 oculimarina Pawl.			
116 pyrenaica Dufour			

117 incisa Buser
118 fissa Günther & Schummel

24 Aphanes L.
  1 arvensis L.   48   Su   Hjelmquist, 1959
  2 microcarpa (Boiss. & Reuter) Rothm. 16   Lu Su Hjelmquist, 1959
  3 cornucopioides Lag.
  4 floribunda (Murb.) Rothm.

25 Cydonia Miller
  1 oblonga Miller   34   It   Bagnoli et al.,
                                                    1963

26 Pyrus L.
  1 cordata Desv.
  2 magyarica Terpó
  3 rossica Danilov
  4 pyraster Burgsd.
  5 caucasica Fedorov
  6 bourgaeana Decne
  7 syriaca Boiss.
  8 amygdaliformis Vill.
  9 salvifolia DC.
 10 elaeagrifolia Pallas
 11 nivalis Jacq.
 12 austriaca A. Kerner
 13 communis L.

27 Malus Miller
  1 trilobata (Labill.) C.K. Schneider
  2 florentina (Zuccagni) C.K. Schneider
  3 sylvestris Miller
  4 dasyphylla Borkh.
  5 praecox (Pallas) Borkh.
  6 domestica Borkh.

28 Sorbus L.
  1 domestica L.
  2 aucuparia L.   34   Fe   V. Sorsa, 1962
    (a) aucuparia
    (b) glabrata (Wimmer & Grab.) Cajander
    (c) sibirica (Hedl.) Krylov
    (d) praemorsa (Guss.) Nyman
    (e) fenenkiana Georgiev & Stoj.
  3 torminalis (L.) Crantz   34   Hu   Baksay, 1956
  4 chamaemespilus (L.) Crantz   34   Au   Liljefors, 1953
  5 aria (L.) Crantz
    leptophylla E.F. Warburg   68   Br   Warburg, 1952
  6 graeca (Spach) Kotschy   34   Hu   Baksay, 1956
    eminens E.F. Warburg   68   Br   Warburg, 1952
    porrigentiformis E.F. Warburg   51   Br   Warburg, 1952
                                      68   Br   Warburg, 1952
  7 rupicola (Syme) Hedl.   68   Europe Liljefors, 1953
  8 umbellata (Desf.) Fritsch
    (a) umbellata
    (b) banatica (Jáv.) Kárpáti
    (c) flabellifolia (Spach) Kárpáti
    (d) koevessii (Penzes) Kárpáti
  9 minima (A. Ley) Hedl.
 10 mougeotii Soyer-Willemet & Godron
 11 austriaca (G. Beck) Hedl.

```
   (a) austriaca
   (b) hazslinszkyana (Soo) Karpati
   (c) croatica Karpati
   (d) serpentini Karpati
       anglica Hedl.                        68      Br      Warburg, 1952
12 dacica Borbas
13 hybrida L.
14 meinichii (Lindeb.) Hedl.
   teodori Liljefors                        51      Su      Liljefors, 1953
15 intermedia (Ehrh.) Pers.                 68      Su      Liljefors, 1955
16 latifolia (Lam.) Pers.
   bristoliensis Wilmott                    51
   semiincisa Borbas                        34      Hu      Baksay, 1956
17 sudetica (Tausch) Fritsch
18 margittaiana (Jav.) Karpati


29 Eriobotrya Lindley
   1 japonica (Thunb.) Lindley

30 Amelanchier Medicus
   1 ovalis Medicus
   2 spicata (Lam.) C. Koch                  68      Su      Lövkvist, 1963
   3 grandiflora Rehder

31 Cotoneaster Medicus
   1 horizontalis Decne
   2 simonsii Baker
   3 integerrimus Medicus
   4 cinnabarinus Juz.
   5 nebrodensis (Guss.) C. Koch
   6 acuminatus Lindley
   7 niger (Thunb.) Fries
   8 granatensis Boiss.
   9 nummularia Fischer & C.A. Meyer
  10 tauricus Pojark.
  11 microphyllus Wallich ex Lindley

32 Pyracantha M.J. Roemer
   1 coccinea M.J. Roemer

33 Mespilus L.
   1 germanica L.

34 Crataegus L.
   1 intricata Lange
   2 crus-galli L.
   3 sanguinea Pallas
   4 altaica (Loudon) Lange
   5 laevigata (Poiret) DC.
    (a) laevigata
    (b) palmstruchii (Lindman) Franco
   6 macrocarpa Hegetschw.
   7 taurica Pojark.
   8 ucrainica Pojark.
   9 calycina Peterm.
    (a) calycina
    (b) curvisepala (Lindman) Franco
  10 microphylla C. Koch
  11 pallasii Griseb.
  12 karadaghensis Pojark.
  13 ambigua C.A. Meyer ex A. Becker
```

14 monogyna Jacq.
  (a) nordica Franco   34   Su   Lövkvist, 1963
  (b) monogyna
  (c) leiomonogyna (Klokov) Franco
  (d) brevispina (G. Kunze) Franco
  (e) aegeica (Pojark.) Franco
  (f) azarella (Griseb.) Franco
15 sphaenophylla Pojark.
16 pentagyna Waldst. & Kit.
17 nigra Waldst. & Kit.
18 schraderana Ledeb.
19 heldreichii Boiss.
20 laciniata Ucria
  (a) laciniata
  (b) pojarkovae (Kossych) Franco
21 azarolus L.
22 pycnoloba Boiss. & Heldr.

35 Prunus L.
  1 persica (L.) Batsch
  2 dulcis (Miller) D.A. Webb
  3 webbii (Spach) Vierh.
  4 tenella Batsch
  5 armeniaca L.
  6 brigantina Vill.
  7 cerasifera Ehrh.
  8 spinosa L.   32   Su   Lövkvist, 1963
  9 ramburii Boiss.
10 domestica L.
  (a) domestica
  (b) insititia (L.) C.K. Schneider
11 cocomilia Ten.
12 prostrata Labill.
13 fruticosa Pallas
14 avium L.   16   Su   Lövkvist, 1963
15 cerasus L.
16 mahaleb L.
17 padus L.   32   Su   Lövkvist, 1963
  (a) padus
  (b) borealis Cajander   32   No   Knaben &
                                                Engelskjon, 1967

18 serotina Ehrh.
19 virginiana L.
20 lusitanica L.
  (a) lusitanica
  (b) azorica (Mouillefert) Franco
21 laurocerasus L.

LXXXI LEGUMINOSAE

1 Cercis L.
  1 siliquastrum L.

2 Ceratonia L.
  1 siliqua L.

3 Gleditsia L.
  1 triacanthos L.

4 Acacia Miller

```
  1 farnesiana (L.) Willd.
  2 karoo Hayne
  3 dealbata Link
  4 mearnsii De Wild.
  5 longifolia (Andrews) Willd.
  6 cyclops A. Cunn. ex G. Don fil.
  7 melanoxylon R. Br.
  8 pycnantha Bentham
  9 cyanophylla Lindley
 10 retinodes Schlecht.
```

5 Sophora L.
```
  1 japonica L.
  2 alopecuroides L.
  3 jaubertii Spach
```

6 Thermopsis R. Br.
```
  1 lanceolata R. Br.
```

7 Anagyris L.
```
  1 foetida L.
```

8 Laburnum Fabr.
```
  1 anagyroides Medicus
  2 alpinum (Miller) Berchtold & J. Presl
```

9 Podocytisus Boiss. & Heldr.
```
  1 caramanicus Boiss. & Heldr.
```

10 Calicotome Link
```
  1 ⨆pᵢₙₒⱼₐ (Lᵢ) Lᵢₙₖ
  2 ᵥᵢᵢᵢₒₛₐ (ᵣₒᵣᵣₑₜ) Lᵢₙₖ              48       ᵢₜ        Larₛₑₙ, 1956a
```

11 Lembotropis Griseb.
```
  1 nigricans (L.) Griseb.
   (a) nigricans
   (b) australis (Freyn ex Wohlf.) J. Holub
```

12 Cytisus L.
```
  1 sessilifolius L.
  2 aeolicus Guss. ex Lindley
  3 villosus Pourret
  4 emeriflorus Reichenb.
  5 purgans (L.) Boiss.
  6 multiflorus (L'Hér.) Sweet
  7 ardoini E. Fourn.
  8 sauzeanus Burnat & Briq.
  9 procumbens (Waldst. & Kit. ex Willd.)
     Sprengel
 10 decumbens (Durande) Spach
 11 agnipilus Velen.
 12 pseudoprocumbens Markgraf
 13 commutatus (Willk.) Briq.
 14 ingramii Blakelock
 15 malacitanus Boiss.
   (a) malacitanus
   (b) catalaunicus (Webb) Heywood
 16 baeticus (Webb) Steudel
 17 striatus (Hill) Rothm.
 18 grandiflorus DC.
 19 cantabricus (Willk.) Reichenb. fil.
```

94

20 scoparius (L.) Link
  (a) scoparius   46   Po   Skalińska et al.,
                                                                 1964
  (b) maritimus (Rouy) Heywood   46   Ga   Bücher & Larsen,
21 reverchonii (Degen & Hervier) Bean   1958b
22 patens L.
23 tribracteolatus Webb

13 Chamaecytisus Link
  1 creticus (Boiss. & Heldr.) Rothm.
  2 subidaeus (Gand.) Rothm.
  3 spinescens (C. Presl) Rothm.
  4 purpureus (Scop.) Link
  5 hirsutus (L.) Link
  6 polytrichus (Bieb.) Rothm.
  7 wulfii (V. Krecz.) A. Klásková
  8 ciliatus (Wahlenb.) Rothm.
  9 glaber (L. fil.) Rothm.
 10 leiocarpus (A. Kerner) Rothm.
 11 borysthenicus (Gruner) A. Klásková
 12 lindemannii (V. Krecz.) A. Klásková
 13 ratisbonensis (Schaeffer) Rothm.
 14 zingeri (Nenukow ex Litv.) A.
    Klaskova
 15 paczoskii (V. Krecz.) A. Klásková
 16 ruthenicus (Fischer ex Woloszczak)
    A. Klásková
 17 graniticus (Rehmann) Rothm.
 18 skrobiszewskii (Pacz.) A. Klásková
 19 supinus (L.) Link
 20 eriocarpus (Boiss.) Rothm.
 21 austriacus (L.) Link
 22 tommasinii (Vis.) Rothm.
 23 heuffelii (Wierzb.) Rothm.
 24 pygmaeus (Willd.) Rothm.
 25 jankae (Velen.) Rothm.
 26 danubialis (Velen.) Rothm.
 27 dorycnioides (Davidov) Frodin &
    Heywood
 28 litwinowii (V. Krecz.) A. Klásková
 29 blockianus (Pawl.) A. Klásková
 30 albus (Hacq.) Rothm.
 31 banaticus (Griseb. & Schenk) Rothm.
 32 rochelii (Wierzb.) Rothm.
 33 podolicus (Blocki) A. Klásková
 34 kovacevii (Velen.) Rothm.
 35 nejceffii (Urum.) Rothm.

14 Chronanthus (DC.) C. Koch
  1 biflorus (Desf.) Frodin & Heywood

15 Teline Medicus
  1 monspessulana (L.) C. Koch
  2 linifolia (L.) Webb & Berth.

16 Genista L.
  1 tinctoria L.   48   Ho   Gadella &
                                                                 Kliphuis, 1966
  2 januensis Viv.
  3 lydia Boiss.

4 ramosissima (Desf.) Poiret
 5 cinerea (Vill.) DC.
　(a) cinerea
　(b) leptoclada (Willk.) O. Bolós &
　　　Molinier
 6 florida L.
 7 valentina (Willd. ex Sprengel) Steudel
 8 obtusiramea Gay ex Spach
 9 pseudopilosa Cosson
10 teretifolia Willk.
11 sakellariadis Boiss. & Orph.
12 sericea Wulfen
13 subcapitata Pančić
14 pulchella Vis.
15 albida Willd.
16 pilosa L.
17 lobelii DC.
18 pumila (Debeaux & Reverchon ex Hervier)
　　Vierh.
19 baetica Spach
20 salzmannii DC.
21 hystrix Lange
　(a) hystrix
　(b) legionensis (Pau) P. Gibbs
22 polyanthos R. de Roemer ex Willk.
23 aspalathoides Lam.

24 scorpius (L.) DC.	40	Hs	Lorenzo-Andreu & Garcia 1950
	48	Ga	Favarger & Contandriopoulos 1961
25 corsica (Loisel.) DC.			
26 carpetana Leresche ex Lange			
27 morisii Colla			
28 anglica L.	42	Ho	Gadella & Kliphuis, 1966

29 falcata Brot.
30 berberidea Lange
31 micrantha Ortega
32 carinalis Griseb.
33 sylvestris Scop.
34 aristata C. Presl
35 hispanica L.
　(a) hispanica
　(b) occidentalis Rouy

36 germanica L.	c.46	Su	Lövkvist, 1963

37 hirsuta Vahl
38 tournefortii Spach
39 lucida Camb.
40 anatolica Boiss.
41 triacanthos Brot.
42 tridens (Cav.) DC.
43 cupanii Guss.
44 radiata (L.) Scop.
45 holopetala (Fleischm. ex Koch) Bald.
46 hassertiana (Bald.) Bald. ex
　　Buchegger

47 dorycnifolia Font Quer	48	Hs	Santos, 1945

48 sessilifolia DC.
49 nissaba Petrović
50 ephedroides DC.
51 aetnensis (Biv.) DC.
52 spartioides Spach

53 haenseleri Boiss.
54 acanthoclada DC.
55 fasselata Decne
56 umbellata (L'Hér.) Poiret

17 Chamaespartium Adanson
 1 sagittale (L.) P. Gibbs
 2 tridentatum (L.) P. Gibbs

18 Echinospartum (Spach) Rothm.
 1 horridum (Vahl) Rothm.   44   Hs   Castro, 1945
 2 boissieri (Spach) Rothm.   44   Hs   Castro, 1945
 3 lusitanicum (L.) Rothm.
  (a) lusitanicum   c.52   Lu   Castro, 1945
  (b) barnadesii (Graells) C. Vicioso   c.52   Hs   Castro, 1945

19 Gonocytisus Spach
 1 angulatus (L.) Spach

20 Lygos Adanson
 1 sphaerocarpa (L.) Heywood
 2 monosperma (L.) Heywood
 3 raetam (Forskål) Heywood

21 Spartium L.
 1 junceum L.

22 Petteria C. Presl
 1 ramentacea (Sieber) C. Presl

23 Erinacea Adanson
 1 anthyllis Link

24 Ulex L.
 1 europaeus L.
  (a) europaeus   96   Su   Lövkvist, 1963
  (b) latebracteatus (Mariz) Rothm.   64   Lu   Castro, 1943
 2 minor Roth   32   Lu   Castro, 1943
 3 gallii   80   Hs   Castro, 1943
 4 densus Welw. ex Webb   64   Lu   Castro, 1945
 5 parviflorus Pourret   32   Lu   Castro, 1943
                                            64   Lu   Castro, 1945
                                            96   Lu   Castro, 1945
  (a) parviflorus
  (b) jussiaei (Webb) D.A. Webb
  (c) funkii (Webb) Guinea
  (d) eriocladus (C. Vicioso) D.A. Webb
 6 micranthus Lange   32   Hs   Castro, 1945
 7 argenteus Welw. ex Webb   64   Lu   Castro, 1941
                                            96   Lu   Castro, 1943
  (a) argenteus
  (b) subsericeus (Coutinho) Rothm.
  (c) erinaceus (Welw. ex Webb) D.A. Webb

25 Stauracanthus Link
 1 boivinii (Webb) Samp.   c.128   Lu   Castro, 1941
 2 genistoides (Brot.) Samp.   48   Lu   Castro, 1943
  (a) genistoides
  (b) spectabilis (Webb) Rothm.

26 Adenocarpus DC.

1 complicatus (L.) Gay
  (a) complicatus
  (b) aureus (Cav.) C. Vicioso
  (c) commutatus (Guss.) Coutinho
2 telonensis (Loisel.) DC.
3 hispanicus (Lam.) DC.
  (a) hispanicus
  (b) argyrophyllus Rivas Goday
4 decorticans Boiss.

27 Lotononis (DC.) Ecklon & Zeyher
   1 lupinifolia (Boiss.) Bentham
   2 genistoides (Fenzl) Bentham

28 Lupinus L.
   1 luteus L.
   2 hispanicus Boiss. & Reuter   52   Lu   Malheiros, 1942
   3 angustifolius L.
   (a) angustifolius
   (b) reticulatus (Desv.) Coutinho
   4 micranthus Guss.
   5 albus L.
   (a) albus
   (b) graecus (Boiss. & Spruner) Franco
       & P. Silva
   6 varius L.
   (a) varius   32   Lu   Malheiros, 1942
   (b) orientalis Franco & P. Silva
   7 perennis L.
   8 nootkatensis Donn ex Sims   48   Br   Maude, 1940
   9 polyphyllus Lindley
   10 arboreus Sims

29 Argyrolobium Ecklon & Zeyher
   1 zanonii (Turra) P.W. Ball   48   Ju   Larsen, 1956a
   2 biebersteinii P.W. Ball

30 Robinia L.
   1 pseudacacia L.
   2 hispida L.

31 Wisteria Nutt.
   1 sinensis (Sims) Sweet

32 Galega L.
   1 officinalis L.
   2 orientalis Lam.

33 Colutea L.
   1 arborescens L.
   (a) arborescens
   (b) gallica Browicz
   2 atlantica Browicz
   3 cilicica Boiss. & Balansa
   4 orientalis Miller

34 Eremosparton Fischer & C.A. Meyer
   1 aphyllum (Pallas) Fischer & C.A. Meyer

35 Halimodendron Fischer ex DC.
   1 halodendron (Pallas) Voss

36 Caragana Fabr.
  1 arborescens Lam.
  2 frutex (L.) C. Koch
  3 grandiflora (Bieb.) DC.

37 Calophaca Fischer
  1 wolgarica (L. fil.) Fischer

38 Astragalus L.
  1 contortuplicatus L.
  2 boeticus L.
  3 striatellus Pallas ex Bieb.
  4 arpilobus Kar. & Kir.
  5 reticulatus Bieb.
  6 ankylotus Fischer & C.A. Meyer
  7 stalinskyi Sirj.
  8 scorpioides Pourret ex Willd.
  9 longidentatus Chater
 10 peregrinus Vahl
 11 pamphylicus Boiss.
 12 haarbachii Spruner ex Boiss.
 13 suberosus Banks & Solander
 14 sinaicus Boiss.
 15 tribuloides Delile
 16 polyactinus Boiss.
 17 stella Gouan
 18 sesameus L.
 19 oxyglottis Steven ex Bieb.
 20 echinatus Murray
 21 epiglottis L.
  (a) epiglottis
  (b) asperulus (Dufour) Nyman
 22 edulis Durieu ex Bunge
 23 cymbicarpos Brot.
 24 hamosus L.
 25 cicer L.
 26 glaux L.

27 danicus Retz.	16	Ga	Guinochet & Logeois, 1962

 28 purpureus Lam.
 29 bourgaeanus Cosson
 30 pseudopurpureus Gusuleac
 31 setosulus Gontsch.
 32 austraegaeus Rech. fil.
 33 turolensis Pau
 34 galegiformis L.
 35 frigidus (L.) A. Gray

(a) frigidus	16	No	Knaben, 1950

  (b) grigorjewii (B. Fedtsch.) Chater
 36 penduliflorus Lam.
 37 umbellatus Bunge
 38 alpinus L.

(a) alpinus	16	He	Favarger, 1965
(b) arcticus Lindman	16	Rs(N)	Sokolovskaja & Strelkova, 1960
39 depressus L.	16	Ga	Favarger, 1965

 40 norvegicus Weber

41 australis (L.) Lam.	32 48	It	Favarger, 1965

 42 lusitanicus Lam.
  (a) lusitanicus
  (b) orientalis Chater & Meikle

43 glycyphyllos L.   16   It   Larsen, 1955
44 glycyphylloides DC.
45 dasyanthus Pallas
46 pubiflorus DC.
47 tanaiticus C. Koch
48 longipetalus Chater
49 exscapus L.
50 huetii Bunge
51 utriger Pallas
52 henningii (Steven) Boriss.
53 wolgensis Bunge
54 nummularius Lam.
55 hellenicus Boiss.
56 anatolicus Boiss.
57 tremolsianus Pau
58 graecus Boiss. & Spruner
59 drupaceus Orph. ex Boiss.
60 creticus Lam.
   (a) creticus
   (b) rumelicus (Bunge) Maire & Petitmengin
61 granatensis Lam.
   (a) granatensis
   (b) siculus (Biv.) Franco & P. Silva
62 arnacantha Bieb.
63 parnassi Boiss.
   (a) parnassi
   (b) calabrus (Fiori) Chater
   (c) cylleneus (Boiss. & Heldr. ex
       Fischer) Hayek
64 thracicus Griseb.
65 troianus Steven ex Fischer
66 physocalyx Fischer   16   Bu   Kuuuluiuv &
                                                           Kuzmanov, 1965b
67 ponticus Pallas
68 vulpinus Willd.
69 alopecurus Pallas
70 alopecuroides L.
71 grossii Pau
72 centralpinus Br.-Bl.
73 ajubensis Bunge
74 clusii Boiss.
75 sempervirens Lam.
   (a) sempervirens
   (b) cephalonicus (C. Presl) Ascherson
       & Graebner
   (c) muticus (Pau) Rivas Goday & Borja
   (d) nevadensis (Boiss.) P. Monts.
76 giennensis Heywood
77 angustifolius Lam.
   (a) angustifolius
   (b) pungens (Willd.) Hayek
78 balearicus Chater
79 massiliensis (Miller) Lam.
80 sirinicus Ten.
   (a) sirinicus
   (b) genargenteus (Moris) Arcangeli
81 odoratus Lam.
82 algarbiensis Cosson ex Bunge
83 falcatus Lam.
84 asper Jacq.
85 roemeri Simonkai

```
 86 austriacus Jacq.
 87 sulcatus L.
 88 clerceanus Iljin & Krasch.
 89 tenuifolius L.
 90 arenarius L.
 91 baionensis Loisel.
 92 tenuifoliosus Maire
 93 mesopterus Griseb.
 94 onobrychis L.
 95 leontinus Wulfen                       32      He      Favarger, 1959a
 96 amarus Pallas
 97 helmii Fischer ex DC.
 98 idaeus Bunge
 99 autranii Bald.
100 agraniotii Orph. ex Boiss.
101 arcuatus Kar. & Kir.
102 reduncus Pallas
103 dolichophyllus Pallas
104 lacteus Heldr. & Sart. ex Boiss.
105 baldaccii Degen
106 wilmottianus Stoj.
107 testiculatus Pallas
108 rupifragus Pallas
109 physodes L.
110 monspessulanus L.
    (a) monspessulanus                     16      Ga      Guinochet &
                                                           Logeois, 1962
    (b) illyricus (Bernh.) Chater
111 spruneri Boiss.
112 incanus L.                             16      Hs      Lorenzo-Andreu &
                                                           Garcia, 1950
    (a) incanus
    (b) incurvus (Desf.) Chater
    (c) nummularioides (Desf. ex DC.) Maire
    (d) macrorhizus (Cav.) Chater
113 sericophyllus Griseb.
114 apollineus Boiss. & Heldr.
115 karelinianus M. Popov
116 subuliformis DC.
117 corniculatus Bieb.
118 muelleri Steudel & Hochst.
119 pugionifer Fischer ex Bunge
120 cornutus Pallas
121 brachylobus DC.
122 varius S.G. Gmelin
123 macropus Bunge
124 pallescens Bieb.
125 glaucus Bieb.
126 zingeri Korsh.
127 vesicarius L.
    (a) vesicarius                         16      Ga      Guinochet &
                                                           Logeois, 1962
    (b) carniolicus (A. Kerner) Chater     32
    (c) pastellianus (Pollini) Arcangeli
128 peterfii Jáv.
129 hispanicus Cosson ex Bunge
130 hegelmaieri Willk.
131 albicaulis DC.
132 medius Schrenk
133 fialae Degen
```

39 Oxytropis DC.
```
  1 lapponica (Wahlenb.) Gay           16    No        Knaben &
                                                       Engelskjön, 1967
  2 deflexa (Pallas) DC.               16    No        Knaben &
                                                       Engelskjön, 1967
  3 jacquinii Bunge
  4 carpatica Uechtr.
  5 gaudinii Bunge
  6 amethystea Arvet-Touvet            16    Ga        Damboldt, 1964
  7 pyrenaica Godron & Gren.           16    It        Favarger, 1959a
  8 triflora Hoppe
  9 mertensiana Turcz.
 10 hippolyti Boriss.
 11 campestris (L.) DC.                48    Fe        Sorsa, 1963
    (a) campestris                     48    Europe    Jalas, 1950
    (b) tiroliensis (Sieber ex Fritsch)
        Leins & Merxm.
    (c) sordida (Willd.) Hartman fil.  48    No        Laane, 1966
 12 prenja (G. Beck) G. Beck
 13 urumovii Ják.
 14 foucaudii Gillot                   32    Ga        Favarger &
                                                       Huynh, 1964
 15 spicata (Pallas) O. & B. Fedtsch.
 16 songorica (Pallas) DC.
 17 ambigua (Pallas) DC.
 18 halleri Bunge ex Koch
    (a) halleri                        32    He        Favarger, 1965
    (b) velutina (Sieber) O. Schwarz   16    Ga        Favarger &
                                                       Huynh, 1964
 19 uralensis (L.) DC.
 20 pilosa (L.) DC.                    16    Hu        Baksay  1958
 21 pallasii Pers.
 22 floribunda (Pallas) DC.
 23 purpurea (Bald.) Markgraf
 24 fetida (Vill.) DC.                 16    He        Favarger, 1959a
```

40 Biserrula L.
  1 pelecinus L.

41 Glycyrrhiza L.
  1 foetida Desf.
  2 glabra L.
  3 korshinskyi Grigoriev
  4 aspera Pallas
  5 echinata L.

42 Amorpha L.
  1 fruticosa L.

43 Psoralea L.
  1 bituminosa L.
  2 americana L.

44 Apios Fabr.
  1 americana Medicus

45 Phaseolus L.
  1 vulgaris L.
  2 coccineus L.

46 Vigna Savi

```
    1 unguiculata (L.) Walpers

47 Glycine Willd.
    1 max (L.) Merr.

48 Cicer L.
    1 arietinum L.
    2 incisum (Willd.) K. Malý
    3 montbretii Jaub. & Spach
    4 graecum Orph. ex Boiss.

49 Vicia L.
    1 orobus DC.
    2 montenegrina Rohlena
    3 sparsiflora Ten.                    12      Hu      Baksay, 1956
    4 pinetorum Boiss. & Spruner
    5 ochroleuca Ten.
      (a) ochroleuca                      12      It      Larsen, 1955b
      (b) dinara (Borbás) K. Malý ex Rohlena
    6 pisiformis L.
    7 argentea Lapeyr.
    8 serinica Uechtr. & Huter
    9 sicula (Rafin.) Guss.
   10 cracca L.                           14      Su      Lövkvist, 1963
                                          27      Ge      Rousi, 1961
                                          28      Fe      V. Sorsa, 1962
                                          30      Da      Rousi, 1961
      oreophila Zertova                   28      Cz      Žertová, 1962
   11 incana Gouan                        12      Cz      Žertová, 1963
   12 tenuifolia Roth                     24      Hu      Baksay, 1954
   13 dalmatica A. Kerner                 12      Hu      Baksay, 1954
   14 sibthorpii Boiss.
   15 cassubica L.                        12      Su      Lövkvist, 1963
   16 sylvatica L.                        14      Su      Lövkvist, 1963
   17 multicaulis Ledeb.
   18 onobrychioides L.
   19 altissima Desf.
   20 dumetorum L.
   21 villosa Roth
      (a) villosa
      (b) varia (Host) Corb.
      (c) eriocarpa (Hausskn.) P.W. Ball
      (d) microphylla (D'Urv.) P.W. Ball
      (e) pseudocracca (Bertol.) P.W. Ball 14      It      Larsen, 1955b
   22 benghalensis L.                     14      Lu      Fernandes &
   23 glauca C. Presl                                     Santos, 1971
   24 biennis L.
   25 cretica Boiss. & Heldr.
      (a) cretica
      (b) aegaea (Halácsy) P.W. Ball
   26 monantha Retz.
      (a) monantha
      (b) triflora (Ten.) B.L. Burtt &
          P. Lewis
   27 articulata Hornem.
   28 ervilia (L.) Willd.
   29 leucantha Biv.
   30 vicioides (Desf.) Coutinho
   31 hirsuta (L.) S.F. Gray
   32 meyeri Boiss.
   33 durandii Boiss.
```

34 disperma DC.
35 tenuissima (Bieb.) Schinz & Thell.
36 tetrasperma (L.) Schreber   14   Ho   Gadella &
                                                     Kliphuis, 1966
37 pubescens (DC.) Link
38 bifoliolata Rodr.
39 oroboides Wulfen   14   Hu   Baksay, 1956
40 truncatula Fischer ex Bieb.
41 pyrenaica Pourret
42 sepium L.   14   Ho   Gadella &
                                                     Kliphuis, 1966
43 grandiflora Scop.
44 barbazitae Ten. & Guss.
45 pannonica Crantz
   (a) pannonica
   (b) striata (Bieb.) Nyman
46 sativa L.
   (a) nigra (L.) Ehrh.   12   It   Larsen, 1956a
                                     14   Su   Lövkvist, 1963
   (b) amphicarpa (Dorthes) Ascherson &
       Graebner
   (c) cordata (Wulfen ex Hoppe)   10   Al Ga Hanelt & Mettin,
       Ascherson & Graebner   Gr It   1966
   (d) incisa (Bieb.) Arcangeli
   (e) sativa   12   It   Larsen, 1956a
   (f) macrocarpa (Moris) Arcangeli
47 lathyroides L.   12   Su   Lövkvist, 1963
48 cuspidata Boiss.
49 peregrina L.
50 melanops Sibth. & Sm.
51 lutea L.
   (a) lutea   14   Su   Lövkvist, 1963
   (b) vestita (Boiss.) Rouy
52 hybrida L
53 bithynica (L.) L.   14   It   Larsen, 1955b
54 narbonensis L.
55 faba L.

50 Lens Miller
   1 nigricans (Bieb.) Godron
   2 culinaris Medicus
   3 orientalis (Boiss.) M. Popov
   4 ervoides (Brign.) Grande

51 Lathyrus L.
   1 vernus (L.) Bernh.   14   Fe   Sorsa, 1962
   2 venetus (Miller) Wohlf.   14   He Ju Rm Brunsberg, 1965
   3 niger (L.) Bernh.
   (a) niger   14   Cz Hu Rm Brunsberg, 1965
   (b) jordanii (Ten.) Arcangeli
   4 japonicus Willd.
   (a) japonicus   14   Is   Löve & Löve, 1956
   (b) maritimus (L.) P.W. Ball   14   Fe Po Su Brunsberg, 1967
   5 laevigatus (Waldst. & Kit.) Gren.
   (a) laevigatus   14   Rm   Brunsberg, 1965
   (b) occidentalis (Fischer & C.A. Meyer)
       Breistr.   14   Ga   Favarger &
                                                     Huynh, 1964
   6 transsilvanicus (Sprengel) Fritsch 14   Rm   Brunsberg, 1965
   7 gmelinii Fritsch
   8 aureus (Steven) Brandza

9 pisiformis L.	14	Hu	Brunsberg, 1967
10 humilis (Ser.) Sprengel			
11 incurvus (Roth) Willd.	14	Ju Su	Brunsberg, 1967
12 filiformis (Lam.) Gay			
13 bauhinii Genty			
14 digitatus (Bieb.) Fiori	14	Rs(K)	Trankovsky, 1940
15 pannonicus (Jacq.) Garcke	14	Hu	Baksay, 1961
(a) pannonicus			
(b) collinus (Ortmann) Soó			
(c) asphodeloides (Gouan) Bassler			
(d) hispanicus (Lacaita) Bassler			
(e) varius (C. Koch) P.W. Ball			
16 pallescens (Bieb.) C. Koch			
17 pancicii (Juriš.) Adamović			
18 alpestris (Waldst. & Kit.) Kit. ex			
Čelak.			
19 montanus Bernh.	14	Ga Ge	Brunsberg, 1965
20 pratensis L.	7	Fe	Simola, 1964
	8	Fe	Simola, 1964
	9	Fe	Simola, 1964
	12	Fe	Simola, 1964
	14	Au Cz Da	Larsen, 1957a
	16	Su	Simola, 1964
	21	Su	Brunsberg, 1965
	28	Be Ga He Ho	Brunsberg, 1965
	42	Ga	Brunsberg, 1965
21 binatus Pančić	14	Su	Brunsberg, 1965
22 hallersteinii Baumg.	14	Rm Su	Brunsberg, 1965
23 palustris L.			
(a) palustris	42	Br Fe Ge Ho	Brunsberg, 1965
(b) nudicaulis (Willk.) P.W. Ball			
24 laxiflorus (Desf.) O. Kuntze	14	Rs(K)	Trankovsky, 1940
25 neurolobus Boiss. & Heldr.	14	Gr	Brunsberg, 1965
26 cirrhosus Ser.			
27 tuberosus L.	14	Cz Ga	Brunsberg, 1965
28 roseus Steven			
29 grandiflorus Sibth. & Sm.			
30 rotundifolius Willd.	14	Rs(K)	Trankovsky, 1940
31 undulatus Boiss.			
32 sylvestris L.	14	Br Cz Su	Brunsberg, 1965
33 latifolius L.	14	Ga Ju	Brunsberg, 1965
34 heterophyllus L.	14	He	Brunsberg, 1965
35 tremolsianus Pau			
36 tingitanus L.			
37 odoratus L.	14	It Tu	Brunsberg, 1965
38 saxatilis (Vent.) Vis.			
39 sphaericus Retz.	14	Da	Larsen, 1954
40 angulatus L.			
41 inconspicuus L.			
42 setifolius L.			
43 cicera L.	14	Ga Gr	Brunsberg, 1965
44 sativus L.	14	Ga Ge Ju	Brunsberg, 1965
45 amphicarpos L.			
46 annuus L.			
47 hierosolymitanus Boiss.			
48 gorgoni Parl.			
49 hirsutus L.	14	Ga Lu	Brunsberg, 1965
50 clymenum L.			
51 articulatus L.			
52 ochrus (L.) DC.	14	Ga	Brunsberg, 1965
53 nissolia L.			

54 aphaca L.   14   Ga Lu   Brunsberg, 1965

52 Pisum L.
  1 sativum L.
   (a) sativum
   (b) elatius (Bieb.) Ascherson & Graebner

53 Ononis L.
  1 rotundifolia L.
  2 tridentata L.   30   Hs   Lorenzo-Andreu, 1951
  3 fruticosa L.
  4 cristata Miller
  5 adenotricha Boiss.
  6 natrix L.
   (a) natrix
   (b) ramosissima (Desf.) Batt.
   (c) hispanica (L. fil.) Coutinho
  7 crispa L.
  8 ornithopodiodes L.
  9 biflora Desf.
 10 maweana Ball
 11 sicula Guss.
 12 reclinata L.
 13 dentata Solander ex Lowe
 14 pendula Desf.
 15 laxiflora Desf.
 16 verae Širj.
 17 pubescens L.
 18 viscosa L.   32   It   Larsen, 1956a
   (a) viscosa
   (b) brachycarpa (DC.) Batt.
   (c) breviflora (DC.) Nyman
   (d) sieberi (Besser ex DC.) Širj.
   (e) subcordata (Cav.) Širj.
   (f) foetida (Schousboe ex DC.) Širj.
 19 crotalarioides Cosson
 20 cintrana Brot.
 21 speciosa Lag.
 22 aragonensis Asso
 23 reuteri Boiss.
 24 pinnata Brot.
 25 leucotricha Cosson
 26 pusilla L.
 27 saxicola Boiss. & Reuter
 28 minutissima L.   30   Ga   Morisset, 1964
 29 striata Gouan
 30 cephalotes Boiss.
 31 hispida Desf.
 32 spinosa L.   30   It   Larsen, 1956a
   (a) spinosa
   (b) antiquorum (L.) Arcangeli
   (c) leiosperma (Boiss.) Širj.
   (d) austriaca (G. Beck) Gams
 33 repens L.   30   Ga   Larsen, 1956a
                                                60   Ho   Gadella &
                                                                Kliphuis, 1967a
 34 arvensis L.   30   Po   Morisset,1964
 35 masquillierii Bertol.   30   It   Morisset,1964
 36 gilivsulid Salzm. ex Boiss.
 37 alba Poiret
 38 oligophylla Ten.

39 variegata L.
40 euphrasiifolia Desf.
41 hirta Poiret
42 cossoniana Boiss.
43 diffusa Ten.
44 serrata Forskål
45 tournefortii
46 subspicata Lag.
47 mitissima L.
48 alopecuroides L.
49 baetica Clemente

54 Melilotus Miller
  1 dentata (Waldst. & Kit.) Pers.   16   Su   Lövkvist, 1963
  2 altissima Thuill.   16   Ge   Scheerer, 1939
  3 alba Medicus   16   Su   Lövkvist, 1963
  4 wolgica Poiret
  5 officinalis (L.) Pallas   16   Hu   Pólya, 1949
  6 polonica (L.) Pallas
  7 taurica (Bieb.) Ser.
  8 italica (L.) Lam.
  9 neapolitana Ten.   16   It   Larsen, 1955b
10 indica (L.) All.
11 elegans Salzm. ex Ser.
12 infesta Guss.
13 sulcata Desf.
14 segetalis (Brot.) Ser.
15 messanensis (L.) All.
16 physocarpa Stefanov

55 Trigonella L.
  1 graeca (Boiss. & Spruner) Boiss.
  2 cretica (L.) Boiss.
  3 maritima Delile ex Poiret
  4 corniculata (L.) L.   16   It   Larsen, 1956a
  5 balansae Boiss. & Reuter
  6 rechingeri Sirj.
  7 grandiflora Bunge
  8 sprunerana Boiss.
  9 smyrnaea Boiss.
10 spinosa L.
11 spicata Sibth. & Sm.
12 striata L. fil.
13 arcuata C.A. Meyer
14 fischerana Ser.
15 polyceratia L.
16 orthoceras Kar. & Kir.
17 monspeliaca L.
18 caerulea (L.) Ser.
19 procumbens (Besser) Reichenb.
20 coerulescens (Bieb.) Halácsy
21 gladiata Steven ex Bieb.
22 cariensis Boiss.
23 foenum-graecum L.   16   Lu   Azevedo Coutinho
                            & Santos, 1943

56 Medicago L.
  1 lupulina L.   16   Hu   Pólya, 1948
  2 secundiflora Durieu
  3 hybrida (Pourret) Trautv.
  4 cretacea Bieb.

5 sativa L.
  (a) sativa   32   Su   Lövkvist, 1963
  (b) ambigua (Trautv.) Tutin
  (c) caerulea (Less. ex Ledeb.) Schmalh. 16   Rs(W)   Sinskaya &
                                                    Maleyeva, 1959
  (d) falcata (L.) Arcangeli   16   Bu   Kozuharov &
                                                  Kuzmanov, 1965a
                                  32   Su   Lövkvist, 1963
  (e) glomerata (Balbis) Tutin
6 prostrata Jacq.   32   Hu   Baksay, 1958
7 rupestris Bieb.
8 saxatilis Bieb.
9 cancellata Bieb.
10 suffruticosa Ramond ex DC.
  (a) suffruticosa
  (b) leiocarpa (Bentham) P. Fourn.
11 arborea L.
12 orbicularis (L.) Bartal.   16   Bu   Kozuharov &
                                                    Kuzmanov, 1965a
13 carstiensis Jacq.
14 intertexta (L.) Miller
15 ciliaris (L.) All.
16 scutellata (L.) Miller
17 blancheana Boiss.
18 rugosa Desr.
19 soleirolii Duby
20 tornata (L.) Miller
21 marina L.
22 truncatula Gaertner
23 rigidula (L.) All.   14   Bu   Kozuharov &
                                                  Kuzmanov, 1965a
                                  10   Bu   Kozuharov &
                                                    Kuzmanov, 1965a
24 pironae Vis.   16   It   Lesins & Lesins,
                                                  1961
25 littoralis Rohde ex Loisel.
26 aculeata Gaertner
27 globosa C. Presl
28 turbinata (L.) All.
29 murex Willd.
30 arabica (L.) Hudson
31 polymorpha L.   14   It   Larsen, 1956a
32 praecox DC.
33 laciniata (L.) Miller
  (a) laciniata
  (b) schimperana P. Fourn.
34 coronata (L.) Bartal.   16   Bu   Kozuharov &
                                                  Kuzmanov, 1965a
35 disciformis DC.
36 tenoreana Ser.
37 minima (L.) Bartal.   16   It   Larsen, 1956a

57 Trifolium L.
  1 ornithopodioides L.   16   Ho   Kliphuis, 1962
  2 lupinaster L.   32   Rs(C)   Iljin &
    ciswolgense Sprygin ex Iljin & Truch. 16   Rs(C)   Trukhaleva,
    litwinowii Iljin   32   Rs(C)   1960
  3 alpinum L.   16   Au   Reese, 1953
  4 pilczii Adamović
  5 strictum L.
  6 nervulosum Boiss. & Heldr.

```
 7 montanum L.                                  16    Hu    Pólya, 1950
   balbisianum Ser.                             16    Ga    Guinochet &
                                                              Logeois, 1962
 8 ambiguum Bieb.
 9 parnassi Boiss. & Spruner
10 repens L.
   (a) repens                                   32    Br    Coombe, 1961
   (b) nevadense (Boiss.) D.E. Coombe
   (c) orbelicum (Velen.) Pawl.
   (d) ochranthum E.I. Nyárády
   (e) orphanideum (Boiss.) D.E. Coombe
   (f) prostratum Nyman
11 occidentale D.E. Coombe                      16    Br    Coombe, 1961
12 pallescens Schreber                          16    Ga    Favarger, 1965b
13 thalii Vill.
14 hybridum L.
   (a) hybridum                                 16    Ho    Gadella &
                                                              Kliphuis, 1966
   (b) elegans (Savi) Ascherson & Graebner
   (c) anatolicum (Boiss.) Hossain
15 bivonae Guss.
16 isthmocarpum Brot.
   (a) isthmocarpum
   (b) jaminianum (Boiss.) Murb.
17 nigrescens Viv.                              16    It    Larsen, 1956a
   (a) nigrescens
   (b) petrisavii (G.C. Clementi)
       Holmboe
18 michelianum Savi
19 angulatum Waldst. & Kit.
20 retusum L.
21 cernuum Brot.
22 glomeratum L.
23 suffocatum L.
24 uniflorum L.
25 spumosum L.
26 vesiculosum Savi
27 mutabile Portenschl.
28 physodes Steven ex Bieb.
29 fragiferum L.
   (a) fragiferum                               16    Hu    Pólya, 1948
   (b) bonannii (C. Presl) Soják
30 resupinatum L.
31 tomentosum L.
32 badium Schreber
33 spadiceum L.
34 speciosum Willd.
35 boissieri Guss. ex Boiss.
36 lagrangei Boiss.
37 brutium Ten.
38 mesogitanum Boiss.
39 aurantiacum Boiss. & Spruner
40 dolopium Heldr. & Hausskn. ex Gibelli
      & Belli
41 patens Schreber
42 aureum Pollich                               16    Su    Lövkvist, 1962
43 velenovskyi Vandas
44 campestre Schreber                           14    Hs    Kliphuis, 1962
45 sebastianii Savi
46 dubium Sibth.                                28    Ho    Kliphuis, 1962
                                                32    Ho    Gadella &
                                                              Kliphuis, 1966
```

47 micranthum Viv.	16	Ho	Kliphuis, 1962
48 striatum L.	14	Ho	Kliphuis, 1962
49 arvense L.	14 Au Da Hu Ju		Bücher & Larsen, 1958

50 affine C. Presl
51 saxatile All.
52 bocconei Savi
53 tenuifolium Ten.
54 trichopterum Pančić
55 phleoides Pourret ex Willd.
56 gemellum Pourret ex Willd.
57 ligusticum Balbis ex Loisel.

58 scabrum L.	10	Ga	Gadella & Kliphuis, 1968

59 dalmaticum Vis.

60 stellatum L.	14	It	Larsen, 1956a

61 dasyurum C. Presl
62 incarnatum L.
  (a) incarnatum
  (b) molinerii (Balbis ex Hornem.) Syme

63 pratense L.	14	Ce	Reese, 1952
var. frigidum Gaudin	14	Au	Reese, 1952

64 pallidum Waldst. & Kit.
65 diffusum Ehrh.
66 noricum Wulfen
67 wettsteinii Dörfler & Hayek
68 ottonis Spruner ex Boiss.
69 lappaceum L.
70 congestum Guss.
71 barbeyi Gibelli & Belli
72 hirtum All.
73 cherleri L.

74 medium L.	c.70	Su	Lövkvist, 1963
	78-80	Su	Lövkvist, 1963
	84	Su	Lövkvist, 1963

  (a) medium
  (b) banaticum (Heuffel) Hendrych
  (c) sarosiense (Hazsl.) Simonkai
  (d) balcanicum Velen.
75 heldreichianum Hausskn.
76 patulum Tausch
77 velebiticum Degen
78 pignantii Fauché & Chaub.
79 alpestre L.
80 rubens L.
81 angustifolium L.
82 purpureum Loisel.
83 desvauxii Boiss. & Blanche
84 smyrnaeum Boiss.
85 ochroleucon Hudson
86 pannonicum Jacq.
87 canescens Willd.
88 alexandrinum L.
89 apertum Bobrov
90 echinatum Bieb.
91 latinum Sebastiani
92 leucanthum Bieb.
93 squamosum L.
94 squarrosum L.
95 obscurum Savi
  (a) obscurum

```
    (b) aequidentatum (Pérez Lara) C.
        Vicioso
96 clypeatum L.
97 subterraneum L.                          16      Ho      Kliphuis, 1962
98 globosum L.
99 pauciflorum D'Urv.


58 Dorycnium Miller
 1 hirsutum (L.) Ser.                        14      Ju      Larsen, 1956a
 2 rectum (L.) Ser.
 3 graecum (L.) Ser.
 4 pentaphyllum Scop.
   (a) pentaphyllum
   (b) germanicum (Gremli) Gams
   (c) gracile (Jordan) Rouy
   (d) herbaceum (Vill.) Rouy


59 Lotus L.
 1 tenuis Waldst. & Kit. ex Willd.          12      Hu      Pólya, 1949
 2 krylovii Schischkin & Serg.
 3 borbasii Ujhelyi                          12      Hu      Ujhelyi, 1960
 4 delortii Timb.-Lagr. ex F.W. Schultz
 5 stenodon (Boiss. & Heldr.) Heldr.
 6 glareosus Boiss. & Reuter
 7 corniculatus L.                           24      Au      Titz, 1964
 8 alpinus (DC.) Schleicher ex Ramond        12      Ga      Favarger &
                                                            Huynh, 1964
 9 uliginosus Schkuhr                        12      Ga Ge Lu  Larsen, 1955a
10 pedunculatus Cav.
11 preslii Ten.
12 palustris Willd.
13 aegaeus (Griseb.) Boiss.
14 parviflorus Desf.                         12      Lu      Larsen, 1956b
15 subbiflorus Lag.
   (a) subbiflorus                           24      Br      Rutland, 1941
   (b) castellanus (Boiss. & Reuter)
       P.W. Ball
16 angustissimus L.                          12      Hu      Larsen, 1955a
17 edulis L.                                 14      It      Larsen, 1955a
18 conimbricensis Brot.                      12      Lu      Larsen, 1956b
19 aduncus (Griseb.) Nyman
20 cytisoides L.                             14      Hs Ju   Larsen, 1958a
21 drepanocarpus Durieu
22 creticus L.                               28      Lu      Larsen, 1958a
23 collinus (Boiss.) Heldr.
24 strictus Fischer & C.A. Meyer
25 halophilus Boiss. & Spruner
26 ornithopodioides L.                       14      It      Larsen, 1955a
27 peregrinus L.
28 azoricus P.W. Ball
29 arenarius Brot.
30 tetraphyllus L.


60 Tetragonolobus Scop.
 1 maritimus (L.) Roth                       14      Su      Lövkvist, 1963
 2 biflorus (Desr.) Ser.
 3 purpureus Moench
 4 conjugatus (L.) Link
 5 requienii (Mauri ex Sanguinetti)
   Sanguinetti
```

61 Hymenocarpos Savi
  1 circinnatus (L.) Savi

62 Securigera DC.
  1 securidaca (L.) Degen & Dörfler   12   It   Larsen, 1955b

63 Anthyllis L.
  1 cytisoides L.
  2 terniflora (Lag.) Pau
  3 hermanniae L.
  4 barba-jovis L.
  5 aegaea Turrill
  6 henoniana Cosson ex Batt.
  7 montana L.   12   Ga   Guinochet &
                                                Logeois, 1962
    (a) montana
    (b) jacquinii (A. Kerner) Hayek
    (c) hispanica (Degen & Hervier) Cullen
  8 aurea Welden
  9 polycephala Desf.
 10 tejedensis Boiss.
 11 rupestris Cosson
 12 ramburii Boiss.
 13 onobrychoides Cav.
 14 gerardi L.
 15 vulneraria L.
    (a) vulneraria
    (b) maritima (Schweigger) Corb.
    (c) polyphylla (DC.) Nyman   12   Hu   Baksay, 1956
    (d) bulgarica (Sagorski) Cullen
    (e) boissieri (Sagorski) Bornm.
    (f) pulchella (Vis.) Bornm.
    (g) arundana (Boiss. & Reuter) Vasc.
    (h) argyrophylla (Rothm.) Cullen
    (i) reuteri Cullen
    (j) atlantis Emberger & Maire
    (k) vulnerarioides (All.) Arcangeli
    (l) forondae (Sennen) Cullen
    (m) pindicola Cullen
    (n) praepropera (A. Kerner) Bornm.
    (o) weldeniana (Reichenb.) Cullen
    (p) maura (G. Beck) Lindb.
    (q) hispidissima (Sagorski) Cullen
    (r) corbierei (Salmon & Travis) Cullen
    (s) carpatica (Pant.) Nyman
    (t) alpestris Ascherson & Graebner   12   Hu   Baksay, 1956
    (u) pyrenaica (G. Beck) Cullen
    (v) iberica (W. Becker) Jalas
    (w) lapponica (Hyl.) Jalas   12   Su   Jalas, 1950
    (x) borealis (Rouy) Jalas
 16 tetraphylla L.
 17 lotoides L.
 18 cornicina L.
 19 hamosa Desf.

64 Ornithopus L.
  1 compressus L.
  2 sativus Brot.
    (a) sativus
    (b) isthmocarpus (Cosson) Dostál
  3 perpusillus L.   14   Ge   Scheerer, 1939

```
  4 pinnatus (Miller) Druce            14      Ga      Ball, 1961

65 Coronilla L.
   1 emerus L.
     (a) emerus                        14      It      Larsen, 1955b
     (b) emeroides (Boiss. & Spruner) Hayek
   2 valentina L.
     (a) valentina
     (b) glauca (L.) Batt.
   3 vaginalis Lam.                     12      Hu      Baksay, 1956
   4 minima L.                          24      Ga      Guinochet &
                                                        Logeois, 1962
   5 juncea L.
   6 coronata L.                        10      Hu      Baksay, 1956
   7 varia L.                           24      Hu      Baksay, 1956
   8 elegans Pančić                     12      Hu      Baksay, 1956
   9 globosa Lam.
  10 cretica L.
  11 rostrata Boiss. & Spruner
  12 scorpioides (L.) Koch              12      It      Larsen, 1955b
  13 repanda (Poiret) Guss.
     (a) repanda
     (b) dura (Cav.) Coutinho

66 Hippocrepis L.
   1 glauca Ten.
   2 scabra DC.
   3 comosa L.                          28      Br      Maude, 1940
   4 squamata (Cav.) Cosson
     (a) squamata
     (b) eriocarpa (Boiss.) Nyman
   5 valentina Boiss.
   6 balearica Jacq.
   7 ciliata Willd.
   8 multisiliquosa L.                  14      Hs      Lorenzo-Andreu &
                                                        Garcia, 1950
   9 salzmanii Boiss. & Reuter
  10 unisiliquosa L.

67 Hammatolobium Fenzl
   1 lotoides Fenzl

68 Scorpiurus L.
   1 muricatus L.                       28      Hs      Dominguez &
   2 vermiculatus L.                                    Galiano, 1974

69 Eversmannia Bunge
   1 subspinosa (Fischer ex DC.) B.
     Fedtsch.

70 Hedysarum L.
   1 coronarium L.                      16      It      Larsen, 1955b
   2 flexuosum L.
   3 spinosissimum L.
   4 glomeratum F.G. Dietrich
   5 hedysaroides (L.) Schinz & Thell.
     (a) hedysaroides                   14      Au      Reese, 1953
     (b) exaltatum (A. Kerner) Žertová
     (c) arcticum (B. Fedtsch.) P.W. Ball
   6 alpinum L.
   7 boutignyanum Alleiz.
```

    8 gmelinii Ledeb.
    9 razoumowianum Helm & Fischer ex DC.
   10 cretaceum Fischer ex DC.
   11 ucrainicum B. Kaschm.
   12 tauricum Pallas ex Willd.
   13 humile L.
   14 varium Willd.
   15 candidum Bieb.
   16 grandiflorum Pallas
   17 biebersteinii Žertová
   18 macedonicum Bornm.   16   Ju   Šopova, 1966

71 Onobrychis Miller
    1 hypargyrea Boiss.
    2 pallasii (Willd.) Bieb.
    3 radiata (Desf.) Bieb.
    4 saxatilis (L.) Lam.
    5 petraea (Bieb. ex Willd.) Fischer
    6 ebenoides Boiss. & Spruner
    7 supina (Chaix) DC.   14   Ga   Guinochet &
                                                             Logeois, 1962
    8 gracilis Besser
    9 pindicola Hausskn.
   10 stenorhiza DC.
   11 alba (Waldst. & Kit.) Desv.
      (a) alba
      (b) laconica (Orph. ex Boiss.) Hayek
      (c) echinata (G. Don fil.) P.W. Ball
      (d) calcarea (Vandas) P.W. Ball
   12 degenii Dörfler
   13 sphaciotica W. Greuter
   14 peduncularis (Cav.) DC.
      (a) peduncularis
      (b) matritensis (Boiss. & Reuter) Maire
   15 argentea Boiss.
      (a) argentea
      (b) hispanica (Širj.) P.W. Ball
   16 reuteri Leresche
   17 montana DC.
      (a) montana   28   It   Larsen, 1960
      (b) scardica (Griseb.) P.W. Ball
      (c) cadmea (Boiss.) P.W. Ball
   18 arenaria (Kit.) DC.
      (a) arenaria   14   He   Favarger, 1953
                                           28   Hu   Baksay, 1956
      (b) taurerica Hand.-Mazz.
      (c) sibirica (Turcz. ex Besser)
          P.W. Ball
      (d) miniata (Steven) P.W. Ball
      (e) tommasinii (Jordan) Ascherson &
          Graebner   14   It   Larsen, 1955b
      (f) lasiostachya (Boiss.) Hayek
      (g) cana (Boiss.) Hayek
   19 oxyodonta Boiss.
   20 inermis Steven
   21 viciifolia Scop.   28   Hu   Pólya, 1950
   22 caput-galli (L.) Lam.   14   It   Larsen, 1956a
   23 aequidentata (Sibth. & Sm.) D'Urv.

72 Ebenus L.
    1 cretica L.

114

2 sibthorpii DC.

73 Alhagi Gagnebin
  1 pseudalhagi (Bieb.) Desv.
  2 graecorum Boiss.

74 Arachis L.
  1 hypogaea L.

LXXXII OXALIDACEAE

1 Oxalis L.

1 corniculata L.	24	Br	Rutland, 1941
2 exilis A. Cunn.			
3 stricta L.	c.24	Po	Skalińska et al., 1959
4 europaea Jordan	24	Ho	Gadella & Kliphuis, 1966
5 articulata Savigny			
6 acetosella L.	22	Br	Marks, 1956
7 corymbosa DC.			
8 latifolia Kunth			
9 tetraphylla Cav.			
10 pes-caprae L.			
11 purpurea L.			
12 incarnata L.			

LXXXIII GERANIACEAE

1 Geranium L.

1 macrorrhizum L.	46	Al	Warburg, 1938
2 dalmaticum (G. Beck) Rech. fil.			
3 cinereum Cav.			
(a) cinereum			
(b) subcaulescens (L'Hér. ex DC.) Hayek			
4 humbertii Beauverd			
5 argenteum L.			
6 sanguineum L.	84	Su	Lövkvist, 1963
	(52-56)	Fe	Sorsa, 1963
7 pratense L.	28	Br	Warburg, 1938
8 sylvaticum L.			
(a) sylvaticum	28	Po	Skalińska et al., 1959
(b) rivulare (Vill.) Rouy	28	He	Favarger, 1965
(c) caeruleatum (Schur) D.A. Webb			
(d) pseudosibiricum (J. Mayer) D.A. Webb & I.K. Ferguson			
9 endressii Gay			
10 versicolor L.			
11 nodosum L.			
12 albiflorum Ledeb.			
13 peloponesiacum Boiss.			
14 ibericum Cav.			
15 phaeum L.	28	Po	Skalińska et al., 1959
16 reflexum L.			
17 aristatum Freyn & Sint.			
18 tuberosum L.			
19 macrostylum Boiss.			

```
20 linearilobum DC.
21 malviflorum Boiss. & Reuter
22 asphodeloides Burm. fil.
23 palustre L.                          28      Hu      Pólya, 1950
24 collinum Stephan ex Willd.
25 divaricatum Ehrh.
26 bohemicum L.                         28      No Su   Dahlgren, 1952
27 lanuginosum Lam.                     48      Su      Dahlgren, 1952
28 sibiricum L.
29 pyrenaicum Burm. fil.                26      Ge      Gauger, 1937
30 rotundifolium L.                     26      It      Larsen, 1956a
31 molle L.                             26      Ho      Gadella & Kliphuis,
                                                        1966
32 brutium Gasparr.
33 pusillum L.                          26      Hu      Pólya, 1950
                                        36              error
34 columbinum L.                        18      Br      Warburg, 1938
35 dissectum L.                         22      Br      Warburg, 1938
36 lucidum L.                           20      Br Lu   Warburg, 1938
37 robertianum L.                       32      Fe      Sorsa, 1962
                                        64      Hu      Pólya, 1950
38 purpureum Vill.                      32      Ga Lu   Bocher & Larsen,
                                                        1955
39 cataractarum Cosson

2 Erodium L'Hér.
  1 gussonii Ten.                       20      It      Guittonneau, 1967
  2 guttatum (Desf.) Willd.
  3 boissieri Cosson                    20      Hs      Guittonneau, 1965
  4 chium (L.) Willd.                   20      Ga      Guittonneau, 1964
    (a) chium
    (b) littoreum (Leman) Ball
  5 laciniatum (Cav.) Willd.            20      Hs      Guittonneau, 1965
  6 malacoides (L.) L'Hér.              40      Lu      Fernandes &
                                                        Queirós, 1971
  7 alnifolium Guss.                    20      Si      Guittonneau, 1967
  8 maritimum (L.) L'Hér.               20      Br      Larsen, 1958
  9 sanguis-christi Sennen              20      Hs      Guittonneau, 1964
 10 corsicum Leman                      20      Co      Contandriopoulos,
                                                        1957
                                        18      Co      Negodi, 1937
 11 reichardii (Murray) DC.             20      Bl      Guittonneau, 1967
 12 ruthenicum Bieb.
 13 hoefftianum C.A. Meyer
 14 botrys (Cav.) Bertol.               40      Hs      Guittonneau, 1965
 15 gruinum (L.) L'Hér.
 16 ciconium (L.) L'Hér.                18      Ga      Guittonneau, 1964
 17 guicciardii Heldr. ex Boiss.
 18 chrysanthum L'Hér. ex DC.
 19 absinthoides Willd.
 20 beketowii Schmalh.
 21 alpinum L'Hér.                       18      It      Guittonneau, 1966
 22 rupestre (Pourret ex Cav.)
    Guittonneau                         20      Hs      Guittonneau, 1965
 23 petraeum (Gouan) Willd.             20      Ga      Guittonneau, 1964
    (a) petraeum
    (b) lucidum (Lapeyr.) D.A. Webb & Chater
    (c) glandulosum (Cav.) Bonnier
    (d) crispum (Lapeyr.) Rouy
    (e) valentinum (Lange) D.A. Webb &
        Chater
```

```
24 rodiei (Br.-Bl.) Poirion
25 cicutarium (L.) L'Hér.
   (a) cicutarium                      20    Ge       Rottgardt, 1956
                                       36    Ge       Rottgardt, 1956
                                       40    Ge       Rottgardt, 1956
                                       48    Ge       Rottgardt, 1956
                                       54    Ge       Rottgardt, 1956
   (b) bipinnatum Tourlet              20
                                       40    Lu       Guittonneau, 1966
   (c) jacquinianum (Fischer, C.A. Meyer
       & Ave-Lall.) Briq.
       subsp. dunense                  40    Br Da    Larsen, 1958
     danicum K. Larsen                 60    Da       Larsen, 1958
26 moschatum (L.) L'Hér.               20    ?Europe  Warburg, 1938
27 acaule (L.) Becherer & Thell.       40    It       Guittonneau, 1966
28 carvifolium Boiss. & Reuter         20
                                       40    Hs       Guittonneau, 1967
29 paui Sennen
30 rupicola Boiss.                     20
31 astragaloides Boiss. & Reuter
32 daucoides Boiss.                    40    Hs       Guittonneau, 1967
                                       60    Hs       Guittonneau, 1967
                                       80    Hs       Guittonneau, 1967
33 manescavi Cosson                    40    Ga       Guittonneau, 1966
34 hirtum (Forskål) Willd.
```

3 Biebersteinia Stephan
  1 orphanidis Boiss.

LXXXIV TROPAEOLACEAE

1 Tropaeolum L.
  1 majus L.

LXXXV ZYGOPHYLLACEAE

```
1 Peganum L.
  1 harmala L.                         24    Hs       Lorenzo-Andreu &
                                                      Garcia, 1950
```

2 Fagonia L.
  1 cretica L.

3 Zygophyllum L.
  1 macropterum C.A. Meyer
  2 fabago L.
  3 ovigerum Fischer & C.A. Meyer ex Bunge
  4 album L. fil.

4 Tribulus L.
  1 terrestris L.

5 Nitraria L.
  1 schoberi L.

6 Tetradiclis Steven ex Bieb.
  1 tenella (Ehrenb.) Litv.

LXXXVI LINACEAE

1 Linum L.
 1 arboreum L.
 2 capitatum Kit. ex Schultes
 3 flavum L.
 4 thracicum Degen
 5 tauricum Willd.
 6 uninerve (Rochel) Jáv.
 7 ucranicum Czern.
 8 campanulatum L.
 9 elegans Spruner ex Boiss.
   dolomiticum Borbás   28   Hu   Baksay, 1956
10 leucanthum Boiss. & Spruner
11 pallasianum Schultes   28   Rm   Ockendon, ined.
12 nodiflorum L.
13 narbonense L.   30   Ga   Ockendon, ined.
14 nervosum Waldst. & Kit.   18   Ockendon, ined.
15 aroanium Boiss. & Orph.
16 hologynum Reichenb.
17 virgultorum Boiss. & Heldr. ex
   Planchon
18 perenne L.
   (a) perenne   18   Ge   Ockendon, 1971
   (b) anglicum (Miller) Ockendon   36   Br   Ockendon, 1971
   (c) alpinum (Jacq.) Ockendon   18   He   Ockendon, 1971
   (d) montanum (DC.) Ockendon   36   Ga   Ockendon, 1971
   (e) extraaxillare (Kit.) Nyman   18   Cz   Ockendon, 1971
19 austriacum L.
   (a) austriacum   18   Au   Ockendon, 1971
   (b) collinum Nyman   18   Ga   Ockendon, 1971
   (c) euxinum (Juz.) Ockendon
20 punctatum C. Presl
21 leonii F.W. Schultz   18   Ga   Ockendon, 1971
22 decumbens Desf.
23 bienne Miller   30   Ga   Ockendon, ined.
24 usitatissimum L.
25 viscosum L.
26 hirsutum L.   16   Cz   Majovsky et al.,
                                                         1967

27 spathulatum (Halácsy & Bald.) Halácsy
28 pubescens Banks & Solander
29 maritumum L.   20   Ga   Ockendon, ined.
30 trigynum L.
31 tenue Desf.
32 tenuifolium L.   16   Hu   Baksay, 1956
                                         18   Cz   Májovský et al.,
                                                         1967
33 suffruticosum L.   72   Hs   Lorenzo-Andreu &
                                                         Garcia, 1950

   (a) suffruticosum
   (b) salsoloides (Lam.) Rouy
34 strictum L.   18   Hs   Lorenzo-Andreu &
                                                         Garcia, 1950

   (a) strictum
   (b) corymbulosum (Reichenb.) Rouy
35 setaceum Brot.   18   Lu   Ockendon, 1964
36 catharticum L.   16   Is   Löve & Löve, 1956

2 Radiola Hill
 1 linoides Roth   18   Da   Hagerup, 1941

LXXXVII EUPHORBIACEAE

118

1 Andrachne L.
  1 telephioides L.

2 Securinega Commerson
  1 tinctoria (L.) Rothm.

3 Chrozophora A. Juss.
  1 tinctoria (L.) A. Juss.
  2 obliqua (Vahl) A. Juss. ex Sprengel

4 Mercurialis L.
  1 annua L.   16   Ho   Gadella & Kliphuis,
                                                    1963
                                  48   Bl   Dahlgren et al.,
                                                    1971
                                  64   Co   Contandriopoulos,
                                                    1962
                          80 96 112   Co   Durand, 1962
    ambigua L. fil.   48   Hs Lu   Durand, 1962
  2 elliptica Lam.
  3 corsica Cosson
  4 reverchonii Rouy
  5 tomentosa L.
  6 perennis L.   42   Hu   Baksay, 1957b
                                  64   Ho   Gadella & Kliphuis,
                                                    1963
                                  66   Ho   Gadella & Kliphuis,
                                                    1963
                                c.80   Su   Lövkvist, 1963
                                  84   Ho   Gadella & Kliphuis,
                                                    1067a
  7 ovata Sternb. & Hoppe   32   Hu   Baksay, 1957b

5 Acalypha L.
  1 virginica L.

6 Ricinus L.
  1 communis L.

7 Euphorbia L.
  1 nutans Lag.
  2 peplis L.
  3 polygonifolia L.
  4 humifusa Willd.
  5 chamaesyce L.
   (a) chamaesyce
   (b) massiliensis (DC.) Thell.
  6 maculata L.
  7 prostrata Aiton
  8 stygiana H.C. Watson
  9 dendroides L.   18   It   D'Amato, 1946
  10 serrata L.
  11 isatidifolia Lam.
  12 villosa Waldst. & Kit. ex Willd.
  13 corallioides L.
  14 lagascae Sprengel
  15 arguta Banks & Solander
  16 microsphaera Boiss.
  17 akenocarpa Guss.
  18 palustris L.   20   Hu   Pólya, 1950
  19 velenovskyi Bornm.

119

20 soongarica Boiss.
21 ceratocarpa Ten.
22 hyberna L.   36   It   Cesca, 1963
  (a) hyberna
  (b) insularis (Boiss.) Briq.
  (c) canuti (Parl.) Tutin
23 gregersenii K. Malý ex G. Beck
24 squamosa Willd.
25 epithymoides L.   16   Hu   Pólya, 1950
26 lingulata Heuffel
27 gasparrinii Boiss.
28 fragifera Jan
29 montenegrina (Bald.) K. Malý ex Rohlena
30 oblongata Griseb.
31 apios L.   12   Bu   Kozuharov &
                                                               Kuzmanov, 1962

32 polygalifolia Boiss. & Reuter
33 uliginosa Welw. ex Boiss.
34 dulcis L.   12   It   Cesca, 1961
                                               24   It   Cesca, 1961
35 angulata Jacq.
36 carniolica Jacq.
37 duvalii Lecoq & Lamotte
38 brittingeri Opiz ex Samp.
39 ruscinonensis Boiss.
40 welwitschii Boiss. & Reuter
41 monchiquensis Franco & P. Silva
42 clementei Boiss.
43 bivonae Steudel
44 squamigera Loisel.
45 spinosa L.   14   It   Cesca, 1963
46 glabriflora Vis.
47 acanthothamnos Heldr. & Sart. ex Boiss.
48 chamaebuxus Bernard ex Gren. & Godron
49 capitulata Reichenb.
50 platyphyllos L.
51 serrulata Thuill.
52 pubescens Vahl   14   It   D'Amato, 1945
53 cuneifolia Guss.
54 pterococca Brot.
55 helioscopia L.   42   Su   Lövkvist, 1963
56 phymatosperma Boiss. & Gaill.
57 myrsinites L.
58 rigida Bieb.   20   It   D'Amato, 1946
59 broteri Daveau
60 lathyris L.
61 aleppica L.
62 medicaginea Boiss.
63 dracunculoides Lam.
64 exigua L.   24   Br   Rutland, 1941
65 falcata L.   16   Hu   Pólya, 1950
                                               36   It   D'Amato, 1939

66 sulcata De Lens ex Loisel.
67 peplus L.   16   Ho   Gadella
                                                               & Kliphuis, 1963

68 ledebourii Boiss.
69 taurinensis All.
70 segetalis L.   16   Ga   D'Amato, 1945
71 pinea L.
72 portlandica L.
73 deflexa Sibth. & Sm.

```
74 petrophila C.A. Meyer
75 transtagana Boiss.
76 matritensis Boiss.
77 biumbellata Poiret
78 boetica Boiss.
79 bupleuroides Desf.
80 nicaeensis All.
   (a) nicaeensis
   (b) glareosa (Pallas ex Bieb.) A.R. Sm.
81 bessarabica Klokov
82 orphanidis Boiss.
83 barrelieri Savi
84 triflora Schott
85 saxatilis Jacq.                       18      Au      Polatschek, 1966
86 valliniana Belli
87 minuta Loscos & Pardo
88 variabilis Cesati
89 gayi Salis
90 maresii Knoche
91 herniariifolia Willd
92 seguierana Necker
   (a) seguierana
   (b) niciciana (Borbás ex Novák)
       Rech. fil.
93 pithyusa L.                           36      It      D'Amato, 1939
   (a) pithyusa
   (b) cupanii (Guss. ex Bertol.)
       A.R. Sm.
94 paralias L.                           16      It      D'Amato, 1939
95 nevadensis Boiss. & Reuter
96 agraria Bieb
97 lucida Waldst. & Kit.                 36      Hu      Pólya, 1950
98 salicifolia Host                      36      Hu      Pólya, 1950
99 esula L.
   (a) esula                             60      Ho      Gadella &
                                                         Kliphuis, 1966
                                         64      Ge      Reese, 1952
   (b) tommasiniana (Bertol.) Nyman      20      Hu      Baksay, 1958
100 undulata Bieb.
101 cyparissias L.                       20      Br      Pritchard, 1957
                                         40      Br      Pritchard, 1957
102 terracina L.
103 amygdaloides L.                      18      It      D'Amato, 1939
   (a) amygdaloides
   (b) semiperfoliata (Viv.) A.R. Sm.
104 heldreichii Orph. ex Boiss.
105 characias L.
   (a) characias                         20      It      Larsen, 1956a
   (b) wulfenii (Hoppe ex Koch) A.R. Sm.


LXXXVIII RUTACEAE

1 Ruta L.
  1  montana (L.) L.
  2  angustifolia Pers.
  3  chalepensis L.
  4  graveolens L.
  5  corsica DC.
```

2 Haplophyllum A. Juss
  1 linifolium (L.) G. Don fil.
  2 suaveolens (DC.) G. Don fil.
  3 thesioides (Fischer ex DC.) G. Don fil.
  4 coronatum Griseb.
  5 balcanicum Vandas
  6 patavinum (L.) G. Don fil.
  7 boissieranum Vis. & Pančić
  8 buxbaumii (Poiret) G. Don fil.

3 Dictamnus L.
  1 albus L.

4 Citrus L.
  1 medica L.
  2 limon (L.) Burm. fil.
  3 limetta Risso
  4 deliciosa Ten.
  5 paradisi Macfadyen
  6 grandis (L.) Osbeck
  7 aurantium L.
  8 sinensis (L.) Osbeck

5 Phellodendron Rupr.
  1 amurense Rupr.

6 Ptelea L.
  1 trifoliata L.

LXXXIX CNEORACEAE

1 Cneorum L.
  1 tricoccon L.

XC SIMAROUBACEAE

1 Ailanthus Desf.
  1 altissima (Miller) Swingle

XCI MELIACEAE

1 Melia L.
  1 azedarach L.

XCII POLYGALACEAE

1 Polygala L.
  1 microphylla L.
  2 chamaebuxus L.   38   Au   Mattick, 1950
                                      c.46   He   Glendinning, 1960
  3 vayredae Costa
  4 myrtifolia L.
  5 rupestris Pourret
  6 exilis DC.
  7 sibirica L.
  8 subuniflora Boiss. & Heldr.
  9 supina Schreber

```
    (a) supina
    (b) hospita (Ileuffel) McNeill
    (c) rhodopea (Velen.) McNeill
10 monspeliaca L.                         c.38      Lu        Glendinning, 1960
11 major Jacq.
12 anatolica Boiss. & Heldr.
13 boissieri Cosson
14 venulosa Sibth. & Sm.
15 preslii Sprengel
16 sardoa Chodat
17 lusitanica Welw. ex Chodat
18 baetica Willk.
19 doerfleri Hayek
20 flavescens DC.
21 nicaeensis Risso ex Koch
    (a) nicaeensis
    (b) mediterranea Chodat
    (c) tomentella (Boiss.) Chodat
    (d) corsica (Boreau) Graebner         34        Co        Contandriopoulos, 1962
    (e) caesalpinii (Bubani) McNeill
    (f) gariodiana (Jordan & Fourr.) Chodat
    (g) carniolica (A. Kerner) Graebner
    (h) forojulensis (A. Kerner) Graebner
22 apiculata Porta
23 comosa Schkuhr                         28-32     Au        Mattick, 1950
                                          34        Su        Lövkvist, 1963
24 vulgaris L.                            28        Au        Mattick, 1950
                                          32        Au        Mattick, 1950
                                          48        Ge        Wulff, 1938
                                          c.56      Au        Mattick, 1950
                                          68        Br He Lu  Glendinning, 1960
                                          c.70      Scand.    Löve & Löve, 1944
25 alpestris Reichenb.
    (a) alpestris                         34        He        Glendinning, 1960
    (b) croatica (Chodat) Hayek
26 serpyllifolia J.A.C. Hose             32        Da No     Björse, 1961
                                          34        Br He     Glendinning, 1960
                                          c.68      Co        Glendinning, 1960
27 carueliana (A.W. Benn) Burnat ex
    Caruel
28 cristagalli Chodat
29 edmundii Chodat
30 calcarea F.W. Schultz                  34        Br He     Glendinning, 1960
31 amara L.
    (a) amara                            28        Hu        Baksay, 1956
    (b) brachyptera (Chodat) Hayek       28        Po        Skalińska et al.,
                                                                 1959
32 amarella Crantz                        34        Fe        Sorsa, 1963
33 alpina (Poiret) Steudel                c.34      He        Glendinning, 1960


XCIII CORIARIACEAE

1 Coriaria L.
  1 myrtifolia L.


XCIV ANACARDIACEAE

1 Rhus L.
  1 coriaria L.
```

2 typhina L.
3 pentaphylla (Jacq.) Desf.
4 tripartita (Ucria) Grande

2 Cotinus Miller
1 coggygria Scop.

3 Pistacia L.
1 terebinthus L.
2 atlantica Desf.
3 vera L.
4 lentiscus L.

4 Schinus L.
1 molle L.
2 terebinthifolia Raddi

XCV ACERACEAE

1 Acer L.
  1 platanoides L.   26   Su   Lövkvist, 1963
  2 lobelii Ten.
  3 campestre L.
  4 tataricum L.
  5 pseudoplatanus L.   52   Su   Lövkvist, 1963
  6 heldreichii Orph. ex Boiss.
   (a) heldreichii
   (b) visianii K. Malý
  7 trautvetteri Medv.
  8 opalus Miller
  0 obtusatum Waldst. & Kit. ex Willd.
 10 granatense Boiss.
 11 hyrcanum Fischer & C.A. Meyer
 12 stevenii Pojark.
 13 monspessulanum L.
 14 sempervirens L.
 15 negundo L.

XCVI SAPINDACEAE

1 Cardiospermum L.
1 halicacabum L.

XCVII HIPPOCASTANACEAE

1 Aesculus L.
1 hippocastanum L.

XCVIII BALSAMINACEAE

1 Impatiens L.
  1 noli-tangere L.   20   Ho   Gadella &
                                        Kliphuis, 1966
                        40   Po   Skalińska et al.,
                                        1959
  2 capensis Meerb.
  3 parviflora DC.   24   Su   Löve & Löve, 1942
                        26   Po   Skalińska et al.,
                                        1959

```
4 glandulifera Royle             18     Br     Jones & Smith,
                                 20     Br       1966
  5 balfourii Hooker fil.
  6 balsamina L.


XCIX AQUIFOLIACEAE

1 Ilex L.
  1 aquifolium L.                40     Br     Maude, 1940
  2 colchica Pojark.
  3 perado Aiton


C CELASTRACEAE

1 Euonymus L.
  1 europaeus L.                 64     Hu     Pólya, 1949
  2 latifolius (L.) Miller
  3 verrucosus Scop.
  4 nanus Bieb.
  5 japonicus L. fil.

2 Maytenus Molina
  1 senegalensis (Lam.) Exell


CI STAPHYLEACEAE

1 Staphylea L.
  1 pinnata L.


CII BUXACEAE

1 Buxus L.
  1 sempervirens L.
  2 balearica Lam.


CIII RHAMNACEAE

1 Paliurus Miller
  1 spina-christi Miller

2 Ziziphus Miller
  1 jujuba Miller
  2 lotus (L.) Lam.

3 Rhamnus L.
  1 alaternus L.
  2 ludovici-salvatoris Chodat
  3 lycioides L.
   (a) lycioides
   (b) velutinus (Boiss.) Tutin
   (c) oleoides (L.) Jahandiez & Maire
   (d) graecus (Boiss. & Reuter)
  4 saxatilis Jacq.
   (a) saxatilis
   (b) tinctorius (Waldst. & Kit.) Nyman
  5 rhodopeus Velen.
```

```
  6 intermedius Steudel & Hochst.
  7 prunifolius Sibth. & Sm.
  8 catharticus L.                        24      Ge      Wulff, 1939
  9 orbiculatus Bornm.
 10 persicifolius Moris
 11 alpinus L.
   (a) alpinus
   (b) fallax (Boiss.) Maire & Petitmengin
   (c) glaucophyllus (Sommier) Tutin
 12 pumilus Turra
 13 sibthorpianus Roemer & Schultes
```

```
4 Frangula Miller
  1 alnus Miller                          20      Br      Rutland, 1941
                                          26      Ge      Wulff, 1937b
  2 azorica Tutin
  3 rupestris (Scop.) Schur
```

CIV VITACEAE

```
1 Vitis L.
  1 vinifera L.
   (a) sylvestris (C.C. Gmelin) Hegi
   (b) vinifera                           38      Rm      Constantinescu
                                                          et al., 1964
```

```
2 Parthenocissus Planchon
  1 quinquefolia (L.) Planchon
  2 inserta (A. Kerner) Fritsch
  3 tricuspidata (Siebold & Zucc.) Planchon
```

CV TILIACEAE

```
1 Tilia L.
  1 tomentosa Moench
  2 dasystyla Steven
  3 platyphyllos Scop.
   (a) platyphyllos
   (b) cordifolia (Besser) C.K. Schneider
   (c) pseudorubra C.K. Schneider
  4 rubra DC.
   (a) rubra
   (b) caucasica (Rupr.) V. Engler
  5 cordata Miller
  6 x vulgaris Hayne
```

CVI MALVACEAE

```
1 Malope L.
  1 malacoides L.
  2 trifida Cav.
```

```
2 Kitaibela Willd.
  1 vitifolia Willd.
```

```
3 Sida L.
  1 rhombifolia L.
```

4 Malvella Jaub. & Spach
  1 sherardiana (L.) Jaub. & Spach

5 Malva L.
  1 hispanica L.
  2 aegyptia L.
  3 stipulacea Cav.
  4 cretica Cav.
   (a) cretica
   (b) althaeoides (Cav.) Dalby
  5 alcea L.   84   Su   Lövkvist, 1963
  6 moschata L.   42   Su   Lövkvist, 1963
  7 tournefortiana L.
  8 sylvestris L.   42   Su   Lövkvist, 1963
  9 nicaeensis All.
 10 parviflora L.
 11 pusilla Sm.
 12 neglecta Wallr.   42   Su   Lövkvist, 1963
 13 verticillata L.

6 Lavatera L.
  1 cretica L.
  2 mauritanica Durieu
  3 arborea L.
  4 maritima Gouan
  5 oblongifolia Boiss.
  6 olbia L.
  7 bryoniifolia Miller
  8 thuringiaca L.
   (a) thuringiaca
   (b) ambigua (DC.) Nyman
  9 punctata All.
 10 trimestris L.
 11 triloba L.
   (a) triloba
   (b) pallescens (Moris) Nyman
   (c) agrigentina (Tineo) R. Fernandes

7 Althaea L.
  1 hirsuta L.
  2 longiflora Boiss. & Reuter
  3 cannabina L.
  4 officinalis L.   42   Hu   Pólya, 1950
  5 armeniaca Ten.

8 Alcea L.
  1 setosa (Boiss.) Alef.
  2 rosea L.
  3 pallida (Willd.) Waldst. & Kit.
   (a) pallida
   (b) cretica (Weinm.) D.A. Webb
  4 heldreichii (Boiss.) Boiss.
  5 lavateriflora (DC.) Boiss.
  6 rugosa Alef.

9 Abutilon Miller
  1 theophrasti Medicus

10 Modiola Moench
  1 caroliniana (L.) G. Don fil.

127

11 Gossypium L.
  1 herbaceum L.
  2 hirsutum L.

12 Hibiscus L.
  1 syriacus L.
  2 palustris L.
  3 trionum L.
  4 cannabinus L.

13 Abelmoschus Medicus
  1 esculentus (L.) Moench

14 Kosteletzkya C. Presl
  1 pentacarpos (L.) Ledeb.

CVII THYMELAEACEAE

1 Daphne L.
  1 mezereum L.   18   Fe   V. Sorsa, 1962
  2 gnidium L.
  3 sophia Kalenicz
  4 laureola L.
   (a) laureola   18   Au   Fuchs, 1938
   (b) philippi (Gren.) Rouy
  5 pontica L.
  6 alpina L.
  7 oleoides Schreber
  8 blagayana Freyer   18   Ju   Newling, 1960
  9 sericea Vahl
 10 cneorum L.   18   Hu   Baksay, 1956
 11 striata Tratt.
 12 arbuscula Čelak
 13 petraea Leybold
 14 rodriguezii Texidor
 15 jasminea Sibth. & Sm.
 16 malyana Blečić
 17 gnidioides Jaub. & Spach

2 Thymelaea Miller
  1 sanamunda All.
  2 pubescens (L.) Meissner
  3 hirsuta (L.) Endl.
  4 tartonraira (L.) All.
   (a) tartonraira
   (b) argentea (Sibth. & Sm.) Holmboe
   (c) thomasii (Duby) Briq.   18   Co   Contandriopoulos,
                                                1964a
  5 nitida (Vahl) Endl.
  6 myrtifolia (Poiret) D.A. Webb
  7 lanuginosa (Lam.) Ceballos & C.
    Vicioso
  8 coridifolia (Lam.) Endl.
  9 procumbens A. & R. Fernandes
 10 broterana Coutinho
 11 calycina (Lapeyr.) Meissner
 12 ruizii Loscos ex Casav.
 13 subrepens Lange
 14 tinctoria (Pourret) Endl.
 15 dioica (Gouan) All.

16 villosa (L.) Endl.
17 passerina (L.) Cosson & Germ.

3 Diarthron Turcz.
  1 vesiculosum (Fischer & C.A. Meyer)
     C.A. Meyer

CVIII ELAEAGNACEAE

1 Hippophae L.
  1 rhamnoides L.   24   Fe Ga Ge   Rousi, 1965a
                                                 He It Po

2 Elaeagnus L.
  1 angustifolia L.

CIX GUTTIFERAE

1 Hypericum L.
   1 calycinum L.
   2 foliosum Aiton
   3 hircinum L.
   4 inodorum Miller
   5 androsaemum L.
   6 balearicum L.
   7 aciferum (W. Greuter) N.K.B. Robson
   8 aegypticum L.
   9 empetrifolium Willd.
  10 amblycalyx Coust. & Cand.
  11 coris L.
  12 ericoides L.
  13 haplophylloides Halácsy & Bald.
  14 hirsutum L.   18   Da   Nielson, 1924
  15 linarioides Bosse
  16 pulchrum L.   18   Da Fa   Bücher, 1940a
  17 nummularium L.
  18 fragile Heldr. & Sart. ex Boiss.
  19 taygeteum Quezel & Contandr.
  20 hyssopifolium Chaix
     (a) hyssopifolium
     (b) elongatum (Ledeb.) Woronow
  21 olympicum L.
  22 cerastoides (Spach) N.K.B. Robson
  23 montanum L.   16   Da   Nielson, 1964
  24 annulatum Moris
  25 delphicum Boiss. & Heldr.
  26 athoum Boiss. & Orph.
  27 atomarium Boiss.
  28 cuisinii W. Barbey
  29 tomentosum L.
  30 pubescens Boiss.
  31 caprifolium Boiss.
  32 elodes L.   32   Br   Robson, unpub.
  33 vesiculosum Griseb.
  34 perfoliatum L.
  35 montbretii Spach
  36 umbellatum A. Kerner
  37 bithynicum Boiss.
  38 richeri Vill.

```
      (a) richeri                              14      He      Favarger, 1959a
      (b) burseri (DC.) Nyman
      (c) grisebachii (Boiss.) Nyman
 39 spruneri Boiss.
 40 rochelii Griseb. & Schenk
 41 barbatum Jacq.
 42 rumeliacum Boiss.
 43 trichocaulon Boiss. & Heldr.
 44 kelleri Bald.
 45 aviculariifolium Jaub. & Spach.
 46 thasium Griseb.
 47 aucheri Jaub. & Spach
 48 linarifolium Vahl                          16      Br      Robson, unpub.
 49 australe Ten.
 50 humifusum L.
 51 tetrapterum Fries                          16      Hu      Pólya, 1950
 52 undulatum Schousboe ex Willd.
 53 maculatum Crantz
      (a) maculatum                            16      Fe      V. Sorsa, 1962
      (b) obtusiusculum (Tourlet) Hayek        32      Br      Robson
 54 perforatum L.
 55 triquetrifolium Turra
 56 elegans Stephan ex Willd.
 57 gymnanthum Engelm. & A. Gray
 58 mutilum L.
 59 majus (A. Gray) Britton
 60 canadense L.                               16      Hb      Moore, 1973
 61 gentianoides (L.) Britton


CX VIOLACEAE

 1 Viola L.
      1 odorata L.                             20      Ge      Schüfer, 1954
      2 suavis Bieb.                           40      Au He It Schmidt, 1961
      3 alba Besser
        (a) alba                               20      Ge      Schüfer, 1954
        (b) scotophylla (Jordan) Nyman         20      Au      Schmidt, 1961
        (c) dehnhardtii (Ten.) W. Becker       20      Bl      Schmidt, 1961
      4 cretica Boiss. & Heldr.                20      Cr      Schmidt, 1961
      5 jaubertiana Marès & Vigineix           20      Bl      Schmidt, 1961
      6 hirta L.                               20      Ge      Schüfer, 1954
      7 collina Besser                         20      Ge      Schüfer, 1954
      8 ambigua Waldst. & Kit.
      9 thomasiana Song. & Perr.               40      Au      Schmidt, 1961
     10 myrenaica Ramond ex DC.                20      Au      Schüfer, 1954
     11 chelmea Boiss. & Heldr.
        (a) chelmea                            20      Gr      Schmidt, 1964b
        (b) vratnikensis Gáyer & Degen
     12 mirabilis L.                           20      Hu      Pólya, 1950
     13 willkommii R. de Roemer
     14 rupestris F.W. Schmidt
        (a) rupestris                          20      Ge      Schüfer, 1954
        (b) relicta Jalas                      20      No      Knaben &
                                                               Engelskjön, 1967
     15 reichenbachiana Jordan ex Boreau       20      Ga      Larsen, 1954
     16 tanaitica Grosset
     17 mauritii Tepl.
     18 riviniana Reichenb.                35 40 45    Ho      Gadella &
                                                               Kliphuis, 1963
```

	46	Ho	Gadella & Kliphuis, 1963
	47	Ho	Gadella & Kliphuis, 1963
19 sieheana W. Becker			
20 canina L.			
(a) canina	40	Ge	Schöfer, 1954
(b) montana (L.) Hartman	40	Ge	Schöfer, 1954
(c) schultzii (Billot) Kirschleger	40	Ge	Schöfer, 1954
21 lactea Sm.	58	Br Lu	Moore, 1958
22 persicifolia Schreber	20	Da	Clausen, 1931
23 pumila Chaix	40	Ge	Schöfer, 1954
24 elatior Fries	40	Ge	Schmidt, 1961
25 jordanii Hanry	40	Ga	Schmidt, 1961
26 uliginosa Besser	20	Ju	Schmidt, 1961
27 palustris L.			
(a) palustris	48	Be Ho	Gadella & Kliphuis, 1963
(b) juressi (Link ex K. Wein) Coutinho			
28 epipsila Ledeb.	24	Fe	Sorsa, 1963
29 jooi Janka	24	Rm	Merxmüller, 1974
30 selkirkii Pursh ex Goldie			
31 pinnata L.			
32 obliqua Hill			
33 biflora L.	12	Fe	Sorsa, 1963
34 crassiuscula Bory			
35 diversifolia (DC.) W. Becker			
36 cenisia L.	20	It	Schmidt, 1961
37 comollia Massara			
38 valderia All.	20	It	Schmidt, 1961
39 magellensis Porta & Rigo ex Strobl	22	It	Schmidt, 1964a
40 brachyphylla W. Becker			
41 perinensis W. Becker			
42 stojanowii W. Becker			
43 fragrans Sieber			
44 poetica Boiss. & Spruner			
45 grisebachiana Vis.	22	Ju	Schmidt, 1964a
46 alpina Jacq.	22	Po	Skalińska et al., 1959
47 nummariifolia Vill.			
48 calcarata L.			
(a) calcarata	40	He It	Schmidt, 1961
(b) zoysii (Wulfen) Merxm.	40	Au	Schmidt, 1961
(c) villarsiana (Roemer & Schultes) Merxm.			
49 bertolonii Pio			
(a) bertolonii	40	It	Schmidt, 1965
(b) messanensis (W. Becker) A. Schmidt	40	It	Schmidt, 1961
heterophylla Bertol.	20	It	Schmidt, 1961
50 splendida W. Becker	40	It	Schmidt, 1965
51 aethnensis Parl.	40	Si	Schmidt, 1964a
52 nebrodensis C. Presl	20	Si	Merxmüller, 1974
53 corsica Nyman			
(a) corsica	52	Sa	Schmidt, 1964
	c.120	Co	Contandriopoulos, 1962
(b) ilvensis (W. Becker) Merxm.	52	It	Merxmüller, 1974
54 munbyana Boiss. & Reuter	52	Si	Schmidt, 1965
55 pseudogracilis Strobl			
(a) pseudogracilis	34	It	Schmidt, 1961
(b) cassinensis (Strobl) Merxm. & A. Schmidt	34	It	Schmidt, 1964a

56 eugeniae Parl.
  (a) eugeniae   34   It   Schmidt, 1965
  (b) levieri (Parl.) A. Schmidt   34   It   Schmidt, 1965

	2n	Region	Reference
56 eugeniae Parl.			
(a) eugeniae	34	It	Schmidt, 1965
(b) levieri (Parl.) A. Schmidt	34	It	Schmidt, 1965
57 oreades Bieb.			
58 eximia Form.			
59 doerfleri Degen			
60 allchariensis G. Beck			
(a) allchariensis			
(b) gostivarensis W. Becker & Bornm.			
61 frondosa (Velen.) Hayek			
62 arsenica G. Beck			
63 cornuta L.	22	Hs	Merxmüller, 1974
64 orphanidis Boiss.			
(a) orphanidis			
(b) nicolai (Pant.) Valentine			
65 dacica Borbás			
66 elegantula Schott			
67 athois W. Becker	20	Gr	Schmidt, 1963
68 gracilis Sibth. & Sm.	20	Ju	Merxmüller, 1974
orbelica Pancic	20	Bu	Griesinger, 1937
69 speciosa Pant.			
70 beckiana Fiala			
71 rhodopeia W. Becker			
72 dubyana Burnat ex Gremli			
73 declinata Waldst. & Kit.			
74 lutea Hudson			
(a) lutea	48	Br	Pettet, 1964
(b) sudetica (Willd.) W. Becker			
75 bubanii Timb.-Lagr.	c.128	Ga Hs	Schmidt, 1964a
76 hispida Lam.			
77 langeana Valentine			
78 tricolor L.			
(a) tricolor	26	Su	Lövkvist, 1963
(b) curtisii (E. Forster) Syme	26	Ho	Gadella & Kliphuis, 1963
(c) macedonica (Boiss. & Heldr.) A. Schmidt	26	Gr	Schmidt, 1963
(d) subalpina Gaudin	26	Po	Skalińska et al., 1966
(e) matutina (Klokov) Valentine			
79 aetolica Boiss. & Heldr.	16	Gr	Schmidt, 1963
80 arvensis Murray	34	Ho	Gadella, 1963
81 kitaibeliana Schultes	16	Sb	Schmidt, 1965
	48	Br	Fothergill, 1944
82 hymettia Boiss. & Heldr.	16	Gr	Schmidt, 1964a
83 parvula Tineo	10	Si	Schmidt, 1962
84 heldreichiana Boiss.			
85 mercurii Orph. ex Halácsy			
86 demetria Prolongo ex Boiss.			
87 occulta Lehm.			
88 arborescens L.	c.140		
89 scorpiuroides Cosson			
90 delphinantha Boiss.	20	Gr	Schmidt, 1964a
91 kosaninii (Degen) Hayek			
92 cazorlensis Gand.			

CXI PASSIFLORACEAE

1 Passiflora L.

1 caerulea L.

CXII CISTACEAE

1 Cistus L.
    1 albidus L.   18   Hs   Löve & Kjellquist,
                                                                1964
    2 crispus L.   18   Lu   Rodrigues, 1954
    3 incanus L.
    4 heterophyllus Desf.
    5 parviflorus Lam.
    6 varius Pourret
    7 monspeliensis L.   18   Hs   Löve & Kjellquist,
                                                                1964
    8 psilosepalus Sweet
    9 albanicus E.F. Warburg ex Heywood
   10 salvifolius L.   18   Lu   Löve & Kjellquist,
                                                                1964
   11 populifolius L.   18   Hs   Löve & Kjellquist,
                                                                1964
      (a) populifolius
      (b) major (Pourret ex Dunal) Heywood
   12 laurifolius L.   18   Hs   Löve & Kjellquist,
                                                                1964
   13 ladanifer L.
   14 palhinhae Ingram
   15 clusii Dunal
   16 libanotis L.

2 Halimium (Dunal) Spach
    1 ocymoides (Lam.) Willk.   18   Lu   Proctor, 1955
    2 alyssoides (Lam.) C. Koch
    3 lasianthum (Lam.) Spach
      (a) lasianthum
      (b) formosum (Curtis) Heywood
    4 atriplicifolium (Lam.) Spach
    5 halimifolium (L.) Willk.   18   Lu   Proctor, 1955
      (a) halimifolium
      (b) multiflorum (Salzm. ex Dunal) Maire
    6 umbellatum (L.) Spach   18   Lu   Proctor, 1955
    7 viscosum (Willk.) P. Silva
    8 verticillatum (Brot.) Sennen
    9 commutatum Pau

3 Tuberaria (Dunal) Spach
    1 lignosa (Sweet) Samp.
    2 globularifolia (Lam.) Willk.
    3 major (Willk.) P. Silva & Rozeira
    4 guttata (L.) Fourr.   36 Br Ga Hb Lu Proctor, 1955
    5 bupleurifolia (Lam.) Willk.
    6 villosissima (Pomel) Grosser
    7 praecox Grosser
    8 acuminata (Viv.) Grosser
    9 macrosepala (Cosson) Willk.
   10 echioides (Lam.) Willk.

4 Helianthemum Miller
    1 lavandulifolium Miller   20   Hs   Coutinho &
                                                                Lorenzo-Andreu, 1948
    2 squamatum (L.) Pers.   10   Hs   Coutinho &
                                                                Lorenzo-Andreu, 1948

```
 3 caput-felis Boiss.
 4 piliferum Boiss.
 5 viscarium Boiss. & Reuter
 6 asperum Lag. ex Dunal
 7 hirtum (L.) Miller
 8 croceum (Desf.) Pers.            20     Hs      Löve & Kjellquist,
                                                    1964
 9 nummularium (L.) Miller          20     Br Ge Hb  Proctor, 1955
   (a) nummularium                  20     He      Rohner, 1954
   (b) tomentosum (Scop.) Schinz. & Thell.
   (c) pyrenaicum (Janchen) Schinz & Thell.
   (d) obscurum (Celak.) J. Holub   20     He      Rohner, 1954
   (e) glabrum (Koch) Wilczek
   (f) semigalbrum (Badaro) M.C.F. Proctor
   (g) berterianum (Bertol.) Breistr.
   (h) grandiflorum (Scop.) Schinz &
       Thell.                       20     He      Rohner, 1954
10 leptophyllum Dunal
11 appeninum (L.) Miller            20     Br Ga   Proctor, 1955
12 pilosum (1.) Pers.
13 virgatum (Desf.) Pers.
14 sessiliflorum (Desf.) Pers.
15 stipulatum (Forskål) C. Chr.
16 villosum Thib.
17 papillare Boiss.
18 ledifolium (L.) Miller           20     Lu      Proctor, 1955
19 salicifolium (L.) Miller         20     Hs      Löve & Kjellquist,
                                                    1964
20 sanguineum (Lag.) Lag. ex Dunal
21 aegyptiacum (L.) Miller
22 oelandicum (L.) DC.
   (a) oelandicum                   22     Su      Proctor, 1955
   (b) alpestre (Jacq.) Breistr.    22     He      Proctor, 1955
   (c) rupifragum (A. Kerner) Breistr.
   (d) italicum (L.) Font Quer & Rothm.  20  Ga    Guinochet &
                                                    Logeois, 1962
   (e) orientale (Grosser) M.C.F. Proctor
23 canum (L.) Baumg.
   (a) canum                        22     Br Ga Hb  Proctor, 1955
   (b) nebrodense (Heldr. ex Guss.)
       Arcangeli
   (c) canescens (Hartman) M.C.F. Proctor
   (d) levigatum M.C.F. Proctor
   (e) piloselloides (Lapeyr.) M.C.F.
       Proctor
   (f) pourretii (Timb.Lagr.) M.C.F.
       Proctor
   (g) stevenii (Rupr. ex Juz. & Pozd.)
       M.C.F. Proctor
24 origanifolium (Lam.) Pers.
   (a) origanifolium
   (b) glabratum (Willk.) Guinea & Heywood
   (c) serrae (Camb.) Guinea & Heywood
   (d) molle (Cav.) Font Quer & Rothm.
25 marifolium (L.) Miller
26 pannosum Boiss.
   (a) pannosum
   (b) frigidulum (Cuatrec.) Font Quer
       & Rothm.
27 cinereum (Cav.) Pers.            20     Hs      Lorenzo-Andreu, 1951
                                    22     Hs      Coutinho &
                                                    Lorenzo-Andreu, 1948
```

```
28 hymettium Boiss. & Heldr.
29 rossmaessleri Willk.
30 viscidulum Boiss.
   (a) viscidulum
   (b) guadiccianum Font Quer & Rothm.
   (c) viscarioides (Debeaux & Reverchon)
        Guinea & Heywood
31 lunulatum (All.) DC.

5 Fumana (Dunal) Spach
  1 arabica (L.) Spach
  2 procumbens (Dunal) Gren. & Godron
  3 ericoides (Cav.) Gand.
  4 scoparia Pomel
  5 bonapartei Maire & Petitmengin
  6 thymifolia (L.) Spach ex Webb          32      Hs       Löve & Kjellquist,
                                                            1964
  7 paradoxa Heywood
  8 laevipes (L.) Spach                    32      Lu       Proctor, 1955
  9 aciphylla Boiss.
```

CXIII TAMARICACEAE

```
1 Reaumuria Hasselq. ex L.
  1 vermiculata L.

2 Tamarix L.
  1 africana Poiret
  2 canariensis Willd.
  3 gallica L.
  4 tetrandra Pallas ex Bieb.
  5 parviflora DC.
  6 hampeana Boiss. & Heldr.
  7 dalmatica Baum.
  8 ramosissima Ledeb.
  9 smyrnensis Bunge
 10 boveana Bunge
 11 meyeri Boiss.
 12 laxa Willd.
 13 gracilis Willd.
 14 hispida Willd.

3 Myricaria Desv.
  1 germanica (L.) Desv.                   24      No       Knaben, 1950
```

CXIV FRANKENIACEAE

```
1 Frankenia L.
  1 pulverulenta L.
  2 boissieri Reuter ex Boiss.
  3 corymbosa Desf.
  4 thymifolia Desf.
  5 laevis L.
  6 hirsuta L.
```

CXV ELATINACEAE

```
1 Bergia L.
```

1 capensis L.

2 Elatine L.
    1 alsinastrum L.
    2 hydropiper L.   c.40   Su   Frisendahl, 1927
    3 macropoda Guss.
    4 hungarica Moesz
    5 triandra Schkuhr   c.40   No Su   Frisendahl, 1927
    6 ambigua Wight
    7 hexandra (Lapierre) DC.   72   Da   Frisendahl, 1927
    8 brochonii Clavaud

CXVI DATISCACEAE

1 Datisca L.
    1 cannabina L.

CXVII CUCURBITACEAE

1 Thladiantha Bunge
    1 dubia Bunge

2 Ecballium A. Richard
    1 elaterium (L.) A. Richard

3 Bryonia L.
    1 alba L.
    2 cretica L.
        (a) cretica
        (b) dioica (Jacq.) Tutin   20   Ho   Gadella & Kliphuis,
                                                                              1966
                                                       40   Co   Contandriopoulos,
                                                                              1964b

        (c) acuta (Desf.) Tutin

4 Citrullus Schrader
    1 lanatus (Thunb.) Mansfeld
    2 colocynthis (L.) Schrader

5 Cucumis L.
    1 melo L.
    2 myriocarpus Naudin
    3 sativus L.

6 Cucurbita L.
    1 pepo L.
    2 moschata Duchesne ex Poiret
    3 maxima Duchesne

7 Echinocystis Torrey & A. Gray
    1 lobata (Michx) Torrey & A. Gray

8 Sicyos L.
    1 angulatus L.

CXVIII CACTACEAE

1 Opuntia Miller

136

```
1 tuna (L.) Miller
2 vulgaris Miller
3 monacantha Haw.
4 stricta (Haw.) Haw.
5 ficus-indica (1.) Miller
6 maxima Miller

2 Cereus Miller
  peruvianus (L.) Miller
```

CXIX LYTHRACEAE

```
1 Lythrum L.
  1 salicaria L.                        60      Ho      Gadella & Kliphuis,
                                                        1966

  2 virgatum L.
  3 junceum Banks & Solander
  4 acutangulum Lag.
  5 castellanum González-Albo ex Borja
  6 flexuosum Lag.
  7 hyssopifolia L.
  8 thymifolia L.
  9 tribracteatum Salzm. ex Sprengel
 10 thesioides Bieb.
 11 borysthenicum (Schrank) Litv.       30      Lu      Cook, 1968
 12 portula (L.) D.A. Webb              10      Hu      Pólya, 1949
 13 volgense D.A. Webb

2 Ammannia L.
  1 auriculata Willd.
  2 coccinea Rottb.
  3 baccifera L.
  4 verticillata (Ard.) Lam.

3 Rotala L.
  1 indica (Willd.) Koehne
  2 filiformis (Bellardi) Hiern
```

CXX TRAPACEAE

```
1 Trapa L.
  1 natans L.                        c.48      Po      Skalińska et al.,
                                                       1959
```

CXXI MYRTACEAE

```
1 Myrtus L.
  1 communis L.

2 Eucalyptus L'Hér.
  1 botryoides Sm.                     22      It      Ruggeri, 1961
  2 robustus Sm.
  3 resinifer Sm.
  4 x trabutii Vilmorin ex Trabut      22      It      Ruggeri, 1961
  5 gomphocephalus DC.
  6 tereticornis Sm.
  7 camaldulensis Dehnh.               22      It      Ruggeri, 1961
  8 rudis Endl.
```

9 globulus Labill.
10 maidenii F. Mueller
11 viminalis Labill.

CXXII PUNICACEAE

1 Punica L.
  1 granatum L.

CXXIII ONAGRACEAE

1 Fuchsia L.			
1 magellanica Lam.	44	Hb	Raven, ined.
2 Circaea L.			
1 lutetiana L.	22	Su	Lövkvist, 1963
2 x intermedia Ehrh.	22	Su	Lövkvist, 1963
3 alpina L.	22	Su	Lövkvist, 1963
3 Oenothera L.			
1 biennis L.	14	Hu	Pólya, 1949
2 rubricaulis Klebahn	14	Rs(C)	Renner, 1951
3 chicagoensis Renner ex Cleland &			
Blakeslee	14	Ga	Renner, 1951
4 strigosa (Rydb.) Mackenzie & Bush	14	Ge	Oehkers, 1926
renneri H. Scholz	14	He	Steiner, 1955
5 erythrosepala Borbás	14	Ho	Gadella & Kliphuis, 1968
conomifica Renner	14	Ci	Rossmann, 1963
6 suaveolens Pers.	14	Ge	Renner, 1951
7 parviflora L.	14	Ge	Renner, 1951
issleri Renner ex Rostanski	14	Ga	Renner, 1951
8 silesiaca Renner	14	Ge	Renner, 1951
atrovirens Shull & Bartlett	14	Ge	Renner, 1937b
9 ammophila Focke	14	Ge	Renner, 1951
syrticola Bartlett	14	Au	Renner, 1951
rubricuspis Renner ex Rostanski	14	Ge	Renner, 1951
10 longiflora L.			
11 affinis Camb. in St.-Hil.	14		
12 stricta Ledeb. ex Link	14	Lu	Rodrigues, 1953
13 rosea L'Hér. ex Aiton			
4 Ludwigia L.			
1 uruguayensis (Camb.) Hara	80	N. Amer.	Raven, 1963
2 peploides (Kunth) P.H. Raven	16	Ga	Raven, ined.
3 palustris (L.) Elliott	16		
5 Epilobium L.			
1 angustifolium L.	36	Br	Mosquin, 1967
2 latifolium L.	72	Is	Löve & Löve, 1956
3 dodonaei Vill.	36	Cz	Májovský et al., 1970
4 fleischeri Hochst.	36	Ca	Zickler, 1967
5 hirsutum L.	36	Br	Raven & Moore, 1964
6 parviflorum Schreber	36	Br	Raven & Moore, 1964
7 duriaei Gay ex Godron			
8 montanum L.	36	Br	Raven & Moore, 1964

9 collinum C.C. Gmelin	36	Au	Griesinger, 1937
10 lanceolatum Sebastiani & Mauri	36	Br	Raven & Moore, 1964
11 alpestre (Jacq.) Krocker	36	Ge	Michaelis, 1925
12 tetragonum L.			
(a) tetragonum	36	Br	Raven & Moore, 1964
(b) lamyi (F.W. Schultz) Nyman	36	Br	Raven & Moore, 1964
(c) tournefortii (Michalet) Léveillé			
13 obscurum Schreber	36	Br	Raven & Moore, 1964
14 roseum Schreber			
(a) roseum	36	Br	Raven & Moore, 1964
(b) subsessile (Boiss.) P.H. Raven			
15 palustre L.	36	Br	Raven & Moore, 1964
16 davuricum Fischer ex Hornem.			
(a) davuricum	36	No	Knaben, 1950
(b) arcticum (Sam.) P.H. Raven			
17 nutans F.W. Schmidt			
18 tundrarum Sam.			
19 atlanticum Litard. & Maire			
20 anagallidifolium Lam.	36	Br	Raven & Moore, 1964
21 hornemanni Reichenb.	36	No	Knaben & Engelskjon, 1967
22 lactiflorum Hausskn.	36	Is	Löve & Löve, 1956
23 alsinifolium Vill.	36	No	Knaben, 1950
24 gemmascens C.A. Meyer			
25 glandulosum Lehm.	36	Non-Europe	Sokolovskaya, 1963
26 adenocaulon Hausskn.	36	Br	Raven & Moore, 1964
27 nerterioides A. Cunn.	36	Br	Raven & Moore, 1964

CXXIV HALORAGACEAE

1 Gunnera L.
  1 tinctoria (Molina) Mirbel

2 Myriophyllum L.			
1 verticillatum L.	28	Ge	Scheerer, 1940
2 spicatum L.			
3 alterniflorum DC.	14	Da	Larsen, 1965
4 brasiliense Camb. in St.-Hil.			
5 heterophyllum Michx			

CXXV THELIGONACEAE

1 Theligonum L.			
1 cynocrambe L.	20	It	Schneider, 1913

CXXVI HIPPURIDACEAE

1 Hippuris L.			
1 vulgaris L.	32	Ho	Gadella & Kliphuis, 1963

CXXVII CORNACEAE

1 Cornus L.
  1 sanguinea L.   22   Hu   Pólya, 1949
    (a) sanguinea
    (b) australis (C.A. Meyer) Jav.
  2 alba L.
  3 sericea L.
  4 mas L.
  5 suecica L.   22   Fe   Sorsa, 1963

CXXVIII ARALIACEAE

1 Hedera L.
  1 helix
    (a) helix   48   Da   Jacobsen, 1954
    (b) poetarum Nyman
    (c) canariensis (Willd.) Coutinho
       hibernica hort.   96 Bot. Gdn.   Jacobsen, 1954
  2 colchica (C. Koch) C. Koch

CXXIX UMBELLIFERAE

1 Hydrocotyle L.
  1 ranunculoides L. fil.
  2 bonariensis Commerson ex Lam.
  3 vulgaris L.   96   Ho   Gadella &
                                        Kliphuis, 1966
  4 sibthorpioides Lam.
  5 moschata G. Forster

2 Bowlesia Ruiz & Pavón
  1 incana Ruiz & Pavón

3 Naufraga Constance & Cannon
  1 balearica Constance & Cannon

4 Sanicula L.
  1 europaea L.   16   Hu   Pólya, 1949
  2 azorica Guthnick ex Seub.

5 Hacquetia DC.
  1 epipactis (Scop.) DC.   16   Po   Skalińska et al.,
                                        1959

6 Astrantia L.
  1 major L.
    (a) major   28   Hu   Baksay, 1957b
    (b) carinthiaca Arcangeli
  2 minor L.   16   It   Favarger &
                                        Huynh, 1964
  3 pauciflora Bertol.
  4 bavarica F.W. Schultz   14   Au   Favarger &
                                        Huynh, 1964
  5 carniolica Jacq.   14   Au   Favarger &
                                        Huynh, 1964

7 Eryngium L.
  1 tenue Lam.

```
 2 barrelieri Boiss.
 3 galioides Lam.
 4 viviparum Gay
 5 duriaei Gay ex Boiss.
 6 ilicifolium Lam.
 7 aquifolium Cav.
 8 maritimum L.                         16     Ge     Wulff, 1937a
 9 alpinum L.
10 planum L.                            16     Po     Skalińska et al.,
                                                      1964
11 dichotomum Desf.
12 creticum Lam.
13 tricuspidatum L.
14 triquetrum Vahl
15 ternatum Poiret
16 serbicum Pančić
17 palmatum Pančić & Vis.
18 glaciale Boiss.
19 dilatatum Lam.
20 spinalba Vill.
21 bourgatii Gouan                      16     He     Burdet & Miége, 1968
22 amorginum Rech. fil.
23 amethystinum L.
24 campestre L.                         14     Po     Skalińska et al.,
                                                      1964
                                        28     Hu     Pólya, 1950
25 corniculatum Lam.
26 pandanifolium Cham. & Schlecht.

 8 Lagoecia L.
   1 cuminoides L.

 9 Petagnia Guss.
   1 saniculifolia Guss.               42     Si     Moore, ined.

10 Echinophora L.
   1 spinosa L.
   2 tenuifolia L.
     (a) tenuifolia
     (b) sibthorpiana (Guss.) Tutin

11 Myrrhoides Heister ex Fabr.
   1 nodosa (L.) Cannon                22     Hu     Baksay, 1956

12 Chaerophyllum L.
   1 aromaticum L.
   2 byzantinum Boiss.
   3 heldreichii Orph. ex Boiss.
   4 coloratum L.
   5 creticum Boiss. & Heldr.
   6 azoricum Trelease
   7 hirsutum L.                       22     He     Rohner, 1954
   8 elegans Gaudin                    22     He     Rohner, 1954
   9 villarsii Koch                    22     He     Rohner, 1954
  10 bulbosum L.
     (a) bulbosum
     (b) prescottii (DC.) Nyman
  11 aureum L.
  12 temulentum L.                     14     Au Da Ga   Böcher & Larsen, 1955
                                       22     Ho     Gadella &
                                                      Kliphuis, 1967a
```

141

13 Anthriscus Pers.
    1 sylvestris (L.) Hoffm.   16   Fe   V. Sorsa, 1962
    2 nitida (Wahlenb.) Garcke
    3 nemorosa (Bieb.) Sprengel
    4 fumarioides (Waldst. & Kit.) Sprengel
    5 cerefolium (L.) Hoffm.
    6 caucalis Bieb.   14   Su   Lövkvist, 1963
    7 tenerrima Boiss. & Spruner

14 Scandix L.
    1 stellata Banks & Solander
    2 australis L.
      (a) microcarpa (Lange) Thell.
      (b) brevirostris (Boiss. & Reuter) Thell.
      (c) grandiflora (L.) Thell.
      (d) australis
    3 pecten-veneris L.
      (a) brachycarpa (Guss.) Thell.
      (b) pecten-veneris   26   Su   Lövkvist, 1963
      (c) macrorhyncha (C.A. Meyer) Rouy &
          Camus

15 Myrrhis Miller
    1 odorata (L.) Scop.

16 Molopospermum Koch
    1 peloponnesiacum (L.) Koch

17 Coriandrum L.
    1 sativum L.   88   Lu   Garde & Malheiros
                                                                               Garde, 1949

18 Bifora Hoffm.
    1 testiculata (L.) Roth
    2 radians Bieb.

19 Scaligeria DC.
    1 cretica (Miller) Boiss.

20 Smyrnium L.
    1 olusatrum L.   22   Br   Rutland, 1941
    2 orphanidis Boiss.
    3 apiifolium Willd.
    4 perfoliatum L.
    5 rotundifolium Miller

21 Bunium L.
    1 bulbocastanum L.
    2 alpinum Waldst. & Kit.
      (a) alpinum
      (b) corydalinum (DC.) Nyman
      (c) montanum (Koch) P.W. Ball
      (d) macuca (Boiss.) P.W. Ball
      (e) petraeum (Ten.) Rouy & Camus
    3 ferulaceum Sibth. & Sm.
    4 pachypodum P.W. Ball

22 Conopodium Koch
    1 majus (Gouan) Loret
    2 pyrenaeum (Loisel.) Miégeville
    3 bourgaei Cosson

4 ramosum Costa
5 capillifolium (Guss.) Boiss.
6 thalictrifolium (Boiss.) Calestani
7 bunioides (Boiss.) Calestani

23 Huetia Boiss.
  1 cynapioides (Guss.) P.W. Ball
   (a) cynapioides
   (b) macrocarpa (Boiss. & Spruner) P.W. Ball
   (c) divaricata (Boiss. & Heldr.) P.W. Ball
  2 cretica (Boiss. & Heldr.) P.W. Ball
  3 pumila (Sibth. & Sm.) Boiss. & Reuter

24 Muretia Boiss.
  1 lutea (Hoffm.) Boiss.

25 Pimpinella L.
  1 rigidula (Boiss. & Orph.) H. Wolff
  2 lutea Desf.
  3 gracilis (Boiss.) H. Wolff
  4 procumbens (Boiss.) H. Wolff
  5 anisum L.
  6 cretica Poiret
  7 peregrina L.
  8 tragium Vill.
   (a) tragium
   (b) lithophila (Schischkin) Tutin
   (c) polyclada (Boiss. & Heldr.) Tutin
   (d) depressa (DC.) Tutin
   (e) titanophila (Woronow) Tutin
  9 pretenderis (Heldr.) Orph. ex Halácsy
 10 villosa Schousboe
 11 anisoides Briganti
 12 serbica (Vis.) Bentham & Hooker fil.
    ex Drude
 13 major (L.) Hudson   20   Hu   Baksay, 1957
 14 siifolia Leresche
 15 bicknellii Briq.
 16 saxifraga L.   40   Su   Lövkvist, 1963

26 Aegopodium L.
  1 podagraria L.   22   Su   Lövkvist, 1963
                                        c.44   Fe   V. Sorsa, 1962

27 Sium L.
  1 latifolium L.   20   Ho   Gadella & Kliphuis,
                                                          1966
  2 sisarum L.

28 Berula Koch
  1 erecta (Hudson) Coville   20   Su   Lövkvist, 1963

29 Crithmum L.
  1 maritimum L.

30 Sclerochorton Boiss.
  1 junceum (Sibth. & Sm.) Boiss.

31 Dethawia Endl.
  1 tenuifolia (Ramond ex DC.) Godron   22   Ga   Favarger & Huynh,
                                                          1964

143

32 Seseli L.
```
 1 libanotis (L.) Koch                            22    Ga    Larsen, 1954
   (a) libanotis
   (b) intermedium (Rupr.) P.W. Ball
 2 sibiricum (L.) Garcke
 3 condensatum (L.) Reichenb. fil.
 4 peucedanoides (Bieb.) Kos.-Pol.               22    Hu    Baksay, 1958
 5 gummiferum Pallas ex Sm.
   (a) gummiferum
   (b) crithmifolium (DC.) P.H. Davis
   (c) aegaeum P.H. Davis
 6 lehmannii Degen
 7 vayrehanum Font Quer
 8 intricatum Boiss.
 9 granatense Willk.
10 peixoteanum Samp.
11 nanum Dufour
12 montanum L.
   (a) montanum
   (b) tommasinii (Reichenb. fil.) Arcangeli
   (c) polyphyllum (Ten.) P.W. Ball
13 cantabricum Lange
14 bocconi Guss.
15 elatum L.
   (a) elatum
   (b) gouanii (Koch) P.W. Ball
   (c) osseum (Crantz) P.W. Ball                  18    Hu    Baksay, 1958
   (d) austriacum (G. Beck) P.W. Ball
16 annuum L.                                      16    Cz    Činčura & Hindakova
   (a) annuum                                                  1963
   (b) carvifolium (Vill.) P. Fourn.
17 campestre Besser
18 arenarium Bieb.
19 tortuosum L.
20 pallasii Besser                                16    Hu    Baksay, 1956
21 glabratum Willd. ex Schultes
22 strictum Ledeb.
23 rigidum Waldst. & Kit.
   (a) rigidum
   (b) peucedanifolium (Besser) Nyman
24 rhodopeum Velen.
25 malyi A. Kerner
26 hippomarathrum Jacq.
   (a) hippomarathrum                             20    Cz    Činčura & Hindakova,
   (b) hebecarpum (DC.) Drude                                  1963
27 dichotomum Pallas ex Bieb.
28 tomentosum Vis.
29 globiferum Vis.
30 leucospermum Waldst. & Kit.                    22    Hu    Baksay, 1956
31 degenii Urum.
32 parnassicum Boiss. & Heldr.
33 bulgaricum P.W. Ball
34 gracile Waldst. & Kit.

33 Oenanthe L.
 1 globulosa L.
   (a) globulosa                                  22    Lu    Cook, 1968
   (b) kunzei (Willk.) Nyman
 2 lisae Moris
 3 fistulosa L.                                   22    Br    Cook, 1968
 4 pimpinelloides L.
```

144

```
  5 millefolia Janka
  6 tenuifolia Boiss. & Orph.
  7 silaifolia Bieb.                          22      Br      Rutland, 1941
  8 peucedanifolia Pollich
  9 lachenalii C.C. Gmelin                    22      Br Lu   Cook, 1968
 10 banatica Heuffel
 11 crocata L.                                22      Br Lu   Cook, 1968
 12 fluviatilis (Bab.) Coleman                22      Br      Cook, 1968
 13 aquatica (L.) Poiret                      22      Br      Cook, 1968

34 Lilaeopsis E.L. Greene
   1 attenuata (Hooker & Arnott) Fernald

35 Aethusa L.
   1 cynapium L.
     (a) cynapium                            20      Au      Titz, 1966
     (b) agrestis (Wallr.) Dostál            20      Su      Løvkvist, 1963
     (c) cynapioides (Bieb) Nyman            20      Su      Løvkvist, 1963

36 Portenschlagiella Tutin
   1 ramosissima (Portenschl.) Tutin

37 Athamanta L.
   1 macedonica (L.) Sprengel
     (a) macedonica
     (b) albanica (Alston & Sandwith) Tutin
     (c) arachnoidea (Boiss. & Orph.) Tutin
         Seseli farinosum Quezel & Contandr. 22     Gr      Quezel &
   2 sicula L.                                                Contandriopoulos, 1965
   3 turbith (L.) Brot.
     (a) turbith
     (b) haynaldii (Borbás & Uechtr.) Tutin
     (c) hungarica (Borbás) Tutin
   4 cretensis L.                             22      It      Favarger, 1965
   5 densa Boiss. & Orph.
   6 cortiana Ferrarini

38 Grafia Reichenb.
   1 golaka (Hacq.) Reichenb.

39 Xatardia Meissner
   1 scabra (Lapeyr.) Meissner

40 Foeniculum Miller
   1 vulgare Miller
     (a) piperitum (Ucria) Coutinho
     (b) vulgare

41 Anethum L.
   1 graveolens L.

42 Kundmannia Scop.
   1 sicula (L.) DC.

43 Silaum Miller
   1 silaus (L.) Schinz & Thell.

44 Trochiscanthes Koch
   1 nodiflora (Vill.) Koch

45 Meum Miller
```

1 athamanticum Jacq.

46 Physospermum Cusson
 1 cornubiense (L.) DC.
 2 verticillatum (Waldst. & Kit.) Vis.

47 Conium L.
 1 maculatum L.

48 Hladnikia Reichenb.
 1 pastinacifolia Reichenb.   22   Ju   Susnik, 1962

49 Pleurospermum Hoffm.
 1 austriacum (L.) Hoffm.   22   Hu   Baksay, 1958
 2 uralense Hoffm.

50 Aulacospermum Ledeb.
 1 isetense (Sprengel) Schischkin

51 Lecokia DC.
 1 cretica (Lam.) DC.

52 Cachrys L.
 1 sicula L.
 2 libanotis L.
 3 pungens Jan ex Guss.
 4 cristata DC.
 5 ferulacea (L.) Calestani
 6 trifida Miller
 7 alpina Bieb.
 8 odontalgica Pallas

53 Heptaptera Margot & Reuter
 1 triquetra (Vent.) Tutin
 2 colladonioides Margot & Reuter
 3 angustifolia (Bertol.) Tutin
 4 macedonica (Bornm.) Tutin

54 Magydaris Koch ex Dc.
 1 pastinacea (Lam.) Paol.
 2 panacifolia (Vahl) Lange

55 Hohenackeria Fischer & C.A. Meyer
 1 exscapa (Steven) Kos.-Pol.
 2 polyodon Cosson & Durieu

56 Bupleurum L.
 1 rotundifolium L.
 2 lancifolium Hornem.
 3 longifolium L.
  (a) longifolium
  (b) aureum (Hoffm.) Soó
 4 stellatum L.   16   Alps   Favarger, 1954
 5 angulosum L.
 6 flavum Forskål
 7 gracile D'Urv.   14   Gr   Snogerup, 1962
 8 aira Snogerup   14   Gr   Snogerup, 1962
 9 apiculatum Friv.
10 glumaceum Sibth. & Sm.   16   Gr   Snogerup, 1962
11 baldense Turra
  (a) baldense

146

```
   (b) gussonei (Arcangeli) Tutin
12 flavicans Boiss. & Heldr.
13 karglii Vis.
14 capillare Boiss. & Heldr.
15 fontanesii Guss. ex Caruel
16 praealtum L.
17 commutatum Boiss. & Balansa
   (a) commutatum
   (b) glaucocarpum (Borbás) Hayek
   (c) aequiradiatum (H. Wolff) Hayek
18 brachiatum C. Koch ex Boiss.
19 gerardi All.
20 trichopodum Boiss. & Spruner
21 affine Sadler
22 asperuloides Heldr. ex Boiss.
23 tenuissimum L.                        16      Hu      Pólya, 1948
24 semicompositum L.
25 petraeum L.                           14      It      Favarger, 1959a
26 ranunculoides L.
   (a) ranunculoides                     42      Ga      Favarger, 1965b
   (b) gramineum (Vill.) Hayek           14      It      Favarger & Huynh,
                                                         1965
27 bourgaei Boiss. & Reuter
28 multinerve DC.
29 falcatum L.
   (a) falcatum                          16      Hu      Baksay, 1956
   (b) cernuum (Ten.) Arcangeli
       subsp. dilatatum                  32      Hu      Baksay, 1956
30 elatum Guss.
31 rigidum L.
   (a) rigidum
   (b) paniculatum (Brot.) H. Wolff
32 spinosum Gouan
33 fruticescens L.
34 dianthifolium Guss.
35 barceloi Cosson ex Willk.
36 acutifolium Boiss.
37 foliosum Salzm. ex DC.
38 gibraltarium Lam.
39 fruticosum L.

57 Trinia Hoffm.
 1 glauca (L.) Dumort.
   (a) glauca
   (b) carniolica (A. Kerner ex Janchen)
 2 multicaulis (Poiret) Schischkin
 3 hispida Hoffm.
 4 ramosissima (Fischer ex Trev.) Koch   20      Hu      Baksay, 1958
 5 kitaibelii Bieb.
 6 muricata Godet
 7 dalechampii (Ten.) Janchen
 8 guicciardii (Boiss. & Heldr.) Drude
 9 crithmifolia (Willd.) H. Wolff

58 Cuminum L.
 1 cyminum L.

59 Apium L.
 1 graveolens L.
 2 nodiflorum (L.) Lag.                   22      Br      Rutland, 1941
 3 repens (Jacq.) Lag.                    22      Hu      Baksay, 1956
```

```
  4 inundatum (L.) Reichenb. fil.          22     Hs     Gadella & Kliphuis,
                                                         1966
  5 crassipes (Koch ex Reichenb.) Reichenb.
    fil.

60 Petroselinum Hill
   1 crispum (Miller) A.W. Hill            22     Br     Rutland, 1941
   2 segetum (L.) Koch

61 Ridolfia Moris
   1 segetum Moris

62 Sison L.
   1 amomum L.

63 Cicuta L.
   1 virosa L.                             22     Hu     Baksay, 1956

64 Cryptotaenia DC.
   1 canadensis (L.) DC.

65 Lereschia Boiss.
   1 thomasii (Ten.) Boiss.                12     It     Grau, ined.

66 Ammi L.
   1 visnaga (L.) Lam.
   2 crinitum Guss.
   3 majus L.
   4 huntii H.C. Watson
   5 trifoliatum (H.C. Watson) Trelease

67 Ptychotis Koch
   1 saxifraga (L.) Loret & Barrandon

68 Ammoides Adanson
   1 pusilla (Brot.) Breistr.

69 Thorella Briq.
   1 verticillatinundata (Thore) Briq.

70 Falcaria Fabr.
   1 vulgaris Bernh.                       22     Su     Löve & Löve, 1942

71 Carum L.
   1 carvi L.                              20     Su     Håkansson, 1953
   2 multiflorum (Sibth. & Sm.) Boiss.
    (a) multiflorum
    (b) strictum (Griseb.) Tutin
   3 verticillatum (L.) Koch
   4 rigidulum (Viv.) Koch ex DC.
   5 heldreichii Boiss.

72 Stefanoffia H. Wolff
   1 daucoides (Boiss.) H. Wolff

73 Brachyapium (Baillon) Maire
   1 dichotomum (L.) Maire

74 Endressia Gay
   1 pyrenaica (Gay ex DC.) Gay
   2 castellana Coincy
```

75 Cnidium Cusson
  1 dubium (Schkuhr) Thell.   20   Hu   Baksay, 1958
  2 silaifolium (Jacq.) Simonkai
    (a) silaifolium
    (b) orientale (Boiss.) Tutin

76 Selinum L.
  1 carvifolia (L.) L.   22   Hu   Pólya, 1949
  2 pyrenaeum (L.) Gouan

77 Ligusticum L.
  1 mutellinoides (Crantz) Vill.   22   He   Favarger, 1954
  2 mutellina (L.) Crantz   22   He   Favarger, 1954
  3 corsicum Gay
  4 albanicum Jáv.
  5 lucidum Miller
    (a) lucidum
    (b) huteri (Porta & Rigo) O. Bolós
  6 ferulaceum All.
  7 scoticum L.

78 Cenolophium Koch ex DC.
  1 denudatum (Hornem.) Tutin

79 Conioselinum Hoffm.
  1 tataricum Hoffm.

80 Angelica L.
  1 palustris (Besser) Hoffm.
  2 sylvestris L.   ??   Hu   Pólya, 1950
  3 archangelica L.
    (a) archangelica   22   Su   Vaarama, 1947
    (b) litoralis (Fries) Thell.   22   Fe   Vaarama, 1947
  4 hererocarpa Lloyd
  5 razulii Gouan
  6 laevis Gay ex Ave-Lall.
  7 angelicastrum (Hoffmanns. & Link)
    Coutinho
  8 pachycarpa Lange

81 Levisticum Hill
  1 officinale Koch

82 Palimbia Besser
  1 rediviva (Pallas) Thell.

83 Bonannia Guss.
  1 graeca (L.) Halácsy

84 Johrenia DC.
  1 distans (Griseb.) Halácsy

85 Capnophyllum Gaertner
  1 peregrinum (L.) Lange

86 Ferula L.
  1 communis L.
    (a) communis
    (b) glauca (L.) Rouy & Camus
  2 tingitana L.
  3 sadlerana Ledeb.   22   Hu   Baksay, 1956

    4 heuffelii Griseb. ex Heuffel
    5 nuda Sprengel
    6 orientalis L.
    7 tatarica Fischer ex Sprengel
    8 caspica Bieb.

87 Ferulago Koch
    1 nodosa (L.) Boiss.
    2 thyrsiflora (Sibth. & Sm.) Koch
    3 asparagifolia Boiss.
    4 campestris (Besser) Grec.
    5 lutea (Poiret) Grande
    6 capillaris (Link ex Sprengel) Coutinho
    7 granatensis Boiss.
    8 sylvatica (Besser) Reichenb.
    9 sartorii Boiss.

88 Eriosynaphe DC.
    1 longifolia (Fischer ex Sprengel) DC.

89 Opopanax Koch
    1 chironium (L.) Koch   22   Hs   Gardé & Malheiros-
                                                              Gardé, 1949

    2 hispidus (Friv.) Griseb.

90 Peucedanum L.
    1 officinale L.   66   Hu   Pólya, 1949
     (a) officinale
     (b) stenocarpum (Boiss. & Reuter) Font
         Quer
    2 longifolium Waldst. & Kit
    3 Lauricum Bieb.
    4 ruthenicum Bieb.
    5 rochelianum Heuffel
    6 paniculatum Loisel.
    7 gallicum Latourr.
    8 aragonense Rouy & Camus
    9 coriaceum Reichenb.
   10 oligophyllum (Griseb.) Vandas
     (a) oligophyllum
     (b) aequiradium (Velen.) Tutin
   11 achaicum Halácsy
   12 carvifolia Vill.
   13 schottii Besser ex DC.   22   It   Favarger, 1959a
   14 vittijugum Boiss.
   15 obtusifolium Sibth. & Sm.
   16 arenarium Waldst. & Kit.
     (a) arenarium
     (b) neumayeri (Vis.) Stoj. & Stefanov
   17 aegopodioides (Boiss.) Vandas
   18 latifolium (Bieb.) DC.
   19 alsaticum L.   22   Hu   Pólya, 1949
   20 venetum (Sprengel) Koch
   21 austriacum (Jacq.) Koch
   22 oreoselinum (L.) Moench
   23 lancifolium Lange
   24 palustre (L.) Moench
   25 cervaria (L.) Lapeyr.
   26 ostruthium (L.) Koch
   27 verticillare (L.) Koch ex DC.
   28 hispanicum (Boiss.) Endl.

29 alpinum (Sieber ex Schultes) B.L.
   Burtt & P.H. Davis

91 Pastinaca L.
   1 sativa L.
     (a) sativa   22   Su   Lövkvist, 1963
     (b) sylvestris (Miller) Rouy & Camus
     (c) urens (Req. ex Godron) Čelak
     (d) divaricata (Desf.) Rouy & Camus
   2 latifolia (Duby) DC.
   3 lucida L.
   4 hirsuta Pančić

92 Heracleum L.
   1 minimum Lam.
   2 austriacum L.   22   Au   Polatschek, 1966
     (a) austriacum
     (b) siifolium (Scop.) Nyman
   3 orphanidis Boiss.
   4 carpaticum Porc.
   5 sphondylium L.   22   Su   Håkansson, 1953
     (a) alpinum (L.) Bonnier & Layens
     (b) pyrenaicum (Lam.) Bonnier & Layens
     (c) transsilvanicum (Schur) Brummitt
     (d) montanum (Schleicher ex Gaudin) Briq.
     (e) sphondylium
     (f) verticillatum (Pančić) Brummitt
     (g) orsinii (Guss.) H. Neumayer
     (h) ternatum (Velen.) Brummitt
     (i) sibiricum (L.) Šimonkai   22   Su   Håkansson, 1953
   6 ligusticifolium Bieb.
   7 stevenii Manden.
   8 pubescens (Hoffm.) Bieb.
   9 mantegazzianum Sommier & Levier

93 Malabaila Hoffm.
   1 aurea (Sibth. & Sm.) Boiss.   22   Gr   Moore, ined.
   2 graveolens (Sprengel) Hoffm.
   3 involucrata Boiss. & Spruner

94 Tordylium L.
   1 maximum L.
   2 officinale L.
   3 pestalozzae Boiss.
   4 apulum L.
   5 byzantinum (Aznav.) Hayek

95 Laser Borkh.
   1 trilobum (L.) Borkh.

96 Elaeoselinum Koch ex DC.
   1 asclepium (L.) Bertol.
     (a) asclepium
     (b) meoides (Desf.) Fiori
   2 tenuifolium (Lag.) Lange
   3 foetidum (L.) Boiss.
   4 gummiferum (Desf.) Tutin

97 Guillonea Cosson
   1 scabra (Cav.) Cosson

98 Laserpitium L.
  1 siler L.
   (a) siler
   (b) garganicum (Ten.) Arcangeli
   (c) zernyi (Hayek) Tutin
  2 latifolium L.   22   Hu   Baksay, 1961
  3 longiradium Boiss.
  4 nestleri Soyer-Willemet
  5 krapfii Crantz
   (a) krapfii
   (b) gaudinii (Moretti) Thell.
  6 peucedanoides L.   22   It   Favarger, 1959a
  7 archangelica Wulfen
  8 nitidum Zanted.   22   It   Favarger, 1959a
  9 halleri Crantz
   (a) halleri
   (b) cynapiifolium (Viv. ex DC.) P. 22   Co   Contandriopoulos,
     Fourn.   1964a
  10 gallicum L.
  11 pseudomeum Orph.,Heldr. & Sart. ex Boiss.
  12 prutenicum L.
   (a) prutenicum
   (b) dufourianum (Rouy & Camus) Tutin
  13 hispidum Bieb.

99 Thapsia L.
  1 villosa L.
  2 maxima Miller
  3 garganica L.

100 Rouya Coincy
  1 polygama (Desf.) Coincy

101 Melanoselinum Hoffm.
  1 decipiens (Schrader & Wendl.) Hoffm.

102 Torilis Adanson
  1 nodosa (L.) Gaertner   22   Lu   Gardé & Malheiros-
                                        Gardé, 1949
  2 arvensis (Hudson) Link   12   Hu   Baksay, 1958
   (a) neglecta (Schultes) Thell.
   (b) arvensis
   (c) elongata (Hoffmanns. & Link) Cannon
   (d) purpurea (Ten.) Hayek
  3 japonica (Houtt.) DC.   16   Su   Lövkvist, 1963
  4 ucranica Sprengel
  5 tenella (Delile) Reichenb. fil.
  6 leptophylla (L.) Reichenb. fil

103 Astrodaucus Drude
  1 orientalis (L.) Drude
  2 littoralis (Bieb.) Drude

104 Turgeniopsis Boiss.
  1 foeniculacea (Fenzl) Boiss.

105 Caucalis L.
  1 platycarpos L.   20   Hs   Lorenzo-Andreu, 1951

106 Turgenia Hoffm.
  1 latifolia (L.) Hoffm.

152

107 Orlaya Hoffm.
```
  1 kochii Heywood                              16      It      Larsen, 1956a
  2 grandiflora (L.) Hoffm.                     20      Hu      Baksay, 1956
  3 daucorlaya Murb.
```

108 Daucus L.
```
  1 durieua Lange
  2 muricatus (L.) L.
  3 broteri Ten.
  4 halophilus Brot.
  5 aureus Desf.
  6 guttatus Sibth. & Sm.
   (a) guttatus
   (b) zahariadii Heywood
  7 involucratus Sibth. & Sm.
  8 carota L.
   (a) carota                                   18      Ho      Gadella & Kliphuis,
                                                                 1966
   (b) maritimus (Lam.) Batt.
   (c) major (Vis.) Arcangeli
   (d) maximus (Desf.) Ball
   (e) sativus (Hoffm.) Arcangeli
   (f) gummifer Hooker fil.
   (g) commutatus (Paol.) Thell.
   (h) hispanicus (Gouan) Thell.
   (i) hispidus (Arcangeli) Heywood
   (j) gadecaei (Rouy & Camus) Heywood
   (k) drepanensis (Arcangeli) Heywood
   (l) rupestris (Guss.) Heywood
  9 setifolius Desf.
 10 crinitus Desf.
```

109 Pseudorlaya (Murb.) Murb.
```
  1 pumila (L.) Grande                          16      It      Larsen, 1956a
  2 minuscula (Pau ex Font Quer) Lainz
```

110 Artedia L.
```
  1 squamata L.
```


CXXX DIAPENSIACEAE

1 Diapensia L.
```
  1 lapponica L.                                12      Is      Löve & Löve, 1956
```


CXXXI PYROLACEAE

1 Pyrola L.
```
  1 minor L.                                    46      Fe      V. Sorsa, 1962
  2 media Swartz                                92      Da      Hagerup, 1941
  3 chlorantha Swartz                           46      Da      Hagerup, 1941
  4 rotundifolia L.
   (a) rotundifolia                             46      Su      Knaben & Engelskjon,
                                                                 1968
   (b) maritima (Kenyon) E.F. Warburg           46      Da No   Knaben & Engelskjon,
                                                                 1968
  5 grandiflora Radius                          46      Is      Löve & Löve, 1956
  6 norvegica Knaben                            46      No      Knaben, 1950
  7 carpatica J. Holub & Křísa                  46      Cz      Holub & Křísa, 1971
```

2 Orthilia Rafin.
  1 secunda (L.) House
    (a) secunda   38   Da   Hagerup, 1941
    (b) obtusata (Turcz.) Bücher

3 Moneses Salisb.
  1 uniflora (L.) A. Gray   26   Da   Hagerup, 1941

4 Chimaphila Pursh
  1 umbellata (L.) W. Barton   26   Da   Knaben & Engelskjon,
                                                        1968
  2 maculata (L.) Pursh

5 Monotropa L.
  1 hypopitys L.   16, 48 Da   Hagerup, 1944

CXXXII ERICACEAE

1 Erica L.
  1 sicula Guss.
  2 ciliaris L.   24   Br   Glanville, ined.
  3 mackaiana Bab.   24   Hb Hs   Glanville, ined.
  4 tetralix L.   24   Da   Hagerup, 1928
  5 terminalis Salisb.
  6 cinerea L.   24   Fa   Hagerup, 1928
  7 australis L.
  8 umbellata L.
  9 arborea L.
  10 lusitanica Rudolphi
  11 vagans L.   24   Br   Maude, 1940
  12 manipuliflora Salisb.
  13 multiflora L.
  14 herbacea L.   24   Da   Hagerup, 1928
  15 erigena R. Ross   24   Hb   Glanville, ined.

2 Bruckenthalia Reichenb.
  1 spiculifolia (Salisb.) Reichenb.

3 Calluna Salisb.
  1 vulgaris (L.) Hull   16   Is   Löve & Löve, 1956b

4 Cassiope D. Don
  1 tetragona (L.) D. Don
  2 hypnoides (L.) D. Don   32   Is   Löve & Löve, 1961

5 Rhododendron L.
  1 luteum Sweet
  2 ponticum L.
    (a) ponticum
    (b) baeticum (Boiss. & Reuter) Hand.-
        Mazz.
  3 ferrugineum L.
  4 myrtifolium Schott & Kotschy
  5 hirsutum L.
  6 lapponicum (L.) Wahlenb.

6 Ledum L.
  1 palustre L.
    (a) palustre   c.52   Fe   V. Sorsa, 1962
    (b) groenlandicum (Oeder) Hultén

154

```
 7 Rhodothamnus Reichenb.
   1 chamaecistus (L.) Reichenb.          24      Au      Polatschek, 1966

 8 Kalmia L.
   1 angustifolia L.
   2 polifolia Wangenh.

 9 Loiseleuria Desv.
   1 procumbens (L.) Desv.               24      Is      Löve & Löve, 1956

10 Phyllodoce Salisb.
   1 caerulea (L.) Bab.                  24      Fe      Sorsa, 1963

11 Daboecia D. Don
   1 cantabrica (Hudson) C. Koch         24      Hb      Maude, 1939
   2 azorica Tutin & E.F. Warburg

12 Arbutus L.
   1 unedo L.                            26      Hb      Sealy & Webb, 1950
   1 andrachne L.

13 Arctostaphylos Adanson
   1 uva-ursi (L.) Sprengel              52      Is      Löve & Löve, 1956
   2 alpinus (1.) Sprengel               26      Rs(N)   Sokolovskaya &
                                                         Strelkova, 1960

14 Gaultheria L.
   1 shallon Pursh

15 Pernettya Gaud.-Beaupré
   1 mucronata (L. fil.) Gaud.-Beaupré
     ex Sprengel

16 Andromeda L.
   1 polifolia L.                        c.48    Fe      Sorsa, 1963

17 Chamaedaphne Moench
   1 calyculata (L.) Moench              22      Fe      Sorsa, 1966

18 Vaccinium L.
   1 oxycoccos (Hill) A. Gray            48      Da      Hagerup, 1940
                                         72      Da      Hagerup, 1940
   2 microcarpum (Turcz. ex Rupr.) Schmalh. 24   Is      Löve & Löve, 1956
   3 macrocarpon Aiton
   4 vitis-idaea L.                      24      Is      Löve & Löve, 1956
     (a) vitis-idaea
     (b) minus (Loddiges) Hultén
   5 uliginosum L.
     (a) uliginosum                      48      Fe      Rousi, 1967
     (b) microphyllum Lange              24      Is      Löve & Löve, 1956
   6 myrtillus L.                        24      Br      I. & O. Hedberg, 1961
   7 arctostaphylos L.
   8 cylindraceum Sm.


CXXXIII EMPETRACEAE

1 Corema D. Don
  1 album (L.) D. Don                    26      Lu      Hagerup, 1941a

2 Empetrum L.
```

1 nigrum L.
  (a) nigrum   26   Fe   Sorsa, 1963
  (b) hermaphroditum (Hagerup) Böcher   52   Is   Löve & Löve, 1956

CXXXIV MYRSINACEAE

1 Myrsine L.
  1 africana L.

CXXXV PRIMULACEAE

1 Primula L.
  1 vulgaris Hudson
    (a) vulgaris   22   Br   Eaton, 1968
    (b) balearica (Willk.) W.W. Sm. &
        Forrest
    (c) sibthorpii (Hoffmanns.) W.W. Sm.
        & Forrest
  2 elatior (L.) Hill
    (a) elatior   22   Br   Eaton, 1968
    (b) leucophylla (Pax) H.-Harrison ex
        W.W. Sm. & Fletcher
    (c) intricata (Gren. & Godron) Ludi
    (d) lofthousei (H.-Harrison) W.W. Sm.
        & Fletcher   22   Hs   Valentine, 1978
    (e) pallasii (Lehm.) W.W. Sm. & Forrest
  3 veris L.
    (a) veris   22   He   Sorsa, 1963
    (b) canescens (Mnch.) Hayek ex Ludi
    (c) columnae (Ten.) Ludi
    (d) macrocalyx (Bunge) Ludi
  4 farinosa L.
    (a) farinosa   18   Br Su   Davies, 1953
                             36   Br   Davies, 1953
    (b) exigua (Velen.) Hayek
  5 frondosa Janka
  6 longiscapa Ledeb.
  7 halleri J.F. Gmelin
  8 scotica Hooker   54   Br   Dovaston, 1955
  9 scandinavica Bruun   72   Br   Dovaston, 1955
  10 stricta Hornem.   126   Is   Löve & Löve, 1956
  11 nutans Georgi   22   Su   Bruun, 1932
  12 egaliksensis Wormsk.
  13 deorum Velen.   66   He   Kress, 1963
  14 glutinosa Wulfen   66   It   Kress, 1963
                             67   It   Kress, 1963
  15 spectabilis Tratt.   66   It   Kress, 1963
  16 glaucescens Moretti   66   It   Kress, 1963
  17 wulfeniana Schott   66   Au   Kress, 1963
  18 clusiana Tausch   c.198   Au   Kress, 1963
  19 minima L.   66 67 68 Au It   Kress, 1963
                   69 70 73
  20 kitaibeliana Schott   66   Ju   Kress, 1963
  21 integrifolia L.   66 c.68 Ga It   Kress, 1963
                     c.70
  22 tyrolensis Schott   66   It   Kress, 1963
  23 allionii Loisel.   66   Ga   Kress, 1963
  24 palinuri Petagna   66   It   Kress, 1963
  25 latifolia Lapeyr. 64 65 66 He   Kress, 1963
                      67

26 marginata Curtis	62 63 ?124	Ga	Favarger & Huynh, 1964
	126 127 128	It	Kress, 1963
27 carniolica Jacq.	62 c.68	He	Kress, 1963
28 auricula L.	62	Po	Skalińska et al., 1959
	63 64	It	Kress, 1963
	65 66		
29 hirsuta All.	62 63 64 67	It	Kress, 1963
30 pedemontana Thomas ex Gaudin	62	It	Kress, 1963
31 apennina Widmer	62	It	Kress, 1963
32 daonensis (Leybold) Leybold	62 63 64	It	Kress, 1963
33 villosa Wulfen	62 63 64	It	Kress, 1963
2 Vitaliana Sesler			
1 primuliflora Bertol.			
(a) primuliflora			
(b) canescens O. Schwarz			
(c) cinerea (Sund.) I.K. Ferguson	40	Ga	Favarger & Küpfer, 1968
(d) assoana Lainz			
(e) praetutiana (Buser ex Sünd.) I.K. Ferguson			
3 Androsace L.			
1 maxima L.			
2 septentrionalis L.			
3 chaixii Gren. & Godron	34	Ga	Favarger, 1965b
4 filiformis Retz.			
5 elongata L.			
6 lactea L.	76	Po He	Favarger, 1962
7 obtusifolia All.	36	He	Favarger, 1962
	38	He	Favarger, 1958
8 hedraeantha Griseb.			
9 carnea L.			
(a) carnea	38	Ga	Favarger, 1958
(b) laggeri (Huet) Nyman	38	Ga	Favarger, 1958
(c) rosea (Jordan & Fourr.) Rouy	38	Ga	Favarger, 1958
(d) brigantiaca (Jordan & Fourr.) I.K. Ferguson	78	He	Favarger, 1958
10 pyrenaica Lam.	38	Ga/Hs	Kress, 1963
11 chamaejasme Wulfen	20	Po	Skalińska, 1959
12 villosa L.	20	He	Favarger, 1958
13 helvetica (L.) All.	40	He	Favarger, 1958
14 vandellii (Turra) Chiov.	40	Ga Hs	Kress, 1963
15 cylindrica DC.	40	Hs	Kress, 1963
16 mathildae Levier	40	It	Kress, 1963
17 pubescens DC.	40	He Hs	Kress, 1963
18 ciliata DC.	c.80	Hs	Kress, 1963
19 hausmannii Leybold	40	It	Kress, 1963
20 alpina (L.) Lam.	40	He	Favarger, 1958
21 wulfeniana Sieber ex Koch	40	Au	Kress, 1963
22 brevis (Hegetschw.) Cesati	40	It	Kress, 1963
4 Cortusa L.			
1 matthioli L.	24	Po	Skalińska et al., 1959
5 Soldanella L.			
1 pusilla Baumg.	40	Rm	Tarnavschi, 1948
2 minima Hoppe			

(a) minima	40	It	Favarger, 1965
(b) samnitica Cristofolini & Pignatti			
3 austriaca Vierh.			
4 alpina L.	40	He	Larsen, 1954
5 carpatica Vierh.	40	Po	Skalińska, 1950
6 pindicola Hausskn.			
7 hungarica Simonkai			
(a) hungarica			
(b) major (Neilr.) S. Pawl.	40	Po	Skalińska, 1950
8 dimoniei Vierh.			
9 montana Willd.	40	Po	Skalińska, 1950
10 villosa Darracq			
6 Hottonia L.			
1 palustris L.	20	Po	Skalińska et al., 1961
7 Cyclamen L.			
1 hederifolium Aiton	24	Bu	Kuzmanov & Kozuharov, ined.
2 graecum Link	78-80	Gr	Glasau, 1939
3 purpurascens Miller	34	Au	Glasau, 1939
4 repandum Sibth. & Sm.			
5 creticum Hildebr.	24	Cr	Glasau, 1939
6 balearicum Willk.	18	Bl	Glasau, 1939
	20	Bl	Kress, 1963
7 coum Miller			
8 persicum Miller			
8 Lysimachia L.			
1 nemorum L.	16	Ho	Gadella & Kliphuis, 1963
	18	Po	Skalińska et al., 1964
	28	Ge	Wulff, 1938
2 serpyllifolia Schreber			
3 vulgaris L.	56 84	Ho	Gadella & Kliphuis, 1966, 1968
4 terrestris (L.) Britton, E.E. Sterns & Poggenb.			
5 ciliata L.			
6 nummularia L.	32 43 45	Ho	Gadella & Kliphuis, 1963, 1967a
	36	Po	Skalińska et al., 1964
7 punctata L.			
8 minoricensis Rodr.			
9 ephemerum L.			
10 clethroides Duby			
11 atropurpurea L.			
12 dubia Aiton			
13 thyrsiflora L.	54	Ho	Gadella & Kliphuis, 1963
9 Trientalis L.			
1 europaea L.	c.160	Is	Löve & Löve, 1956
10 Asterolinon Hoffmanns. & Link			
1 linum-stellatum (L.) Duby			
11 Glaux L.			
1 maritima L.	30	Is	Löve & Löve, 1956

```
12 Anagallis L.
   1 minima (L.) E.H.L. Krause              22      Da      Hagerup, 1941b
   2 crassifolia Thore
   3 tenella (L.) L.                         22      Au      Haffner, 1950
   4 arvensis L.                             40      Gr      Kollman & Feinbrun,
                                                            1968
   5 foemina Miller                          40      ?Br     Marsden-Jones &
                                                            Weiss, 1960
   6 monelli L.                              22      Hs      D.M. Moore, ined.

13 Samolus L.
   1 valerandi L.                           ?24      Ge      Wulff, 1937a
                                            26       Ho      Gadella & Kliphuis,
                                                            1968


14 Coris L.
   1 monspeliensis L.                        18      It      Kress, 1963
                                             56      Hs      Gadella et al., 1966
   2 hispanica Lange


CXXXVI PLUMBAGINACEAE

1 Plumbago L.
  1 europaea L.                              12      Rm      Tarnavschi, 1938

2 Ceratostigma Bunge
  1 plumbaginoides Bunge

3 Acantholimon Boiss.
  1 androsaceum (Jaub. & Spach) Boiss.

4 Armeria Willd.
  1 maritima (Miller) Willd.
    (a) maritima                             18      Is      Löve & Löve, 1956
    (b) sibirica (Turcz. ex Boiss.) Nyman
    (c) miscella (Merino) Malagarriga
    (d) elongata (Hoffm.) Bonnier            18      Br      Baker, 1959
    (e) halleri (Wallr.) Rothm.              18      Ge      Griesinger, 1937
    (f) purpurea (Koch) A. & D. Löve         18      Ge      Griesinger, 1937
    (g) alpina (Willd.) P. Silva             18      Ju      Lovka et al., 1971
                                             36      Europe  Donadille, 1967
    (h) barcensis (Simonkai) P. Silva
  2 canescens (Host) Boiss.
    (a) canescens                            18      Europe  Donadille, 1967
    (b) nebrodensis (Guss.) P. Silva         18      Europe  Donadille, 1967
  3 undulata (Bory) Boiss.
  4 rumelica Boiss.                          18      Europe  Donadille, 1967
  5 sancta Janka
  6 vandasii Hayek
  7 denticulata (Bertol.) DC.
  8 alliacea (Cav.) Hoffmans. & Link         18      Europe  Donadille, 1967
  9 ruscinonensis Girard                     18      Ga      Donadille, 1967
 10 pseudarmeria (Murray) Mansfeld           18      Lu      Donadille, 1967
 11 vestita Willk.
 12 langei Boiss. ex Lange
    (a) langei
    (b) daveaui (Coutinho) P. Silva
 13 villosa Girard                           18      Hs      Phillips, 1938
 14 transmontana (Samp.) Lawrence
```

15 morisii Boiss.			
16 colorata Pau			
17 macropoda Boiss.			
18 filicaulis (Boiss.) Boiss.	18	Ga/Hs	Donadille, 1967
19 trachyphylla Lange			
20 littoralis Willd.	18	Europe	Donadille, 1967
21 duriaei Boiss.			
22 arcuata Welw. ex Boiss. & Reuter			
23 girardii (Bernis) Litard.	18	Europe	Donadille, 1967
24 eriophylla Willk.			
25 humilis (Link) Schultes			
(a) humilis	18	Lu	Fernandes, 1950
(b) odorata (Samp.) P. Silva			
26 sardoa Sprengel	18	Sa	Donadille, 1967
27 leucocephala Salzm. ex Koch	18	Co	Contandriopoulos,
28 multiceps Wallr.	18	Co	1962
29 soleirolii (Duby) Godron	18	Co	Donadille, 1967
30 juniperifolia (Vahl) Hoffmanns. & Link	18	Hs	Sugiura, 1938
31 splendens (Lag. & Rodr.) Webb			
(a) splendens	18	Hs	Donadille, 1967
(b) bigerrensis (C. Vicioso & Beltran)			
P. Silva			
32 pubigera (Desf.) Boiss.			
33 berlengensis Daveau			
34 welwitschii Boiss.	18	Lu	Donadille, 1967
35 pungens (Link) Hoffmanns. & Link	18	Europe	Donadille, 1967
36 pinifolia (Brot.) Hoffmanns. & Link	18	Lu	Donadille, 1967
37 rouyana Daveau	18	Europe	Donadille, 1967
38 macrophylla Boiss. & Reuter	18	Hs/Lu	Sugiura, 1944
39 velutina Welw. ex Boiss. & Reuter			
40 gaditana Boiss.			
41 hirta Willd.			
42 hispalensis Pau			
43 cariensis Boiss.			
5 Limonium Miller			
1 sinuatum (L.) Miller			
2 bonduellei (Lestib.) O. Kuntze			
3 thouinii (Viv.) O. Kuntze			
4 brassicifolium (Webb & Berth.) O.			
Kuntze			
5 ferulaceum (L.) O. Kuntze			
6 diffusum (Pourret) O. Kuntze			
7 caesium (Girard) O. Kuntze	16		
8 insigne (Cosson) O. Kuntze	18	Hs	Erben, 1978
(a) insigne			
(b) carthaginiense Pignatti			
9 latifolium (Sm.) O. Kuntze			
10 asterotrichum (Salmon) Salmon			
11 sareptanum (A. Becker) Gams			
12 bungei (Claus) Gamajun.			
13 gmelinii (Willd.) O. Kuntze	18	Hu	Borhidi, 1968
	27	Rs(E)	Aleskoorskij, 1929
	36	Hu	Polya, 1948
14 meyeri (Boiss.) O. Kuntze			
15 hirsuticalyx Pignatti			
16 tomentellum (Boiss.) O. Kuntze			
17 vulgare Miller			
(a) vulgare	36	Lu	Rodrigues, 1953
(b) serotinum (Reichenb.) Gams			
18 humile Miller	36	Br	Choudhuri, 1942

```
19 bellidifolium   (Gouan) Dumort.          18    Rs(E)     Aleskoovskij, 1929
20 cancellatum (Bernh. ex Bertol.) O.
   Kuntze
21 vestitum (Salmon) Salmon
22 johannis Pignatti
23 cordatum (L.) Miller
24 furfuraceum (Lag.) O. Kuntze            18    Hs        Erben, 1978
25 lucentinum Pignatti & Freitag
26 dichotomum (Cav.) O. Kuntze
27 calcarae (Tod. ex Janka) Pignatti
28 catalaunicum (Willk. & Costa) Pignatti
   (a) catalaunicum
   (b) procerum (Willk.) Pignatti
   (c) viciosoi (Pau) Pignatti
29 articulatum (Loisel.) O. Kuntze
30 japygicum (Groves) Pignatti
31 minutum (L.) Fourr.
32 caprariense (Font Quer & Marcos)
      Pignatti
   (a) caprariense
   (b) multiflorum Pignatti
33 parvifolium (Tineo) Pignatti
34 acutifolium (Reichenb.) Salmon
35 aragonense (Debeaux) Pignatti
36 cosyrense (Guss.) O. Kuntze
37 pontium Pignatti
38 multiforme (U. Martelli) Pignatti
39 bocconei (Lojac.) Litard.
40 anfractum (Salmon) Salmon
41 melium (Nyman) Pignatti
42 tremolsii (Rouy) P. Fourn.              18    Hs        Erben, 1978
43 calaminare Pignatti ex Pignatti        18    Hs        Erben, 1978
44 tenuiculum (Tineo ex Guss.) Pignatti
45 hermaeum (Pignatti) Pignatti
46 remotispiculum (Lacaita) Pignatti
47 minutiflorum (Guss.) O. Kuntze
   (a) minutiflorum
   (b) balearicum Pignatti
48 inarimense (Guss.) Pignatti
   (a) inarimense
   (b) ebusitanum (Font Quer) Pignatti
49 tenoreanum (Guss.) Pignatti
50 laetum (Nyman) Pignatti
51 gougetianum (Girard) O. Kuntze
52 albidum (Guss.) Pignatti
53 frederici (W. Barbey) Rech. fil.
54 panormitanum (Tod.) Pignatti
55 graecum (Poiret) Rech. fil.
   (a) graecum
   (b) divaricatum (Rouy) Pignatti
56 oleifolium Miller
   (a) oleifolium
   (b) dictyocladum Arcangeli
   (c) sardoum (Pignatti) Pignatti)
   (d) pseudodictyocladum (Pignatti) Pignatti
57 carpathum (Rech. fil.) Rech. fil.
58 emarginatum (Willd.) O. Kuntze
59 duriusculum (Girard) Fourr.
60 ramosissimum (Poiret) Maire
   (a) confusum (Gren. & Godron) Pignatti
   (b) provinciale (Pignatti) Pignatti
```

    (c) tommasinii (Pignatti) Pignatti
    (d) siculum Pignatti
    (e) doerfleri (Halácsy) Pignatti
61 ocymifolium (Poiret) O. Kuntze
62 densissimum (Pignatti) Pignatti
63 parvibracteatum Pignatti
64 girardianum (Guss.) Fourr.   35   Ga   Baker, 1955, 1971
65 densiflorum (Guss.) O. Kuntze   26   Si   Dolcher &
66 dufourei (Girard) O. Kuntze   Pignatti, 1971
67 binervosum (G.E. Sm.) Salmon   34-36   Br   Baker, 1955, 1971
68 recurvum Salmon   27   Br   Baker, 1955, 1971
69 transwallianum (Pugsley) Pugsley   27   Br   Baker, 1955, 1971
                                            35   Hb   Baker, 1955, 1971
70 paradoxum Pugsley   33   Br   Baker, 1955, 1971
71 salmonis (Sennen & Elias) Pignatti
72 auriculae-ursifolium (Pourret) Druce   25   Ga   Baker, 1955, 1971
                                            26
    (a) auriculae-ursifolium
    (b) lusitanicum Pignatti
    (c) multiflorum (Pignatti) Pignatti
73 majoricum Pignatti
74 ovalifolium (Poiret) O. Kuntze   16   Hs   Erben, 1978
    (a) gallicum Pignatti
    (b) lusitanicum Pignatti
75 biflorum (Pignatti) Pignatti
76 lausianum Pignatti
77 delicatulum (Girard) O. Kuntze   27   Hs   Pignatti, ined.
    (a) delicatulum
    (b) tournefortii Pignatti
    (c) valentinum Pignatti
78 gibertii (Sennen) Sennen   08 L)   Hs   Pignatti, ined.
79 costae (Willk.) Pignatti
80 sibthorpianum (Guss.) O. Kuntze
81 eugeniae Sennen   16   Hs   Pignatti, ined.
82 coincyi Sennen   16   Hs   Pignatti, ined.
83 album (Coincy) Sennen   18   Hs   Pignatti, ined.
84 cymuliferum (Boiss.) Sauvage & Vindt   16   Hs   Erben, 1978
85 supinum (Girard) Pignatti   26-27   Hs
86 suffruticosum (L.) O. Kuntze
87 echioides (L.) Miller   18   Ga Hs   Erben, 1978

6 Goniolimon Boiss.
    1 elatum (Fischer ex Sprengel) Boiss.
    2 speciosum (L.) Boiss.
    3 heldreichii Halácsy
    4 dalmaticum (C. Presl) Reichenb. fil.
    5 tataricum (L.) Boiss.
    6 tauricum Klokov
    7 rubellum (S.G. Gmelin) Klokov & Grossh.
    8 graminifolium (Aiton) Boiss.
    9 collinum (Griseb.) Boiss.
    10 besseranum (Schultes ex Reichenb.)   36   Rm   Tarnavschi, 1938
        Kusn.
    11 sartorii Boiss.

7 Limoniastrum Heister ex Fabr.
    1 monopetalum (L.) Boiss.

8 Psylliostachys (Jaub. & Spach) Nevski
    1 spicata (Willd.) Nevski

CXXXVII EBENACEAE

1 Diospyros L.
  1 lotus L.

CXXXVIII STYRACACEAE

1 Styrax L.
  1 officinalis L.

CXXXIX OLEACEAE

1 Jasminum L.
  1 nudiflorum Lindley
  2 officinale L.
  3 fruticans L.
  4 humile L.

2 Fontanesia Labill.
  1 philliraeoides Labill.

3 Forsythia Vahl
  1 europaea Degen & Bald.

4 Fraxinus L.
  1 ornus L.
  2 pennsylvanica Marshall
  3 excelsior L.   46   Ho   Gadella & Kliphuis,
    (a) excelsior   46   Ho   1963
    (b) coriariifolia (Scheele) E. Murray
  4 angustifolia Vahl
    (a) angustifolia
    (b) oxycarpa (Bieb. ex Willd.) Franco &
        Rocha Afonso
  5 pallisiae Wilmott

5 Syringa L.
  1 vulgaris L.
  2 josikaea Jacq. fil. ex Reichenb.

6 Ligustrum L.
  1 vulgare L.   46   Ho   Gadella & Kliphuis,
                                                    1966

7 Olea L.
  1 europaea L.   46   It   Breviglieri &
                                                    Battaglia, 1954

8 Phillyrea L.
  1 angustifolia L.
  2 latifolia L.   46   It   Armenise, 1958

9 Picconia DC.
  1 azorica (Tutin) Knobl.

CXL GENTIANACEAE

1 Cicendia Adanson

163

1 filiformis (L.) Delarbre	26	Ga	Favarger, 1960

2 **Exaculum** Caruel

1 pusillum (Lam.) Caruel	20	Ga	Favarger, 1969

3 **Blackstonia** Hudson
 1 perfoliata (L.) Hudson

(a) perfoliata	40	He Gr Ga It	Zeltner, 1966
(b) serotina (Koch ex Reichenb.)			
Vollmann	40	Ga Gr Hs Cz	Zeltner, 1966
(c) imperfoliata (L. fil.) Franco &			
Rocha Afonso	20	Ga	Zeltner, 1962
(d) grandiflora (Viv.) Maire			

4 **Centaurium** Hill

1 scilloides (L. fil.) Samp.	20	Br Lu Hs	Zeltner, 1970
2 erythraea Rafn			
(a) erythraea	40	Br Ga Hs It	Zeltner, 1962,
		Si	1970
(b) turcicum (Velen.) Melderis			
(c) rumelicum (Velen.) Melderis	20	Ga It Gr Sa	Zeltner, 1970
		Co	
(d) grandiflorum (Biv.) Melderis			
(e) rhodense (Boiss. & Reuter)			
Melderis	40	Si It Sa	Zeltner, 1970
(f) majus (Hoffmanns. & Link)			
Melderis	20	Hs	Zeltner, 1963
		Lu It Ga Sa	Zeltner, 1970
3 suffruticosum (Griseb.) Ronniger	20	Hs	Zeltner, 1970
4 chloodes (Brot.) Samp.	40	Lu	Zeltner, 1970
5 littorale (D. Turner) Gilmour			
(a) littorale	38	Ge	Wulff, 1937a
	40	Ge No	Zeltner, 1970
	c.56	Br	Warburg, 1939
(b) uliginosum (Waldst. & Kit.)			
Melderis	40	Be Br Ho Su	Zeltner, 1970
6 favargeri Zeltner	20	Ga	Zeltner, 1966
7 triphyllum (W.L.E. Schmidt) Melderis	20	Hs	Zeltner, 1967
8 rigualii Esteve			
9 linariifolium (Lam.) G. Beck	20	Hs	Zeltner, 1963
10 microcalyx (Boiss. & Reuter) Ronniger			
11 pulchellum (Swartz) Druce	36	Hs Ga Gr Ju	Zeltner, 1962,
		Au Br Be Ho	1963, 1966
12 tenuiflorum (Hoffmanns. & Link)			
Fritsch			
(a) tenuiflorum	40	Gr Lu Sa	Zeltner, 1966
(b) acutiflorum (Schott) Zeltner	20	Hs Ga It Co	Zeltner, 1961
		Si	
13 spicatum (L.) Fritsch	22	Hs Si Sa Ga	Zeltner, 1970
		Lu	
14 maritimum (L.) Fritsch	20	Co Ga Sa	Zeltner, 1970

5 **Gentiana** L.
 1 lutea L.

(a) lutea	40	Ga	Favarger, 1949b
(b) symphyandra (Murb.) Hayek			
2 punctata L.	40	Po	Skalińska, 1952
3 pannonica Scop.	40	He	Favarger, 1965b
4 purpurea L.	40	No	Knaben & Engelskjon,
			1967
5 burseri Lapeyr.			

(a) burseri	40	Ga	Favarger & Küpfer, 1968
(b) villarsii (Griseb.) Rouy	c.40	Ga	Favarger & Küpfer, 1968
6 asclepiadea L.	36	Po	Skalińska, 1950
7 pneumonanthe L.	26	Hu	Pólya, 1950
8 frigida Haenke	24	Po	Skalińska, 1950
9 froelichii Jan ex Reichenb.	42	Ju	Favarger, 1965b
10 cruciata L.	52	Po	Skalińska, 1952
(a) cruciata			
(b) phlogifolia (Schott & Kotschy) Tutin			
11 decumbens L. fil.			
12 prostrata Haenke	c.36	He	Favarger, 1952
13 pyrenaica L.	26	Ge	Favarger & Küpfer, 1968
14 boryi Boiss.			
15 clusii Perr. & Song.	36	Po	Skalińska, 1950
16 occidentalis Jakowatz			
17 ligustica R. de Vilmorin & Chopinet			
18 acaulis L.	36	Ju	Lovka et al., 1971
19 alpina Vill.	36	Ga	Favarger & Küpfer, 1968
20 dinarica G. Beck			
21 angustifolia Vill.	36	Ga	Favarger, 1965
22 verna L.			
(a) verna	28	Po	Skalińska, 1950
(b) pontica (Soltok.) Hayek			
(c) tergestina (G. Beck) Hayek			
23 brachyphylla Vill.			
(a) brachyphylla	28	He	Favarger, 1965
(b) favratii (Rittener) Tutin	28 30 + 2B	He	Favarger, 1965
	32	Ga	Favarger, 1965
24 pumila Jacq.	20	Au	Favarger, 1965
(a) pumila			
(b) delphinensis (Beauverd) P. Fourn.			
25 bavarica L.	30	It	Favarger, 1965
26 rostanii Reuter ex Verlot	30	Ga	Favarger, 1969
27 terglouensis Hacq.	40	It	Favarger, 1965
(a) tergloensis			
(b) schleicheri (Vacc.) Tutin			
28 nivalis L.	14	Po	Skalińska et al., 1964
29 utriculosa L.			
6 Gentianella Moench			
1 tenella (Rottb.) Börner	10	No	Knaben, 1950
2 nana (Wulfen) Pritchard	30	Au	Favarger, 1965
3 ciliata (L.) Borkh.			
(a) ciliata	44	Po	Skalińska, 1959
(b) doluchanovii (Grossh.) Pritchard			
4 detonsa (Rottb.) G. Don fil.	44	Is	Löve & Löve, 1956
5 barbata (Froelich) Berchtold & J. Presl			
6 crispata (Vis.) J. Holub			
7 columnae (Ten.) J. Holub			
8 campestris (L.) Börner			
(a) campestris	36	Is	Löve & Löve, 1956
(b) baltica (Murb.) Tutin			
9 hypericifolia (Murb.) Pritchard	36	Ga	Favarger & Küpfer, 1968
10 amarella (L.) Börner	36	Is	Löve & Löve, 1956

(a) amarella
(b) septentrionalis (Druce) Pritchard
11 uliginosa (Willd.) Börner   c.54   Da   Holmen, 1961
12 anglica (Pugsley) E.F. Warburg
  (a) anglica
  (b) cornubiensis Pritchard
13 bulgarica (Velen.) J. Holub
14 ramosa (Hegetschw.) J. Holub
15 pilosa (Wettst.) J. Holub
16 engadinensis (Wettst.) J. Holub   36   He   Favarger &
                                                Huynh, 1964
17 anisodonta (Borbás) Á. & D. Löve   36   It   Favarger, 1965b
18 aspera (Hegetschw. & Heer) Dostál ex
  Skalický
19 germanica (Willd.) E.F. Warburg   36   Ge   Reese, 1961
20 austriaca (A. & J. Kerner) J. Holub
21 lutescens (Velen.) J. Holub   36   Po   Skalińska, 1950
22 aurea (L.) H. Sm.   36   Is   Löve & Löve, 1956

7 Lomatogonium A. Braun
  1 carinthiacum (Wulfen) Reichenb.   40   Au   Fürnkranz, 1965
  2 rotatum (L.) Fries ex Fernald   10   Is   Löve & Löve, 1956

8 Swertia L.
  1 perennis L.   28   Po   Skalińska,et al.,
                                              1959

9 Halenia Borkh.
  1 corniculata (L.) Cornaz

CXLI MENYANTHACEAE

1 Menyanthes L.
  1 trifoliata L.   54   Su   Palmgren, 1943

2 Nymphoides Séguier
  1 peltata (S.G. Gmelin) O. Kuntze   54   Ho   Gadella & Kliphuis,
                                              1965

CXLII APOCYNACEAE

1 Nerium L.
  1 oleander L.

2 Trachomitum Woodson
  1 venetum (L.) Woodson
  2 sarmatiense Woodson
  3 tauricum (Pobed.) Pobed.

3 Rhazya Decne
  1 orientalis (Decne) A.DC.

4 Vinca L.
  1 minor L.   46   Br   Rutland, 1941
  2 herbacea Waldst. & Kit.
  3 difformis Pourret   c.46   Lu   K. Jones, unpub.
    (a) difformis
    (b) sardoa Stearn
  4 balcanica Pénzes

5 major L. 92 Br Rutland, 1941

CXLIII ASCLEPIADACEAE

1 Periploca L.
  1 graeca L. 24 It Lopane, 1951
  2 laevigata Aiton

2 Araujia Brot.
  1 sericifera Brot.

3 Gomphocarpus R. Br.
  1 fruticosus (L.) Aiton 16 Hs Björkqvist et al.,
                                                      1969

4 Asclepias L.
  1 syriaca L.

5 Cynanchum L.
  1 acutum L. 18 It Francini, 1927

6 Vincetoxicum N.M. Wolf
  1 canescens (Willd.) Decne
  2 speciosum Boiss. & Spruner
  3 nigrum (L.) Moench
  4 scandens Sommier & Levier
  5 huteri Vis. & Ascherson
  6 juzepczukii (Pobed.) Privalova
  7 rossicum (Kleopow) Barbarich
  8 schmalhausennii (Kusn.) Markgraf
  9 fuscatum (Hornem.) Reichenb. fil.
 10 pannonicum (Borhidi) J. Holub 44 Hu Borhidi, 1968
 11 hirundinaria Medicus 22 Cz Da He Böcher & Larsen,
                                            Hu It Sa 1955

    (a) lusitanicum Markgraf
    (b) intermedium (Loret & Barrandon)
        Markgraf
    (c) nivale (Boiss. & Heldr.) Markgraf
    (d) adriaticum (G. Beck) Markgraf
    (e) stepposum (Pobed.) Markgraf
    (f) cretaceum (Pobed.) Markgraf
    (g) jailicola (Juz.) Markgraf
    (h) contiguum (Koch) Markgraf
    (i) hirundinaria

7 Cionura Griseb.
  1 erecta (L.) Griseb.

8 Caralluma R. Br.
  1 europaea (Guss.) N.E. Br.
  2 munbyana (Decne) N.E. Br.

CXLIV RUBIACEAE

1 Putoria Pers.
  1 calabrica (L. fil.) DC. 22 Al Strid, 1971

2 Sherardia L.
  1 arvensis L. 22 Po Skalińska et al.,
                                           1964

3 Crucianella L.
  1 maritima L.   22   Lu   Rodrigues, 1953
  2 macrostachya Boiss.
  3 bithynica Boiss.
  4 graeca Boiss.   22   Gr   Ehrendorfer, ined.
  5 angustifolia L.   22   Bu Hs   Ehrendorfer, ined.
  6 imbricata Boiss.
  7 latifolia L.   44   Bl   Dahlgren et al.,
                                             1971

  8 patula L.

4 Asperula L.
  1 aristata L. fil.
    (a) scabra (J. & C. Presl) Nyman   20
                                      40
    (b) nestia (Rech. fil.) Ehrend. &
        Krendl   40
    (c) thessala (Boiss. & Heldr.) Hayek
    (d) oreophila (Briq.) Hayek   20
    (e) condensata (Heldr. ex Boiss.)
        Ehrend. & Krendl   20
  2 wettsteinii Adamović
  3 tenella Heuffel ex Degen
  4 crassifolia L.
  5 calabra (Fiori) Ehrend. & Krendl
  6 staliana Vis.
  7 garganica Huter, Porta & Rigo ex
    Ehrend. & Krendl
  8 suffruticosa Boiss. & Heldr.
  9 ophiolithica Ehrend.
 10 idaea Halácsy
 11 suberosa Sibth. & Sm.
 12 gussonii Boiss.
 13 pumila Moris
 14 oetaea (Boiss.) Heldr. ex Halácsy
 15 nitida Sibth. & Sm.
 16 lutea Sibth. & Sm.
    (a) lutea
    (b) rigidula (Halácsy) Ehrend.
    (c) euboea Ehrend.
    (d) mungieri (Boiss. & Heldr.) Ehrend.
       & Krendl
 17 abbreviata (Halácsy) Rech. fil.
 18 pulvinaris (Boiss.) Heldr. ex Boiss.
 19 boissieri Heldr. ex Boiss.   22   Gr   Faure & Pietrera,
                                           1969
 20 neilreichii G. Beck   20   Rs(K)   Fagerlind, 1934
 21 beckiana Degen
 22 neglecta Guss.
 23 rupicola Jordan
 24 pyrenaica L.
 25 cretacea Willd.
 26 supina Bieb.   20   Rs(K)   Ehrendorfer, ined.
 27 tephrocarpa Czern. ex M. Popov &
    Chrshan.
 28 exasperata V. Krecz. ex Klokov
 29 petraea V. Krecz. ex Klokov
 30 occidentalis Rouy
 31 cynanchica L.   20   Cz   Mičieta, 1981
                                     40
 32 rumelica Boiss.

33 danilewskiana Basiner			
34 laevissima Klokov			
35 graveolens Bieb. ex Schultes & Schultes fil.			
36 diminuta Klokov			
37 leiograveolens M. Popov & Chrshan.			
38 savranica Klokov			
39 setulosa Boiss.			
40 littoralis Sibth. & Sm.			
41 incana Sibth. & Sm.	44	Cr	Ehrendorfer, ined.
42 taygetea Boiss. & Heldr.			
43 rupestris Tineo			
44 hirsuta Desf.	22 44	Hs	Ehrendorfer & Ernet, ined.
45 taurica Pacz.			
46 arcadiensis Sims	22	Gr	Faure & Pietrera, 1969
47 doerfleri Wettst.	22	Ju	Ehrendorfer & Ernet, ined.
48 hirta Ramond	22	Ga	Favarger & Küpfer, 1968
49 hercegovina Degen			
50 capitata Kit. ex Schultes	22	Rm	Ehrendorfer &
51 hexaphylla All.	22	Ga	Ernet, ined.
52 taurina L.	22	Hu	Baksay, 1956
(a) taurina			
(b) leucanthera (G. Beck) Hayek			
53 involucrata Wahlenb.			
54 laevigata L.	44		error
55 tinctoria L.	22	Cz	Holub et al., 1970
	44	Po	Skalińska et al., 1959
56 arvensis L.			
57 rigida Sibth. & Sm.	22	Cr	Ehrendorfer, ined.
58 tournefortii Sieber ex Sprengel	22	Cr	Faure & Pietrera, 1969
59 muscosa Boiss. & Heldr.	22	Gr	Faure & Pietrera, 1969
60 baenitzii Heldr. ex Boiss.			
61 chlorantha Boiss. & Heldr.	22	Gr	Faure & Pietrera, 1969
62 scutellaris Vis.	22	Ju	Ehrendorfer &
63 baldaccii (Halácsy) Ehrend.			Ernet, ined.
64 saxicola Ehrend.			
65 boryana (Walpers) Ehrend.			
66 purpurea (L.) Ehrend.			
(a) purpurea	22	It	Krendl &
(b) apiculata (Sibth. & Sm.) Ehrend.	22	Gr	Ehrendorfer, ined.
5 Galium L.			
1 paradoxum Maxim.			
2 rotundifolium L.	22	Po	Skalińska et al., 1964
3 scabrum L.	22	Co	Ehrendorfer
4 baillonii Brandza	22	Rm	&
5 broterianum Boiss. & Reuter	22	Lu	Krendl, ined.
6 boreale L.	44	Po	Skalińska et al., 1961
	55	Po	Skalińska et al., 1961
	66	Po	Skalińska et al., 1961

7 rubioides L.	66		error
8 fruticosum Willd.	22	Cr	Ehrendorfer, ined.
9 ephedroides Willk.			
10 odoratum (L.) Scop.	44	Po	Piotrowicz, 1958
11 triflorum Michx			
12 rivale (Sibth. & Sm.) Griseb.	66	Ju	Ehrendorfer, ined.
13 uliginosum L.	22	Ho	Gadella & Kliphuis, 1963
	44	Ho	Gadella & Kliphuis, 1963
14 viridiflorum Boiss. & Reuter			
15 debile Desv.	24	Bu	Ančev, 1974
16 palustre L.	24	Po	Skalińska et al., 1961
	48	Ju Rm	Teppner et al., 1976
17 elongatum C. Presl	96	Au Da Ge	Teppner et al., 1976
	144	Au	Teppner et al., 1976
18 trifidum L.	24	Au	Puff, 1976
brevipes	24	Is	Löve & Löve, 1956
19 saturejifolium Trev.			
20 boissieranum Ehrend. & Krendl			
21 baeticum (Rouy) Ehrend. & Krendl			
22 concatenatum Cosson			
23 humifusum Bieb.			
24 maritimum L.			
25 tunetanum Lam.			
26 verum L.			
(a) verum	22	Bu	Ančev, 1974
	44	Cz	Májovský et al 1970
(b) wirtgenii (F.W. Schultz) Oborny	22	Ju	Fritsch, 1973
27 x pomeranicum Retz.			
28 thymifolium Boiss. & Heldr.			
29 kerneri Degen & Dörfler	22	Bu	Markova &
30 degenii Bald. ex Degen			Ančev, 1973
31 erythrorrhizon Boiss. & Reuter			
32 pulvinatum Boiss.			
33 litorale Guss.			
34 arenarium Loisel.	66	Ga	Krendl, ined.
35 firmum Tausch	22	Ju	Ehrendorfer & Krendl, ined.
36 heldreichii Halácsy	22	Gr	Ehrendorfer, 1974
37 protopycnotrichum Ehrend. & Krendl	22	Bu Ju Tu	Krendl, ined.
38 mollugo L.	22	Au It Ju	Krendl, 1967
39 album Miller			
(a) album	44	Bu	Ančev, 1974
(b) pycnotrichum (H. Braun) Krendl	44	Bu	Ančev, 1974
(c) prusense (C. Koch) Ehrend. & Krendl	44	Gr Tu	Ehrendorfer &
40 reiseri Halacsy	22	Gr	Krendl, ined.
41 fruticescens Cav.			
42 corrudifolium Vill.	22	Ju It	Krendl, 1967
43 truniacum (Ronniger) Ronniger	22	Au	Krendl, 1967
44 montis-arerae Merxm. & Ehrend.	22	It	Krendl, 1967
45 meliodorum (G. Beck) Fritsch	44	Au	Krendl, 1967
46 lucidum All.	44	Bu	Ančev, 1974
47 bernardii Gren. & Godron			
48 aetnicum Biv.	44	It Si	Krendl, 1976
49 cinereum All.	44	Ga	Krendl, 1976
50 crespianum J.J. Rodr.			
51 mirum Rech. fil.	22	Bu	Ančev, 1974
52 scabrifolium (Boiss.) Hausskn.	22 44	Bu	Ančev, 1974

170

53 peloponnesiacum Ehrend. & Krendl	44	Gr	Krendl, 1974
54 asparagifolium Boiss. & Heldr.			
55 melanantherum Boiss.	22	Gr	Krendl, ined.
56 rhodopeum Velen.	22 44	Bu	Ančev, 1974
57 incurvum Sibth. & Sm.	44	Cr	Ehrendorfer, ined.
58 flavescens Borbas	44	Bu	Ančev, 1974
59 pruinosum Boiss.			
60 glaucophyllum E. Schmid			
61 murcicum Boiss. & Reuter			
62 octonarium (Klokov) Pobed.	22	Bu	Krendl, ined.
63 glaucum L.	22	Au	Ernet &
	44	Au Cz Ga	Ehrendorfer, ined.
64 biebersteinii Ehrend.			
65 xeroticum (Klokov) Pobed.			
66 volhynicum Pobed.			
67 moldavicum (Dobrescu) Franco			
68 kitaibelianum Schultes & Schultes fil.	22	Rm	Ehrendorfer, 1975
69 pseudaristatum Schur	22	Bu	Ančev, 1974
70 aristatum L.	22	Ga It	Ehrendorfer, 1975
71 abaujense Borbas	44	Cz	Ehrendorfer, 1975
72 polonicum Blocki	44	Po	Piotrowicz, 1958
73 laconicum Boiss. & Heldr.	22	Gr	Faure & Pietrera, 1969
74 procurrens Ehrend.	22	Bu	Ehrendorfer & Ančev, 1975
75 laevigatum L.	44	Ju	Ehrendorfer, 1975
76 schultesii Vest	44	Ju	Ehrendorfer, 1975
	66	Cz	Majovsky et al., 1970
77 sylvaticum L.	22	Ho	Gadella & Kliphuis, 1963
78 longifolium (Sibth. & Sm.) Griseb.	22	Tu	Ehrendorfer, 1975
79 bulgaricum Velen.	22	Bu	Ančev, ined.
80 saxosum (Chaix) Breistr.	22	Ga	Ehrendorfer, ined.
81 cometerhizon Lapeyr.			
82 incanum Sibth. & Sm.			
(a) incanum	44	Gr	Faure & Pietrera, 1969
(b) creticum Ehrend.			
83 cyllenium Boiss. & Heldr.	22	Gr	Faure & Pietrera, 1969
84 palaeoitalicum Ehrend.	20	It	Ehrendorfer, 1974
85 pyrenaicum Gouan	22	Ga Hs	Ehrendorfer, ined.
	44	Ga	Ehrendorfer, ined.
86 corsicum Sprengel	22 44	Co	Ehrendorfer, ined.
87 obliquum Vill.	22	Ga	Ehrendorfer, 1955
	44	It Ga	Ehrendorfer, 1955
88 rubrum L.	88	It	Ehrendorfer, 1955
89 x centroniae Cariot	88	It	Ehrendorfer, 1955
90 x carmineum Beauverd			
91 balearicum Briq.			
92 valentinum Lange			
93 rosellum (Boiss.) Boiss. & Reuter	22	Hs	Küpfer, 1968
94 helodes Hoffmanns. & Link	22	Lu	Ehrendorfer, ined.
95 rivulare Boiss. & Reuter	22	Hs	Ehrendorfer, ined.
96 asturiocantabricum Ehrend.	88	Hs	Ehrendorfer, ined.
97 papillosum Lapeyr.	22	Hs	Ehrendorfer, ined.
	44	Ga Hs	Ehrendorfer, ined.
98 pinetorum Ehrend.	22 44	Hs	Ehrendorfer, ined.
99 marchandii Roemer & Schultes	88	Ga Hs	Ehrendorfer, ined.

171

100 nevadense Boiss. & Reuter	22	Hs	Küpfer, 1969
101 timeroyi Jordan	22	Ga	Ehrendorfer, 1962
102 fleurotii Jordan	44	Ga	Ehrendorfer, 1962
	88	Ga	Ehrendorfer, 1962
103 pumilum Murray	66	Europe	Ehrendorfer, 1962
	88	Ge	Ehrendorfer, 1962
104 valdepilosum H. Braun	22	Au	Ehrendorfer, 1962
	44	Da	Ehrendorfer, 1962
105 suecicum (Sterner) Ehrend.	22	Su	Ehrendorfer, 1962
106 oelandicum (Sterner & Hyl.) Ehrend.	22	Su	Ehrendorfer, 1962
107 cracoviense Ehrend.	22	Po	Piotrowicz, 1958
108 sudeticum Tausch			
109 sterneri Ehrend.	22	Br	Ehrendorfer, 1962
	44	Da No	Ehrendorfer, 1962
110 normanii O.C. Dahl	44	Is	Löve & Löve, 1956
111 austriacum Jacq.	22	Au	Titz, 1964
	44	Au	Titz, 1964
112 anisophyllon Vill.	22	Gr	Faure & Pietrara, 1969
	44	Po	Skalińska et al., 1964
	66	Au Ge	Ehrendorfer, 1963
	88	He It	Ehrendorfer, 1963
	110	It	Ehrendorfer, 1963
113 pseudohelveticum Ehrend.	44	Ca	Ehrendorfer, ined.
114 megalospermum All.	22	Au Ga He	Ehrendorfer, ined.
	44	Ca	Ehrendorfer, ined.
115 pusillum L.	22 88	Ga	Ehrendorfer, ined.
116 brockmannii Briq.	22	Hs	Ehrendorfer, ined.
117 ˙labudut (Pau ex Deheaux) Dou ˙˙ Ehrend.	22	Hs	Ehrendorfer, ined.
118 saxatile L.	22	Lu	Kliphuis, 1967
	44	Ho	Kliphuis, 1967
119 tendae Reichenb. fil.	22	Ga	Ehrendorfer, ined.
120 magellense Ten.	22	It	Ehrendorfer, 1963
121 margaritaceum A. Kerner	22	It	Ehrendorfer, ined.
122 baldense Sprengel	22	It	Ehrendorfer, 1953
123 noricum Ehrend.	44	Au	Ehrendorfer, 1953
124 demissum Boiss.	22	Bu	Ančev, 1974
125 stojanovii Degen	22	Bu	Ančev, 1974
126 cespitosum Lam.	22	Ga	Ehrendorfer, ined.
127 graecum L.			
(a) graecum	22	Cr	Ehrendorfer, ined.
(b) pseudocanum Ehrend.			
128 canum Req. ex DC.			
129 setaceum Lam.	22	Cr	Ehrendorfer, ined.
(a) setaceum			
(b) decaisnei (Boiss.) Ehrend.			
130 monachinii Boiss. & Heldr.	22	Cr	Ehrendorfer, ined.
131 spurium L.	20	Bu	Ančev, 1974
132 aparine L.	42	Au	Ehrendorfer, ined.
	44	Ho	Kliphuis, 1962b
	48	Au	Ehrendorfer, ined.
	62	Au	Ehrendorfer, ined.
	66	Bu	Ančev, 1974
	68	Au	Ehrendorfer, ined.
133 tricornutum Dandy	44	Au Ga It	Ehrendorfer, ined.
134 verrucosum Hudson	22	Ju	Ehrendorfer, ined.
135 intricatum Margot & Reuter			
136 capitatum Bory & Chaub.			
137 incrassatum Halácsy			

138 viscosum Vahl
139 parisiense L.   44   Ga Hs Ju   Ehrendorfer, ined.

138 viscosum Vahl			
139 parisiense L.	44	Ga Hs Ju	Ehrendorfer, ined.
	66	Bu	Ancev, 1974
140 divaricatum Pourret ex Lam.	44	Ga Lu	Ehrendorfer, ined.
141 tenuissimum Bieb.	22	Asia	
	44	Bu	Ančev, 1974
142 minutulum Jordan			
143 recurvum Req. ex DC.			
144 verticillatum Danth.	22	Bu	Ančev, 1974
	44	Gr	Ehrendorfer, ined.
145 murale (L.) All.	44 Bot. Gard.		Fagerlind, 1934

6 Callipeltis Steven
  1 cucullaris (L.) Rothm.

7 Cruciata Miller			
1 laevipes Opiz	22	Cz	Májovský et al., 1970
2 taurica (Pallas ex Willd.) Ehrend.			
(a) euboea (Ehrend.) Ehrend.			
(b) taurica	44	Rs(K)	Ehrendorfer, ined.
3 glabra (L.) Ehrend.	22	It	Ehrendorfer, ined.
	44	Bu	Ančev, 1976
4 balcanica Ehrend.			
5 pedemontana (Bellardi) Ehrend.	18	Cz	Májovský et al., 1970

8 Valantia L.			
1 aprica (Sibth. & Sm.) Boiss. & Heldr.	22	Gr	Faure & Pietrera, 1969
2 hispida L.	18	Bl	Dahlgren et al., 1971
3 muralis L.	18	Bl	Dahlgren et al., 1971

9 Rubia L.			
1 peregrina L.	44	Bl	Dahlgren et al., 1971
angustifolia L.	66	Bl	Cardona, 1973
2 tenuifolia D'Urv.			
3 tatarica (Trev.) Friedrich Schmidt Petrop.			
4 tinctorum L.			

CXLV POLEMONIACEAE

1 Polemonium L.			
1 caeruleum L.	18	No	Knaben & Engelskjon, 1967
2 acutiflorum Willd.	18	No	Knaben & Engelskjon, 1967
3 boreale Adams	18	No	Laane, 1966

173

2 Collomia Nutt.
  1 grandiflora Douglas ex Lindley

CXLVI CONVOLVULACEAE

1 Cuscuta L.
  1 australis R. Br.
   (a) tinei (Insenga) Feinbrun
   (b) cesattiana (Bertol.) Feinbrun
  2 campestris Yuncker
  3 suaveolens Ser.
  4 gronovii Willd.
  5 pedicellata Ledeb.
  6 europaea L.
  7 pellucida Butkov
  8 epilinum Weihe
  9 palaestina Boiss.
 10 atrans Feinbrun
 11 triumvirati Lange
 12 epithymum (L.) L.   14   Su   Ehrenberg, 1945
   (a) epithymum
   (b) kotschyi (Desmoulins) Arcangeli
 13 brevistyla A. Braun ex A. Richard
 14 planiflora Ten.
 15 approximata Bab.
   (a) approximata
   (b) macranthera (Heldr. & Sart. ex Boiss.)
       Feinbrun & W. Greuter
   (c) episonchum (Webb & Berth.) Feinbrun
 16 lupuliformis Krocker
 17 monogyna Vahl

2 Cressa L.
  1 cretica L.

3 Dichondra J.R. & G. Forster
  1 micrantha Urban

4 Calystegia R. Br.
  1 soldanella (L.) R. Br.   22   Lu   Rodriguez, 1953
  2 sepium (L.) R. Br.   22   Ho   Gadella & Kliphuis,
                                                          1967
   (a) sepium   22   Cz   Holub et al., 1970
   (b) roseata Brummitt   22   Br   Brummitt, 1973
   (c) americana (Sims) Brummitt
   (d) spectabilis Brummitt
  3 silvatica (Kit.) Griseb.   22   Br   Brummitt, 1973
  4 pulchra Brummitt & Heywood   22   Cz   Holub et al., 1970

5 Convolvulus L.
  1 persicus L.
  2 boissieri Steudel
   (a) boissieri
   (b) compactus (Boiss.) Stace
  3 libanoticus Boiss.
  4 dorycnium L.

174

```
  5 cneorum L.
  6 oleifolius Desr.
  7 lanuginosus Desr.
  8 calvertii Boiss.
  9 holosericeus Bieb.
 10 lineatus L.
 11 cantabrica L.                            30      Hu      Baksay, 1958
 12 humilis Jacq.
 13 tricolor L.
    (a) tricolor
    (b) cupanianus (Sa'ad) Stace
 14 meonanthus Hoffmanns. & Link
 15 pentapetaloides L.
 16 siculus L.
    (a) siculus                             44      Bl      Dahlgren et al.,
                                                            1971
    (b) agrestis (Schweinf.) Verdcourt
 17 sabatius Viv.
 18 valentinus Cav.
 19 arvensis L.                              50      Da      Hagerup, 1941b
 20 farinosus L.
 21 betonicifolius Miller
 22 scammonia L.
 23 althaeoides L.
    (a) althaeoides
    (b) tenuissimus (Sibth. & Sm.) Stace

 6 Ipomoea L.
    1 stolonifera (Cyr.) J.F. Gmelin
    2 sagittata Poiret
    3 acuminata (Vahl) Roemer & Schultes
    4 purpurea Roth
    5 batatas (L.) Lam.
```

CXLVII HYDROPHYLLACEAE

```
1 Nemophila Nutt.
  1 menziesii Hooker & Arnott

2 Phacelia Juss.
  1 tanacetifolia Bentham

3 Wigandia Kunth
  1 caracasana Kunth
```

CXLVIII BORAGINACEAE

```
1 Argusia Boehmer
  1 sibirica (L.) Dandy                      26      Rm      Tarnavschi, 1948

2 Heliotropium L.
  1 europaeum L.
  2 ellipticum Ledeb.
  3 lasiocarpum Fischer & C.A. Meyer
  4 dolosum De Not.
  5 suaveolens Bieb.
   (a) suaveolens
   (b) bocconei (Guss.) Brummitt
  6 halacsyi Reidl
```

```
     7 hirsutissimum Grauer
     8 micranthos (Pallas) Bunge
     9 arguzioides Kar. & Kir.
    10 curassavicum L.
    11 amplexicaule Vahl
    12 supinum L.

3 Arnebia Forskål
     1 decumbens (Vent.) Cosson & Kralik

4 Lithospermum L.
     1 officinale L.                      28       Ho      Gadella & Kliphuis,
                                                            1966

5 Neatostema I.M. Johnston
     1 apulum (L.) I.M. Johnston          28       Bl      Dahlgren et al.,
                                                            1971

6 Buglossoides Moench
     1 purpurocaerulea (L.) I.M. Johnston  16      Ge      Reese, 1952
     2 gastonii (Bentham) I.M. Johnston
     3 calabra (Ten.) I.M. Johnston        20      It      Grau, 1968b
     4 arvensis (L.) I.M. Johnston       28 42     Ga      Grau, 1968b
                                                   It      Gadella & Kliphuis,
                                                            1970
       (a) arvensis                       28       Scand.  Löve & Löve, 1944

       (b) permixta (Jordan ex F.W. Schultz)
             R. Fernandes                ⸗ 28               Grau, 1968b
       (c) gasparrinii (Heldr. ex Guss.)
             R. Fernandes
       (d) sibthorpiana (Griseb.) R. Fernandes
     5 minima (Moris) R. Fernandes
     6 glandulosa (Velen.) R. Fernandes
     7 tenuiflora (L. fil.) I.M. Johnston

7 Lithodora Griseb.
     1 rosmarinifolia (Ten.) I.M. Johnston  26     Si      Fürnkranz, 1967
     2 fruticosa (L.) Griseb.               28     Hs      Lorenzo-Andreu, 1951
     3 hispidula (Sibth. & Sm.) Griseb.
     4 zahnii (Heldr. ex Halácsy) I.M. Johnston
     5 oleifolia (Lapeyr.) Griseb.
     6 nitida (Ern) R. Fernandes            50     Hs      Grau, 1971
     7 diffusa (Lag.) I.M. Johnston
       (a) diffusa                          24     Hs      Grau, 1968b
       (b) lusitanica (Samp.) P. Silva &
             Rozeira

8 Macrotomia DC.
     1 densiflora (Ledeb. ex Nordm.) Macbride

9 Onosma L.
     1 frutescens Lam.                      14     Gr      Grau, 1964a
     2 graeca Boiss.                        14     Gr      Grau, 1968b
     3 taygetea Boiss. & Heldr.
     4 visianii G.C. Clementi               18     Au      Teppner, 1971
     5 rhodopea Velen.
     6 setosa Ledeb.
     7 tricerosperma Lag.
       (a) tricerosperma
       (b) hispanica (Degen & Hervier) P.W. Ball
```

(c) granatensis (Dubeaux & Degen) Stroh
8 simplicissima L.
9 polyphylla Ledeb.
10 tinctoria Bieb.
11 propontica Aznav.
12 bubanii Stroh
13 fastigiata (Br.-Bl.) Br.-Bl. ex Lacaita

14 vaudensis Gremli	20	He	Teppner, 1971
15 arenaria Waldst. & Kit.	12	Hu	Baksay, 1957b
		Hu	Pólya, 1950
(a) arenaria	20	Au It	Teppner, 1971
(b) fallax (Borbás) Jáv.	26	Sb Ju	Teppner, 1971
16 austriaca (G. Beck) Fritsch	26	Au	Teppner, 1971
17 pseudarenaria Schur			
18 tridentina Wettst.	28	It	Grau, 1964a
19 helvetica (A. DC.) Boiss.	26 28	It	Grau, 1964a
		Ga It	Favarger, 1971
20 lucana Lacaita			
21 thracica Velen.			
22 echioides L.	14	It	Grau, 1968b
23 tornensis Jáv.	14	Cz	Baksay, 1957b
24 heterophylla Griseb.			
25 elegantissima Rech. fil. & Goulimy			
26 euboica Rech. fil.			
27 mattirolii Bald.			
28 montana Sibth. & Sm.	14	Gr	Grau, 1968b
29 erecta Sibth. & Sm.			
30 taurica Pallas ex Willd.	14	Rs(K)	Teppner, 1971
31 leptantha Heldr.			
32 stellulata Waldst. & Kit.	22	Ju	Teppner, 1971
33 spruneri Boiss.			

10 Cerinthe L.
  1 minor L.
  2 glabra Miller
    (a) minor

(b) tenuiflora (Bertol.) Domac	16	Co	Contandriopoulos, 1964a
(c) smithiae (A. Kerner) Domac			
3 major L.	16	Hs	Gadella et al., 1966

  4 retorta Sibth. & Sm.

11 Moltkia Lehm.

1 suffruticosa (L.) Brand	16	It	Grau, 1968b
2 petraea (Tratt.) Griseb.			
3 doerfleri Wettst.			

12 Alkanna Tausch

1 orientalis (L.) Boiss.	28	Gr	Grau, 1968b
2 graeca Boiss. & Spruner	30	Gr	Grau, 1968b
(a) graeca			
(b) baeotica (DC.) Nyman			
3 calliensis Heldr. ex Boiss.			
4 primuliflora Griseb.			
5 stribrnyi Velen.			
6 methanaea Hausskn.			
7 tinctoria (L.) Tausch	30	It	Grau, 1968b
8 corcyrensis Hayek			
9 lutea DC.	28	Bl	Dahlgren et al., 1971

```
  10 pelia (Halacsy) Rech. fil.
  11 pindicola Hausskn.
  12 sandwithii Rech. fil.
  13 noneiformis Griseb.
  14 pulmonaria Griseb.
  15 scardica Griseb.
  16 sieberi DC.
  17 sartoriana Boiss. & Heldr.

13 Halacsya Dörfler
   1 sendtneri (Boiss.) Dörfler          22    Ju    Sušnik, 1967

14 Echium L.
   1 albicans Lag. & Rodr.
   2 angustifolium Miller
   3 humile Desf.
   4 asperrimum Lam.                      14    Ga    Delay, 1969
   5 italicum L.
                                          32    Hs    Gadella et al.,
                                                      1966
                                          16    Cz    Májovský et al.,
                                                      1970

   6 lusitanicum L.
     (a) lusitanicum
     (b) polycaulon (Boiss.) P. Gibbs
   7 flavum Desf.
   8 boissieri Steudel
   9 russicum J.F. Gmelin                 24    Hu    Pólya, 1950
  10 vulgare L.                        16 32    Ho    Gadella & Kliphuis,
                                                      1966
  11 plantagineum L.                      16    Co    Litardière, 1943
  12 sabulicola Pomel
  13 creticum L.
     (a) creticum
     (b) coincyanum (Lacaita) R. Fernandes
     (c) algarbiense R. Fernandes         16    Lu    Fernandes, 1969
  14 tuberculatum Hoffmanns. & Link       16    Lu    Rodriguez, 1953
  15 gaditanum Boiss.
  16 rosulatum Lange
  17 parviflorum Moench                   16    Bl    Dahlgren, et al.,
                                                      1971
  18 arenarium Guss.

15 Pulmonaria L.
   1 obscura Dumort.                      14    Hu       Pólya, 1949
   2 officinalis L.                       16    Ge Ju Au Merzmuller &
                                                   It    Grau, 1969
                                          14    Ga       Larsen, 1954

   3 rubra Schott                         14    Rm       Merxmüller &
                                                         Grau, 1969
   4 filarszkyana Jáv.                    14    Rm       Tarnavschi, 1935
   5 stiriaca A. Kerner                   18    Au Ju    Wolkinger, 1966
   6 mollis Wulfen ex Hornem.             18    Au       Merxmüller &
                                                         Grau, 1969
                                          14
                                          28    Rm       Tarnavschi, 1935
   7 vallarsae A. Kerner                  22    It       Merxmüller &
                                                         Grau, 1969
   8 affinis Jordan                       22    Ga It    Merxmüller &
                                                         Grau, 1969
```

9 saccharata Miller	22	It	Merxmüller & Grau, 1969
10 montana Lej.	22	Ge	Merxmüller & Grau, 1969
	18	Ge	Merxmüller & Grau, 1969
11 longifolia (Bast.) Boreau	14	Ga	Merxmüller & Grau, 1969
12 angustifolia L.	14	Ga It	Merxmüller & Grau, 1969
13 visianii Degen & Lengyel	20	Au He It Ju	Merxmüller & Grau, 1969
14 kerneri Wettst.	26	Au	Merxmüller, 1970

16 Nonea Medicus

1 lutea (Desr.) DC.	14	Rm	Gusuleac & Tarnavschi, 1935
2 obtusifolia (Willd.) DC.			
3 pulla (L.) DC.	18	Rm	Tarnavschi & Lungeneau, 1970
	20	Au	Furnkranz, 1967
4 pallens Petrović			
5 caspica (Willd.) G. Don fil.			
6 versicolor (Steven) Sweet			
7 ventricosa (Sibth. & Sm.) Griseb.			
8 micrantha Boiss. & Reuter			
9 vesicaria (L.) Reichenb.			

17 Elizaldia Willk.
  1 calycina (Roemer & Schultes) Maire

18 Symphytum L.

1 officinale L.	24 + 4B	It	Gadella & Kliphuis, 1970
	24 40 48 26 36	Ho	Gadella & Kliphuis 1967
	54	Bu	Markova & Ivanova, 1970
(a) officinale			
(b) uliginosum (A. Kerner) Nyman			
2 asperum Lepechin			
3 x uplandicum Nyman	36	Fe	Vaarama, 1948
4 tuberosum L.			
(a) tuberosum	44	It	Grau, 1968b
(b) nodosum (Schur) Soó	96	Ge	Grau, 1968b
	100	Ju	
	18	Rm	Tarnavschi, 1948
	72	Bu	Markova & Ivanova, 1970
5 gussonei F.W. Schultz			
6 cordatum Waldst. & Kit. ex Willd.	18	Rm	Tarnavschi, 1948
7 naxicola Pawl.	30	Gr	Runemark, 1967
8 ibiricum Steven			
9 davisii Wickens			
10 cycladense Pawl.	30	Gr	Pawlowski, unpub.
11 tauricum Willd.	18	Rm	Tarnavschi, 1948
12 orientale L.			
13 bulbosum C. Schimper			
14 ottomanum Friv.	20	Bu	Markova & Ivanova, 1970

19 Procopiania Guşuleac
  1 cretica (Willd.) Guşuleac
  2 insularis Pawl.   28   Cr   Runemark, 1967a
  3 circinalis (Runemark) Pawl.   28   Cr   Runemark, 1967a

20 Brunnera Steven
  1 macrophylla (Adams) I.M. Johnston

21 Anchusa L.
  1 leptophylla Roemer & Schultes
  2 gmelinii Ledeb.
  3 ochroleuca Bieb.
  4 cespitosa Lam.
  5 undulata L.
    (a) undulata
    (b) hybrida (Ten.) Coutinho
  6 sartorii Heldr. ex Guşuleac
  7 granatensis Boiss.
  8 calcarea Boiss.
  9 crispa Viv.   16   Co   Contandriopoulos,
                                                 1962
  10 officinalis L.   16 + 0-2B   Au   Furnkranz. 1967

  11 azurea Miller
  12 aggregata Lehm.
  13 thessala Boiss. & Spruner   24   Rm   Tarnavschi, 1948
  14 pusilla Guşuleac
  15 macrosyrinx Rech. fil.
  16 macedonica Degen & Dörfler
  17 stylosa Bieb.
  18 spruneri Boiss.
  19 barrelieri (All.) Vitman
  20 serpentinicola Rech. fil.
  21 aegyptiaca (L.) DC.
  22 arvensis (L.) Bieb.   48   It   Gadella & Kliphuis,
                                                 1970
    (a) arvensis   48   Su   Lövkvist, 1963
    (b) orientalis (L.) Nordh.
  23 cretica Miller   16   It   Grau, 1968b
  24 variegata (L.) Lehm.   16   Gr   Grau, 1968b

22 Pentaglottis Tausch
  1 sempervirens (L.) Tausch ex L.H.
    Bailey

23 Borago L.
  1 officinalis L.   16   Rm   Tarnavschi, 1948
  2 pygmaea (DC.) Chater & W. Greuter   32   Co   Contrandriopoulos,
                                                 1962

24 Trachystemon D. Don
  1 orientalis (L.) G. Don fil.   56   Bu   Markova, 1970

25 Mertensia Roth
  1 maritima (L.) S.F. Gray   24   Is   Löve & Löve, 1956

26 Amsinckia Lehm.
  1 lycopsoides (Lehm.) Lehm.
  2 calycina (Moris) Chater

27 Asperugo L.
  1 procumbens L.   48   Cz   Májovský et al.,
                                                                          1970

28 Rochelia Reichenb.
  1 disperma (L. fil.) C. Koch
    (a) disperma
    (b) retorta (Pallas) E. Kotejowa

29 Myosotis L.
  1 incrassata Guss.   24   Ga   Merxmüller &
                                                                          Grau, 1963
  2 cadmea Boiss.   24   Gr   Grau, 1968a
  3 ucrainica Czern.
  4 pusilla Loisel.   24   Co Ga   Grau, 1968a
  5 litoralis Steven ex Bieb.
  6 arvensis (L.) Hill
    (a) arvensis   52   Ho   Gadella &
                                                                          Kliphuis, 1967a
    (b) umbrata (Rouy) O. Schwarz   66   Ga Ge Ju No Grau, 1968a
                                                               Su
  7 ramosissima Rochel
    (a) ramosissima   48   Is   Löve & Löve, 1956
    (b) globularis (Samp.) Grau
  8 ruscinonensis Rouy   ?36 48   Ga   Blaise, 1966
  9 discolor Pers.
    (a) discolor   72   Ga Ge Hs Lu Grau, 1968a
                                                      64   Ho   Gadella &
                                                                          Kliphuis, 1968
    (b) dubia (Arrondeau) Blaise   24   Ga   Blaise, 1970
  10 congesta R.J. Shuttlew. ex Albert
     & Reyn.   24 ?32   Gr   Grau, 1970
                                                      48   Gr   Blaise, 1970
  11 balbisiana Jordan
  12 persoonii Rouy   48   Hs   Grau, 1968a
  13 stricta Link ex Roemer & Schultes   36   Ga   Blaise, 1966
                                                      48   Bu Ga Ge Ju Grau, 1968a
  14 minutiflora Boiss. & Reuter   48   Hs   Grau, 1968a
  15 speluncicola (Boiss.) Rouy   24   Ga   Merxmüller &
                                                                          Grau, 1963
  16 refracta Boiss.
     (a) refracta   44   Cr Gr Hs   Grau, 1968a
     (b) paucipilosa Grau   20   Gr   Grau, 1968a
     (c) aegagrophila W. Grueter & Grau
  17 sylvatica Hoffm.
     (a) sylvatica   18 20   Au Ge   Grau, 1964b
     (b) subarvensis Grau   18   It Ju   Grau, 1964b
     (c) cyanea (Boiss. & Heldr.)
         Vestergren   18 20   Gr   Grau, 1964b
     (d) elongata (Strobl) Grau   20 22   Si   Grau, 1964b
  18 soleirolii Gren. & Godron   18   Co   Blaise, 1969
  19 decumbens Host Host
     (a) decumbens   32   Au Ga He It Grau, 1964b
                                                               Su
     (b) teresiana (Sennen) Grau   32 34   Ga Hs   Grau, 1964b
     (c) variabilis (M. Angelis) Grau   32   Au   Grau, 1964b
     (d) kerneri (Dalla Torre & Sarnth.)
         Grau   32   Au   Grau, 1964b
     (e) florentina Grau   28   It   Grau, 1970
  20 latifolia Poiret
  21 alpestris F.W. Schmidt   24 48 72   Au Ga Ge   Grau, 1964b
                                                      70   He It

181

```
22 gallica Vestergren                    24      Ga      Merxmüller &
                                                         Grau, 1963
23 ambigens (Béguinot) Grau              24      It      Grau, 1964
24 stenophylla Knaf                      48      Au      Merxmüller &
                                                         Grau, 1963
25 asiatica (Vestergren) Schischkin &
   Serg.
26 alpina Lapeyr.                        24      Ga Hs   Grau, 1964b
27 corsicana (Fiori) Grau                24      Co      Contandriopoulos,
                                                         1966
28 suaveolens Waldst. & Kit. ex Willd.   24 48   Gr      Grau, 1964b
29 lithospermifolia (Willd.) Hornem.
30 secunda A. Murray                     24 48   Lu      Fernandes & Leitao,
                                                         1969
31 welwitschii Boiss. & Reuter           24      Lu      Grau, 1967a
32 stolonifera (DC.) Gay ex Leresche     24      Hs      Merxmüller &
   & Levier                                              Grau, 1963
33 sicula Guss.                          46 46+2  Ga Sa   Grau, 1967a
34 debilis Pomel                         48      Lu      Grau, 1967a
35 laxa Lehm.
   (a) caespitosa (C.F. Schultz) Hyl.
       ex Nordh.
   (b) baltica (Sam.) Hyl. ex Nordh.     ?84 88  Su Fe   Merxmüller &
                                                         Grau, 1963
36 scorpioides L.                        66      Au Ge   Merxmüller &
                                                         Grau, 1963
                                         ?64     Hu      Pólya, 1950
37 rehsteineri Wartm.                    22      Ge      Merxmüller &
                                                         Grau, 1963
38 nemorosa Besser                       80 84   Cz      Holub et al,
39 ramottiana (Br.-Bl) ex Chassagni)                     1971
   Grau                                  44      Ga Hs   Grau, 1970
40 azorica H.C. Watson
41 sparsiflora Mikan ex Polh             18      Au      Merxmüller &
                                                         Grau, 1963


30 Eritrichium Schrader ex Gaudin
   1 pectinatum (Pallas) DC.
   2 nanum (L.) Schrader ex Gaudin
     (a) nanum                           46      He      Favarger, 1965b
     (b) jankae (Simonkai) Jáv.
   3 villosum (Ledeb.) Bunge
   4 aretioides (Cham.) DC.

31 Lappula Gilib.
   1 deflexa (Wahlenb.) Garcke           24      No      Knaben & Engelskjon,
                                                         1967
   2 spinocarpos (Forskål) Ascherson ex
     O. Kuntze
   3 echinophora (Pallas) O. Kuntze
   4 marginata (Bieb.) Gurke
   5 squarrosa (Retz.) Dumort            48      Hu      Baksay, 1956
     (a) squarrosa
     (b) heteracantha (Ledeb.) Chater
   6 barbata (Bieb.) Gurke

32 Trigonotis Steven
   1 peduncularis (Trev.) Bentham ex Baker

33 Omphalodes Miller
   1 scorpioides (Haenke) Schrank        24      Ge      Grau, 1967b
```

```
  2 nitida Hoffmanns. & Link          24      Lu      Grau, 1967b
  3 verna Moench                       48      It Lu   Grau, 1967b
  4 luciliae Boiss.
  5 linifolia (L.) Moench              28      Lu      Grau, 1967b
  6 kuzinskyanae Willk.                28      Lu      Grau, 1967b
  7 littoralis Lehm.
  8 brassicifolia (Lag.) Sweet

34 Cynoglossum L.
   1 officinale L.                     24      Cz      Májovský et al.,
                                                         1970
                                       28      Cz      Murín & Váchová,
                                                         1967


   2 dioscoridis Vill.
   3 columnae Ten.
   4 germanicum Jacq.                  24      Cz      Májovský et al.,
                                                         1970
   5 creticum Miller                   24      Ga      Delay, 1970
   6 sphacioticum Boiss. & Heldr.
   7 hungaricum Simonkai               24      Hu      Baksay, 1958
   8 nebrodense Guss.
   9 cheirifolium L.
  10 magellense Ten.
  11 clandestinum Desf.

35 Solenanthus Ledeb.
   1 reverchonii Degen.
   2 biebersteinii DC.
   3 apenninus (L.) Fischer & C.A. Meyer
   4 scardicus Bornm.
   5 stamineus (Desf.) Wettst.
   6 albanicus (Degen & Bald.) Degen & Bald.

36 Mattiastrum (Boiss.) Brand
   1 lithospermifolium (Lam.) Brand

37 Rindera Pallas
   1 tetraspis Pallas
   2 umbellata (Waldst. & Kit.) Bunge
   3 graeca (A. DC.) Boiss. & Heldr.


CXLIX VERBENACEAE

1 Vitex L.
  1 agnus-castus L.

2 Lantana L.
  1 camara L.

3 Verbena L.
  1 rigida Sprengel
  2 bonariensis L.
  3 officinalis L.                     14      Hs      Björkqvist et al.,
                                                         1969
  4 supina L.
```

4 Lippia L.
  1 nodiflora (L.) Michx
  2 canescens Kunth
  3 triphylla (L'Hér.) O. Kuntze

CL CALLITRICHACEAE

1 Callitriche L.
  1 hermaphroditica L.   6   Is   Löve & Löve, 1956
  2 truncata Guss.
   (a) truncata
   (b) fimbriata Schotsman
   (c) occidentalis (Rouy) Schotsman   6   Lu   Schotsman, 1961a
  3 pulchra Schotsman   8   Cr   Schotsman, 1969
  4 lusitanica Schotsman   8   Lu   Schotsman, 1967
  5 stagnalis Scop.   10   Br Ga He Lu Schotsman, 1961b
  6 obtusangula Le Gall   10   Co Ga Ge Lu Schotsman, 1961b
  7 cophocarpa Sendtner   10   He   Schotsman, 1961b
  8 platycarpa Kütz.   20   He   Schotsman, 1961b
  9 palustris L.   20   Is   Löve & Löve, 1956
  10 hamulata Kütz. ex Koch   38   He   Schotsman, 1961
                                            ?40   Is   Löve & Löve, 1956
  11 brutia Petagna   28   Lu   Schotsman, 1961c
  12 peploides Nutt.
  13 terrestris Rafin.

CLI LABIATAE

1 Ajuga L.
  1 triuncatus L.   32   Gr   Strid, 1965
  2 genevensis L.   32   Ge   Scheerer, 1940
  3 pyramidalis L.   32   Fe   Sorsa, 1963
  4 reptans L.   32   Bu   Markova & Ivanova,
                                                              1970
  5 tenorii C. Presl
  6 salicifolia (L.) Schreber
   (a) salicifolia
   (b) bassarabica (Savul. & Zahar.)
       P.W. Ball
  7 laxmannii (L.) Bentham   62   Bu   Markova & Ivanova,
                                                              1971
  8 piskoi Degen & Bald.
  9 iva (L.) Schreber   c.86   Bl   Dahlgren et al.,
                                                              1971
  10 chamaepitys (L.) Schreber
   (a) chamaepitys   28   Br   Rutland, 1941
   (b) chia (Schreber) Arcangeli   30   Gr   Strid, 1965

2 Teucrium L.
  1 fruticans L.
  2 brevifolium Schreber   30   Gr   Strid, 1965
  3 aroanium Orph. ex Boiss.
  4 pseudochamaepitys L.
  5 campanulatum L.
  6 arduini L.
  7 lamifolium D'Urv.
  8 halacsyanum Heldr.   32   Gr   Quezel & Contand-
                                                              riopoulos, 1966
  9 francisci-werneri Rech. fil.

```
10 heliotropifolium W. Barbey
11 scorodonia L.
   (a) scorodonia                        32      Ho      Gadella & Kliphuis,
                                                           1963
                                        ?34      Br      Rutland, 1941
   (b) euganeum (Vis.) Arcangeli
   (c) baeticum (Boiss. & Reuter) Tutin
12 massiliense L.                        32      Co      Contandriopoulos,
                                                           1962
13 salviastrum Schreber
14 asiaticum L.
15 scordium L.
   (a) scordium
   (b) scordioides (Schreber) Maire &
       Petitmengin                       32      Rm      Tarnavschi, 1938
16 spinosum L.
17 resupinatum Desf.
18 botrys L.                             32      Cz      Májovský et al.,
                                                           1970
19 chamaedrys L.                         60      Hu      Baksay, 1958
20 webbianum Boiss.
21 krymense Juz.
22 lucidum L.
23 divaricatum Sieber ex Boiss.       62 64      Gr      Strid, 1965
24 flavum L.
   (a) flavum
   (b) glaucum (Jordan & Fourr.) Ronniger
   (c) hellenicum Rech. fil.
   (d) gymnocalyx Rech. fil
25 intricatum Lange
26 fragile Boiss.
27 microphyllum Desf.
28 marum L.
29 subspinosum Pourret ex Willd.
30 compactum Clemente ex Lag.
31 pyrenaicum L.                         26      Ga      Küpfer, 1969
32 rotundifolium Schreber               26      Hs      Küpfer, 1969
33 buxifolium Schreber
34 freynii Reverchon ex Willk.
35 cuneifolium Sibth. & Sm.
36 alpestre Sibth. & Sm.
   (a) alpestre                          78      Cr      Strid, 1965
   (b) gracile (W. Barbey & Major) D. Wood
37 montanum L.                           16      Ga      Guinochet &
                                                           Logeois, 1962
                                         26      Ga      Küpfer, 1969
                                         30      Hu      Baksay, 1956
                                         60      Cz      Murín & Váchová,
                                                           1967
38 thymifolium Schreber
39 cossonii D. Wood
40 libanitis Schreber
41 carthaginense Lange
42 aragonense Loscos & Pardo
43 pumilum L.
   (a) pumilum
   (b) carolipaui (C. Vicioso ex Pau) D. Wood
44 turredanum Losa & Rivas Goday
45 polium L.                             26      Hs      Lorenzo-Andreu &
                                                           Garcia, 1950
                                         78      Ge      Strid, 1965
                                         52      Ga      Pueck, 1968
```

(a) polium
(b) aureum (Schreber) Arcangeli
(c) capitatum (L.) Arcangeli
(d) pii-fontii Palau
(e) vincentinum (Rouy) D. Wood
46 gnaphalodes L'Hér.
47 eriocephalum Willk.
48 haenseleri Boiss.
49 charidemi Sandwith

3 Scutellaria L.
1 orientalis L.
2 alpina L.
(a) alpina   22   Ga   Favarger, 1959a
(b) supina (L.) I.B.K. Richardson
3 columnae All.   34   Hu   Baksay, 1958
(a) columnae
(b) gussonii (Ten.) Rech. fil.
4 altissima L.
5 rubicunda Hornem.
(a) linnaeana (Caruel) Rech. fil.
(b) rubicunda   34   Gr   Bothmer, 1969
(c) rupestris (Boiss. & Heldr.) I.B.K.
    Richardson
(d) geraniana (Tuntas ex Halácsy)
    I.B.K. Richardson
6 albida L.
7 velenovskyi Rech. fil.
  naxensis Bothmer   34   Gr   Bothmer, 1969
8 sieberi Bentham
9 hirta Sibth. & Sm.
10 bulgarica Barc.
11 galericulata L.   31 32   Ho   Gadella & Kliphuis,
                                                         1968

12 hastifolia L.
13 minor Hudson

4 Prasium L
1 majus L.   34   Cr Gr   Bothmer, 1970

5 Marrubium L.
1 thessalum Boiss. & Heldr.
2 cylleneum Boiss. & Heldr.
3 incanum Desr.
4 supinum L.
5 velutinum Sibth. & Sm.
6 friwaldskyanum Boiss.
7 leonuroides Desr.
8 alysson L.
9 peregrinum L.   34   Cz   Murín & Váchová,
                                                         1967
10 vulgare L.   34   Br   Rutland, 1941
11 pestalozzae Boiss.
12 alternidens Rech. fil.

6 Sideritis L.
1 grandiflora Salzm. ex Bentham
2 foetens Clemente ex Lag.
3 arborescens Salzm. ex Bentham
(a) arborescens

(b) luteola (Font Quer) P.W. Ball ex
   Heywood
(c) paulii (Pau) P.W. Ball ex Heywood
4 hirsuta L.
5 endressii Willk.
   (a) endressii
   (b) laxespicata (Degen & Debeaux) Heywood
6 scordioides L.
   (a) scordioides
   (b) cavanillesii (Lag.) P.W. Ball ex
      Heywood

7 glacialis Boiss.	34	Hs	Küpfer & Favarger, 1967
8 hyssopifolia L.	32+6B	Ge He	Favarger, 1965b
	34		
	30	Ga	Favarger & Küpfer, 1968

   (a) hyssopifolia
   (b) guillonii (Timb.-Lagr.) Rouy
9 ovata Cav.
10 linearifolia Lam.
11 angustifolia Lag.
12 reverchonii Willk.
13 javalambrensis Pau

14 spinulosa Barnades ex Asso	22 28 30	Hs	Coutinho & Lorenzo-Andreu, 1948

15 serrata Cav. ex Lag.
16 ilicifolia Willd.
17 leucantha Cav.
18 stachydioides Willk.
19 incana L.
   (a) incana
   (b) virgata (Desf.) Malagarriga
   (c) sericea (Pers.) P.W. Ball ex
      Heywood
   (d) glauca (Cav.) Malagarriga
20 lacaitae Font Quer

21 syriaca L.	24	Rs(K)	Borhidi, 1968
22 clandenstina (Bory & Chaub.) Hayek			
23 scardica Griseb.	32	Bu	Kozuharov & Kuzmanov, 1965b

24 perfoliata L.

25 montana L.			
(a) montana	16	Cz	Májovský et al., 1970

   (b) remota (D'Urv.) Heywood
   (c) ebracteata (Asso) Murb.
26 lanata L.

27 romana L.			
(a) romana	28	Hs	Björkqvist et al., 1969
(b) purpurea (Talbot ex Bentham)			
28 curvidens Stapf	28	Cr Gr	Strid, 1965

7 Melittis L.

1 melissophyllum L.	30	Hu	Baksay, 1958

   (a) melissophyllum
   (b) carpatica (Klokov) P.W. Ball
   (c) albida (Guss.) P.W. Ball

8 Eremostachys Bunge
  1 tuberosa (Pallas) Bunge

9 Phlomis L.
  1 tuberosa L.   22   Bu   Markova, 1970
  2 samia L.
  3 herba-venti L.
   (a) herba-venti
   (b) pungens (Willd.) Maire ex
       DeFilipps
  4 purpurea L.
  5 italica L.
  6 floccosa D. Don
  7 cretica C. Presl
  8 ferruginea Ten.
  9 lychnitis L.   20   Hs   Lorenzo-Andreu &
                                            Garcia, 1950

 10 crinita Cav.
 11 lanata Willd.   20   Cr   Strid, 1965
 12 fruticosa L.   20   Gr   Strid, 1965

10 Galeopsis L.
  1 segetum Necker   16   Rm   Muntzing, 1929
  2 pyrenaica Bartl.   16   Ga   Muntzing, 1929
  3 ladanum L.   16   Su   Muntzing, 1929
  4 angustifolia Ehrh. ex Hoffm. 16   Su   Muntzing, 1929
  5 reuteri Reichenb. fil.
  6 speciosa Miller   18   Au Su Muntzing, 1929
  7 pubescens Besser   16 Au Cz Hu Rm Muntzing, 1929
  8 tetrahit L.   32 Br Be Su Muntzing, 1929
  9 bifida Boenn.   32 Au Rm Su Muntzing, 1929

11 Lamium L.
  1 orvala L.
  2 garganicum L.   18   Co   Contandriopoulos,
                                      1962
   (a) garganicum
   (b) laevigatum Arcangeli
   (c) pictum (Boiss. & Heldr.) P.W. Ball
  3 corsicum Gren. & Godron 18   Co   Contandriopoulos,
                                      1962
  4 glaberrimum Taliev
  5 flexuosum Ten.
  6 maculatum L.   18   Au   Mattick, 1950
  7 album L.   18   Hu   Pólya, 1949
  8 moschatum Miller   18   Gr   Strid, 1965
  9 bifidum Cyr.
   (a) bifidum
   (b) balcanicum Velen.
   (c) albimontanum Rech. fil.
 10 purpureum L.   18   Fe   Sorsa, 1963
 11 hybridum Vill.   36   Is   Löve & Löve, 1956
 12 moluccellifolium Fries   36   Is   Löve & Löve, 1956
 13 amplexicaule L.
   (a) amplexicaule   18   Hu   Pólya, 1950
   (b) orientale Pacz.

12 Lamiastrum Heister ex Fabr.

1 galeobdolon (L.) Ehrend. & Polatschek
  (a) galeobdolon   18   Ge   Dersch, 1964
  (b) montanum (Pers.) Ehrend. &
      Polatschek   36   Au   Polatschek, 1966
  (c) flavidum (F. Hermann) Ehrend. &
      Polatschek   18   Au   Polatschek, 1966

13 Leonurus L.
  1 cardiaca L.   18   Hu   Pólya, 1949
  2 marrubiastrum L.   24   Rm   Tarnavschi, 1948

14 Moluccella L.
  1 spinosa L.

15 Ballota L.
  1 frutescens (L.) J. Woods
  2 acetabulosa (L.) Bentham   28   Gr   Strid, 1965
  3 pseudodictamnus (L.) Bentham
  4 hirsuta Bentham
  5 rupestris (Biv.) Vis.
  6 macedonica Vandas
  7 nigra L.   22   Br   Rutland, 1941
    (a) nigra
    (b) sericea (Vandas) Patzak
    (c) foetida Hayek
    (d) velutina (Pospichal) Patzak
    (e) uncinata (Fiori & Béguinot) Patzak
    (f) anatolica P.H. Davis

16 Stachys L.
  1 alopecuros (L.) Bentham
  2 monieri (Gouan) P.W. Ball
  3 officinalis (L.) Trevisan   16   Hu   Pólya, 1949
  4 balcanica P.W. Ball
  5 scardica (Griseb.) Hayek
  6 balansae Boiss. & Kotschy
  7 alpina L.
  8 germanica L.   30   Cz   Májovský et al.,
                                                             1970
    (a) germanica
    (b) heldreichii (Boiss.) Hayek
    (c) lusitanica (Hoffmanns. & Link)
        Coutinho
  9 tymphaea Hausskn.
  10 tournefortii Poiret
  11 sericophylla Halácsy
  12 thirkei C. Koch
  13 cassia (Boiss.) Boiss.
  14 cretica L.
    (a) cretica
    (b) salviifolia (Ten.) Rech. fil.
    (c) bulgarica Rech. fil
  15 byzantina C. Koch
  16 heraclea All.
  17 acutifolia Bory & Chaub.
  18 obliqua Waldst. & Kit.
  19 decumbens Pers.
  20 canescens Bory & Chaub.
  21 anisochila Vis. & Pančić
  22 methifolia Vis.
  23 plumosa Griseb.

24 sylvatica L.	66	Hu	Pólya, 1950
25 palustris L.	102	Su	Löve, 1954a
	64	Cz	Májovský et al., 1970
26 recta L.			
(a) recta	34	Ga	Májovský et al., 1970
	32	Hu	Pólya, 1949
(b) labiosa (Bertol.) Briq.	34	It	Favarger, 1959a

(c) subcrenata (Vis.) Briq.
27 beckeana Dörfler & Hayek
28 parolinii Vis.
29 atherocalyx C. Koch
30 iberica Bieb.
31 virgata Bory & Chaub.
32 angustifolia Bieb.
33 tetragona Boiss. & Heldr.
34 leucoglossa Griseb.

35 mucronata Sieber ex Sprengel	30	Cr	Strid, 1965

36 spruneri Boiss.
37 swainsonii Bentham

(a) swainsonii	34	Gr	Strid, 1965
(b) argolica (Boiss.) Phitos & Damboldt	34	Gr	Phitos & Damboldt, 1969

(c) scyronica (Boiss.) Phitos & Damboldt
38 ionica Halácsy

39 euboica Rech. fil	34	Gr	Strid, 1965

40 circinata L'Hér.

41 glutinosa L.	32	Co	Contandriopoulos, 1962

42 spinosa L.
43 iva Griseb.
44 chrysantha Boiss. & Heldr.
45 candida Bory & Chaub.
46 spreitzenhoferi Heldr.
47 maritima Gouan
48 pubescens Ten.
49 arenaria Vahl

50 corsica Pers.	16	Co	Contandriopoulos, 1962
51 ocymastrum (L.) Briq.	18	Hs	Björkqvist et al., 1969
52 marrubiifolia Viv.	16	Co	Contandriopoulos, 1962

53 annua (L.) L.
54 spinulosa Sibth. & Sm.
55 milanii Petrović

56 arvensis (L.) L.	10	Cr Gr	Strid, 1965

57 brachyclada De Noë ex Cosson
58 serbica Pančić

17 Cedronella Moench
  1 canariensis (L.) Webb & Berth.

18 Nepeta L.
  1 multibracteata Desf.
  2 tuberosa L.
    (a) tuberosa

```
  (b) reticulata (Desf.) Maire
  (c) gienensis (Degen & Hervier) Heywood
 3 apuleii Ucria
 4 scordotis L.
 5 granatensis Boiss.
 6 sibthorpii Bentham
 7 parnassica Heldr. & Sart.
 8 spruneri Boiss.
 9 dirphya (Boiss.) Heldr. ex Halácsy
10 camphorata Boiss. & Heldr.
11 heldreichii Halácsy
12 italica L.
13 cataria L.                             36      Cz      Májovský et al.,
                                                          1970
14 nepetella L.                           34      He      Contandriopoulos,
                                                          1962
15 agrestis Loisel.                       18      Co      Contandriopoulos,
                                                          1962
16 foliosa Moris
17 grandiflora Bieb.
18 melissifolia Lam.                      18      Cr      Strid, 1965
19 latifolia DC.
20 nuda L.
   (a) nuda
   (b) albiflora Gams
21 ucranica L.
22 parviflora Bieb.
23 hispanica Boiss. & Reuter
24 beltranii Pau


19 Glechoma L.
 1 hederacea L.                           18      Fe      Sorsa, 1963


                                        18 36     Po      Skalińska et al., 1959
 2 hirsuta Waldst. & Kit.

20 Dracocephalum L.
 1 nutans L.
 2 thymiflorum L.                        c.14     Fe      V. Sorsa, 1962
 3 ruyschiana L.                          14      Su      Löve & Löve, 1944
 4 austriacum L.                          14      Hu      Baksay, 1958
 5 moldavica L.

21 Prunella L.
 1 laciniata (L.) L.                      28      Bu      Kozuharov &
                                                          Kuzmanov, 1969
                                                  Cz
                                          32      Cz      Májovský et al.,
                                                          1970
 2 grandiflora (L.) Scholler             28      Cz      Hrubý, 1932
 3 vulgaris L.                            28      Cz      Hrubý, 1932
 4 hyssopifolia L.

22 Cleonia L.
 1 lusitanica (L.) L.

23 Melissa L.
 1 officinalis L.
   (a) officinalis                        32      Ga      Litardière, 1945
```

          (b) altissima (Sibth. & Sm.) Arcangeli   64   Co   Litardière, 1945

24 Ziziphora L.
   1 capitata L.
   2 hispanica L.
   3 acinoides L.
   4 tenuior L.
   5 taurica Bieb.
   6 persica Bunge

25 Satureja L.
   1 thymbra L.
   2 salzmannii P.W. Ball
   3 spinosa L.
   4 rumelica Velen.
   5 montana L.
    (a) montana
    (b) variegata (Host) P.W. Ball
    (c) illyrica Nyman
    (d) kitaibelii (Wierzb.) P.W. Ball   30   Ju   Lovka et al.,
                                                          1971
    (e) taurica (Velen.) P.W. Ball
   6 parnassica Heldr. & Sart. ex Boiss.
   7 pilosa Velen.
   8 athoa K. Maly
   9 cuneifolia Ten.
  10 obovata Lag.
  11 coerulea Janka
  12 hortensis L.

26 Acinos Miller
   1 corsicus (Pers.) Getliffe
   2 alpinus (L.) Moench.
    (a) alpinus   18   Au   Reese, 1953
    (b) majoranifolius (Miller) P.W. Ball
    (c) meridionalis (Nyman) P.W. Ball   18   Ga   Favarger & Küpfer,
                                                          1968
   3 suaveolens (Sibth. & Sm.) G. Don fil.
   4 arvensis (Lam.) Dandy   18   Ho   Gadella & Kliphuis,
                                                          1966
   5 rotundifolius Pers.

27 Calamintha Miller
   1 grandiflora (L.) Moench   22   He   Wieffering, 1969
   2 sylvatica Bromf.
    (a) sylvatica   24   Hu   Baksay, 1958
    (b) ascendens (Jordan) P.W. Ball   48   Bl   Nilsson & Lassen,
                                                          1971
   3 nepeta (L.) Savi
    (a) nepeta   24   Hs   Bjorkqvist et al.,
                                                          1969
    (b) glandulosa (Req.) P.W. Ball   20   Co   Contandriopoulos,
                                                          1964a
                                          24   Cz   Holub et al.,
                                                          1970
   4 cretica (L.) Lam.
   5 incana (Sibth. & Sm.) Boiss.

28 Clinopodium L.
   1 vulgare L.
    (a) vulgare   20   Ge   Scheerer, 1939

                              192

(b) arundanum (Boiss.) Nyman	20	Gr	Bothmer, 1967

29 Micromeria Bentham
  1 fruticosa (L.) Druce
    (a) fruticosa
    (b) serpyllifolia (Bieb.) P.H. Davis
  2 thymifolia (Scop.) Fritsch
  3 taygetea P.H. Davis
  4 dalmatica Bentham
  5 pulegium (Rochel) Bentham
  6 frivaldszkyana (Degen) Velen.

7 filiformis (Aiton) Bentham	30	Bl	Dahlgren et al., 1971

  8 microphylla (D'Urv.) Bentham
  9 acropolitana Halácsy
 10 hispida Boiss. & Heldr. ex Bentham
 11 marginata (Sm.) Chater
 12 croatica (Pers.) Schott

13 nervosa (Desf.) Bentham	30	Gr	Strid, 1965
14 juliana (L.) Bentham ex Reichenb.	30	Gr	Strid, 1965

 15 myrtifolia Boiss. & Hohen.
 16 cristata (Hampe) Griseb.
 17 cremnophila Boiss. & Heldr.
 18 kerneri Murb.
 19 parviflora (Vis.) Reichenb.
 20 graeca (L.) Bentham ex Reichenb.

(a) graeca	20	Hs	Björkqvist et al., 1969

    (b) imperica Chater
    (c) garganica (Briq.) Guinea
    (d) longiflora (C. Presl) Nyman
    (e) tenuifolia (Ten.) Nyman
    (f) consentina (Ten.) Guinea
    (g) fruticulosa (Bertol.) Guinea
 21 inodora (Desf.) Bentham

30 Thymbra L.
  1 spicata L.
  2 calostachya (Rech. fil.) Rech. fil.

31 Hyssopus L.
  1 officinalis L.
    (a) canescens (DC.) Briq.
    (b) montanus (Jordan & Fourr.) Briq.
    (c) aristatus (Godron) Briq.
    (d) officinalis

32 Origanum L.
  1 compactum Bentham

2 heracleoticum L.	30	Gr	Bothmer, 1970
3 vulgare L.	30	Ho	Gadella & Kliphuis, 1963
	?32	Ge	Scheerer, 1940
4 virens Hoffmanns. & Link	30	Lu	Fernandes, ined.
5 majorana L.			
6 onites L.	30	Gr	Bothmer, 1970
7 microphyllum (Bentham) Boiss.			

```
 8 majoricum Camb.
 9 dictamnus L.
10 tournefortii Aiton                     30      Gr      Bothmer, 1970
11 scabrum Boiss. & Heldr.
   (a) scabrum
   (b) pulchrum (Boiss. & Heldr.)
       P.H. Davis
12 lirium Heldr. ex Halácsy
13 vetteri Briq. & W. Barbey

33 Thymus L.
 1 capitatus (L.) Hoffmanns. & Link        30      Si      Sibiho, 1961
 2 mastichina L.                           56      Hs      Kaleva, 1969
 3 tomentosus Willd.
 4 caespititius Brot.
 5 piperella L.
 6 teucrioides Boiss. & Spruner
 7 cephalotos L.
 8 villosus L.
 9 longiflorus Boiss.
10 membranaceus Boiss.
11 antoninae Rouy & Coincy
12 mastigophorus Lacaita
13 dolopicus Form.
14 cherlerioides Vis.                      28      Rs(K)   Gogina &
                                                           Suetozacova, 1968

15 parnassicus Halácsy
16 capitellatus Hoffmanns. & Link.
17 camphoratus Hoffmanns. & Link.
18 carnosus Boiss.
19 vulgaris L.                             30      Fe      Jalas, 1948
20 hyemalis Lange
21 zygis L.                                60      Lu      Jalas & Pohjo,
                                                           1965

22 baeticus Boiss. ex Lacaita
23 hirtus Willd.
24 loscosii Willk.
   (a) loscosii
   (b) fontqueri Jalas                     56      Hs      Kaleva, 1969
25 serpylloides Bory
   (a) serpylloides
   (b) gadorensis (Pau) Jalas
26 holosericeus Celak
27 laconicus Jalas
28 bracteosus Vis. ex Bentham
29 granatensis Boiss.
30 aranjuezii Jalas                        28      Hs      Kaleva, 1969
31 bracteatus Lange ex Cutanda             56      Hs      Kaleva, 1969
32 leptophyllus Lange
33 atticus Celak.
34 plasonii Adamović
35 striatus Vahl                        26 54      Ju      Jalas & Kaleva,
                                                           1966

36 spinulosus Ten.
37 aznavourii Velen.
38 kirgisorum Dubjanski
39 zygioides Griseb.
40 comptus Friv.
41 sibthorpii Bentham
42 pannonicus All.                         28      Cz      Jalas & Kaleva,
                                                           1967

43 glabrescens Willd.
```

(a) glabrescens	28 32 56 58	Po	Trela-Sawicha, 1968
(b) decipiens (H. Braun) Domin	52	Ga	Jalas & Kaleva, 1967
(c) urumovii (Velen.) Jalas	56	Rs(K)	Gogina & Svetozocova, 1968

44 longedentatus (Degen & Urum.) Ronniger
45 pallasianus H. Braun
  (a) pallasianus
  (b) brachyodon (Borbas) Jalas

46 herba-barona Loisel	56	Co	Contandriopoulos, 1962
47 nitens Lamotte	28	Ga	Bonnet, 1961

48 willkommii Ronniger
49 richardii Pers.

(a) richardii	28	Ju	Kaleva, 1969

  (b) nitidus (Guss.) Jalas
  (c) ebusitanus (Font Quer) Jalas
50 guberlinensis Iljin
51 binervulatus Klokov & Schost.
52 ocheus Heldr. & Sart. ex Boiss.
53 thracicus Velen.
54 stojanovii Degen

55 longicaulis C. Presl	30	Rm Su	Jalas & Kaleva, 1966
dolomiticus Coste	28	Ga	Bonnet, 1958
embergeri Roussine	48	Ga	Shimoya, 1952

56 praecox Opiz

(a) praecox	c.50 c.54	He Hu	Jalas & Pohjo, 1965
	56	He	
	58	Cz	Jalas & Kaleva, 1967

  (b) skorpilii (Velen.) Jalas

(c) polytrichus (A. Kerner ex	28	Ga	Bonnet, 1966
Borbas) Jalas	c.50 54	It	Jalas & Kaleva, 1966
	55 56	He	Jalas & Pohjo, 1965

  (d) zygiformis (H. Braun) Jalas

(e) arcticus (E. Durand) Jalas	c.50 Br Ho Hs Ga 51 54		Pigott, 1954
57 nervosus Gay ex Willk.	28	Ga	Bonnet, 1966
58 pulcherrimus Schur	56 60	Po	Trela-Sawicka, 1968
59 comosus Heuffel ex Griseb.	28	Rm	Kaleva, 1969

60 bihoriensis Jalas

61 pulegioides L.	28 30	Po	Trela-Sawicka, 1968
62 alpestris Tausch ex A. Kerner	28	Po	Trela-Sawicka, 1968

63 oehmianus Ronniger & Soska
64 alternans Klokov
65 serpyllum L.

(a) serpyllum	24	Po	Trela-Sawicka, 1968
(b) tanaensis (Hyl.) Jalas	24	Fe	Jalas, 1948
66 talijevii Klokov & Schost.			

34 Lycopus L.
1 europaeus L.	22	Su	Ehrenberg, 1945
2 exaltatus L. fil.			

35 Mentha L.
1 requienii Bentham	18	Co	Contandriopoulos, 1962
2 pulegium L.	20	Rm	Tarnavschi, 1948
3 micrantha (Bentham) Schost.			
4 cervina L.			
5 arvensis L.	72	Su	Olsson, 1967
	24 90	Ho	Ouweneel, 1968
x muellerana F.W. Schultz	60	Br	Morton, 1956
	42	Br	Morton, 1956
6 x verticillata L.	78 84 90	Su	Olsson, 1967
	120 132		
	96	Ho	Ouweneel, 1968
7 x gentilis L.	54 60 84 96	Br	Morton, 1956
	108 120		
8 x smithiana R.A. Graham	120	Br	Morton, 1956
9 aquatica L.	96	Br	Morton, 1956
10 x piperita L.	66 72	Br	Morton, 1956
x piperita nm. citrata (Ehrh.) Boivin	84 120	Br	Morton, 1956
x maximilianea F.W. Schultz	60 72-78 120	Br	Morton, 1956
x pyramidalis Ten.	c.72	It	Harley, 1977
11 suaveolens Ehrh.	24	Br	Harley, 1977
insularis Req.	24	Bl	Harley, 1977
x rotundifolia (L.) Hudson	24	Ho	Ouweneel, 1968
x villosa Hudson	36	Ho	Ouweneel, 1968
x villosa Hudson nm. alopecuroides (Hull)	36	Br	Morton, 1956
12 longifolia (L.) Hudson	24	Au Ga Gr He Hs	Harley, 1977
13 microphylla C. Koch	48	Gr	Harley, 1977
14 spicata L.	48	Br	Morton, 1956
x villosonervata auct., ?an Opiz	38	Br	Harley, 1977

36 Perilla L.
1 frutescens (L.) Britton

37 Rosmarinus L.
1 officinalis L.	24	Bl	Nilsson & Lassen, 1971
2 eriocalix Jordan & Fourr.			

38 Lavandula L.
1 stoechas L.	30	Gr	Bothmer, 1970
(a) stoechas			
(b) pedunculata (Miller) Samp. ex Rozeira			
(c) lusitanica (Chaytor) Rozeira			
(d) luisieri (Rozeira) Rozeira			
(e) sampaiana Rozeira			

196

(f) cariensis (Boiss.) Rozeira
2 viridis L'Hér.
3 dentata L.   44   Bl   Nilsson & Lassen, 1971

4 angustifolia Miller
  (a) angustifolia
  (b) pyrenaica (DC.) Guinea
5 latifolia Medicus
6 lanata Boiss.   50   Hs   Küpfer, 1969
7 multifida L.

39 Horminum L.
1 pyrenaicum L.   12   He   Favarger, 1953

40 Salvia L.
1 officinalis L.
2 lavandulifolia Vahl
3 grandiflora Etlinger   16   Bu   Markova, 1970
4 eichlerana Heldr. ex Halácsy
5 triloba L. fil.
6 candelabrum Boiss.
7 blancoana Webb & Heldr.
8 brachyodon Vandas
9 ringens Sibth. & Sm.
10 pinnata L.
11 scabiosifolia Lam.
12 pomifera L.   14   Gr   Bothmer, 1970
13 phlomoides
14 sclarea L.   22   Bu   Markova, 1970
15 argentea L.   20   Bu   Markova, 1970
16 aethiopis L.   24   Hu   Felfüldy, 1947
17 candidissima Vahl
18 glutinosa L.   16   Hu   Pólya, 1949
19 forskaohlei L.
20 bicolor Lam.
21 pratensis L.   18   Cz   Májovský et al., 1970

22 dumetorum Andrz. ex Besser
23 teddii Turrill
24 transsylvanica (Schur ex Griseb.) Schur
25 nemorosa L.   12   Cz   Májovský et al., 1970

  (a) nemorosa
  (b) tesquicola (Klokov & Pobed.) Soó
26 amplexicaulis Lam.
27 valentina Vahl
28 sclareoides Brot.
29 verbenaca L.   42   Cr   Bothmer, 1970
                        59 64 Ga Hs   Gadella et al., 1966
                        64   Ga   van Loon et al., 1971
                        60   Bl   Dahlgren et al., 1971
30 nutans L.
31 austriaca Jacq.
32 jurisicii Kosanin
33 virgata Jacq.   18   Bu   Markova, 1970
34 viridis L.   16   Cr Gr   Bothmer, 1970
35 verticillata L.   16   Cz   Májovský et al., 1970

36 napifolia Jacq.

41 Elsholtzia Willd.
   1 ciliata (Thunb.) Hyl.

CLII SOLANACEAE

1 Nicandra Adanson
  1 physalodes (L.) Gaertner

2 Lycium L.
  1 europaeum L.
  2 intricatum Boiss.   24   Lu   Fernandes &
  3 ruthenicum Murray   Queiros, 1970/71
  4 barbarum L.   24   Ga   Delay, 1967
  5 chinense Miller
  6 afrum L.

3 Atropa L.
  1 bella-donna L.
  2 baetica Willk.

4 Scopolia Jacq.
  1 carniolica Jacq.   48   Hu   Pólya, 1950

5 Hyoscyamus L.
  1 niger L.   34   Au   Griesinger 1937
  2 reticulatus L.
  3 albus L.   68   Bl   Dahlgren et al.,
                                                        1971
  4 aureus L.
  5 pusillus L.

6 Withania Pauquy
  1 somnifera (L.) Dunal
  2 frutescens (L.) Pauquy

7 Physalis L.
  1 alkekengi L.   24   Cz   Májovský et al.,
  2 peruviana L.   1970
  3 pubescens L.
  4 philadelphica Lam.   24   Lu   R. Fernandes, 1970

8 Salpichroa Miers
  1 origanifolia (Lam.) Baillon

9 Capsicum L.
  1 annuum L.

10 Solanum L.
   1 nigrum L.
     (a) nigrum   72   Br   Edmonds, 1977
     (b) schultesii (Opiz) Wessely   72   Ju   Edmonds, 1977
   2 sublobatum Willd. ex Roemer & Schultes
   3 luteum Miller
     (a) luteum   48   Ju   Edmonds, 1977
     (b) alatum (Moench) Dostál   48   Rs(K)   Edmonds, 1977
   4 sarrachoides Sendtner   24   Br   Edmonds, 1977
   5 triflorum Nutt.
   6 pseudocapsicum L.
   7 dulcamara L.   24   Ga   Delay, 1967
   8 tuberosum L.   48   Br   Hawkes, 1956

```
   9 laciniatum Aiton
  10 melongena L.
  11 bonariense L.
  12 sodomeum L.
  13 mauritianum Scop.
  14 cornutum Lam.
  15 elaeagnifolium Cav.

11 Lycopersicon Miller
   1 esculentum Miller                24      Br      Haskell, 1964

12 Mandragora L.
   1 officinarum L.
   2 autumnalis Bertol.               84      Hs      Tamayo & Grande,
                                                      1967

13 Triguera Cav.
   1 ambrosiaca Cav.
   2 osbeckii (L.) Willk.

14 Datura L.
   1 stramonium L.                    24      Hu      Pólya, 1949
   2 ferox L.
   3 innoxia Miller

15 Cestrum L.
   1 parqui L'Hér.

16 Nicotiana L.
   1 glauca R.C. Graham
   2 rustica L.
   3 tabacum L.
   4 alata Link & Otto


CLIII BUDDLEJACEAE

1 Buddleja L.
   1 davidii Franchet
   2 albiflora Hemsley
   3 japonica Hemsley
   4 lindleyana Fortune ex Lindley


CLIV SCROPHULARIACEAE

1 Gratiola L.
   1 officinalis L.                   32      Rm      Tarnavschi, 1948
   2 linifolia Vahl
   3 neglecta Torrey

2 Lindernia All.
   1 procumbens (Krocker) Philcox
   2 dubia (L.) Pennell

3 Bacopa Aublet
   1 monnieri (L.) Pennell

4 Limosella L.
   1 aquatica L.                      40      Is      Löve & Löve, 1956
   2 tenella Quezel & Contandr.       40      Gr      Quézel &
                                                      Contandriopoulos,1965
```

```
3 australis R. Br.                    20      Br      Blackburn, 1939

5 Mimulus L.
  1 guttatus DC.                      48      Br      Maude, 1939
  2 luteus L.
  3 moschatus Douglas ex Lindley

6 Dodartia L.
  1 orientalis L.

7 Verbascum L.
  1 spinosum L.
  2 ovalifolium Donn ex Sims
   (a) ovalifolium
   (b) thracicum (Velen.) Murb.
  3 purpureum (Janka) Huber-Morath
  4 adrianopolitanum Podp.
  5 blattaria L.
  6 siculum Tod. ex Lojac.
  7 virgatum Stokes
  8 daenzeri (Fauche & Chaub.) O. Kuntze
  9 rupestre (Davidov) I.K. Ferguson
 10 roripifolium (Halácsy) I.K.
      Ferguson
 11 boissieri (Heldr. & Sart. ex Boiss.)
      O. Kuntze
 12 barnadesii Vahl.
 13 laciniatum (Poiret) O. Kuntze
 14 creticum (L.) Cav.
 15 arcturus L.
 16 levanticum I K Fomguian
 17 bueulifolium Iam
 18 spectabile Bieb.
 19 hervieri Degen
 20 phoeniceum L.                     32      Cz      Májovský et al.,
                                                       1970
                                      36      Hu      Baksay, 1961
 21 xanthophoeniceum Griseb.
 22 cylleneum (Boiss. & Heldr.) O. Kuntze
 23 acaule (Bory & Chaub.) O. Kuntze
 24 pyramidatum Bieb.
 25 zuccarinii (Boiss.) I.K. Ferguson
 26 orientale (L.) All.
 27 phlomoides L.
 28 densiflorum Bertol.
 29 samniticum Ten.
 30 guicciardii Heldr. ex Boiss.
 31 vandasii (Rohlena) Rohlena
 32 niveum Ten.
   (a) niveum
   (b) visianianum (Reichenb.) Murb.
   (c) garganicum (Ten.) Murb.
 33 macrurum Ten.
 34 thapsus L.                        32      Lu      Fernandes, 1950
                                      36      Ho      Gadella & Kliphuis,
                                                       1966

   (a) thapsus
   (b) crassifolium (Lam.) Murb.
   (c) giganteum (Willk.) Nyman
 35 litigiosum Samp.
 36 longifolium Ten.
```

37 boerhavii L.
38 argenteum Ten.
39 nevadense Boiss.
40 lagurus Fischer & C.A. Meyer
41 georgicum Bentham
42 nicolai Rohlena
43 anisophyllum Murb.
44 eriophorum Godron
45 baldaccii Degen
46 epixanthinum Boiss. & Heldr.
47 euboicum Murb. & Rech. fil.
48 pelium Halácsy
49 dieckianum Borbás & Degen
50 macedonicum Kosanin & Murb.
51 botuliforme Murb.
52 dentifolium Delile
53 speciosum Schrader
  (a) speciosum
  (b) megaphlomos (Boiss. & Heldr.)
      Nyman
54 adeliae Heldr. ex Boiss.
55 lasianthum Boiss. ex Bentham
56 undulatum Lam.
57 pentelicum Murb.
58 sinuatum L.   30   It   Mori, 1957
59 halacsyanum Sint. & Bornm. ex Halácsy
60 pinnatifidum Vahl
61 leucophyllum Griseb.
62 cylindrocarpum Griseb.
63 pseudonobile Stoj. & Stefanov
64 humile Janka
65 adenanthum Bornm.
66 nobile Velen.
67 graecum Heldr. & Sart. ex Boiss.
68 herzogii Bornm.
69 mallophorum Boiss. & Heldr.
70 decorum Velen.
71 mucronatum Lam.
72 pulverulentum Vill.
73 gnaphalodes Bieb.
74 glandulosum Delile
75 davidoffii Murb.
76 jankaeanum Pančić
77 durmitoreum Rohlena
78 rotundifolium Ten.
  (a) rotundifolium
  (b) haenseleri (Boiss.) Murb.
  (c) conocarpum (Moris) I.K. Ferguson
79 reiseri Halácsy
80 delphicum Boiss. & Heldr.
81 banaticum Schrader
82 lychnitis L.
83 chaixii Vill.
  (a) chaixii
  (b) austriacum (Schott ex Roemer &
      Schultes) Hayek
  (c) orientale Hayek
84 bithynicum Boiss.
85 nigrum L.
  (a) nigrum
  (b) abietinum (Borbás) I.K. Ferguson

201

```
 86 lanatum Schrader
 87 glabratum Friv.
    (a) glabratum
    (b) brandzae (Franchet ex Brandza) Murb.
    (c) bosnense (K. Maly) Murb.

 8 Scrophularia L.
    1 vernalis L.                          40       Ge Po      Vaarama & Leikas,
                                                               1970

    2 bosniaca G. Beck
    3 aestivalis Griseb.                   44       Bu         Grau, 1976
    4 divaricata Ledeb.
    5 peregrina L.                         36       Ga         Vaarama & Leikas,
                                                               1970

    6 arguta Aiton
    7 pyrenaica Bentham                    58       Ga         Grau, 1976
    8 scorodonia L.                        58       Br Hs Lu   Grau, 1976
                                          60       Ga Lu      Vaarama & Leikas,
                                                               1970
    9 grandiflora DC.
      (a) grandiflora                      60       Lu         Vaarama & Leikas,
                                                               1970

      (b) reuteri (Daveau) I.B.K. Richardson  58   Hs         Grau, 1976
   10 sambucifolia L.                      58       Hs         Grau, 1976
   11 trifoliata L.                      c.84       Co         Contandriopoulos,
                                                               1962

   12 sciophila Willk.                     58       Hs         Grau, 1976
   13 sublyrata Brot.
      schousboei Lange                     60       Lu         Grau, 1976
   14 laevigata Vahl
   15 scopolii Hoppe                       26       Bu         Vaarama & Leikas,
                                                               1970

   16 alpestris Gay ex Bentham            68       Hs         Grau, 1976
                                          72       Ga         Vaarama & Leikas,
                                                               1970
   17 herminii Hoffmanns. & Link        c.52       Lu         Vaarama & Hiirsalmi,
                                                               1967
   18 nodosa L.                            36       It         Mori, 1957
   19 auriculata L.                  78 84 86       Ho         Gadella & Kliphuis,
                                                               1966
                                          80       Ga It Lu   Vaarama & Leikas,
                                                               1970


      lyrata Willd.                        58       Hs         Grau, 1976
   20 umbrosa Dumor .                      52       Cz         Májovský et al.,
                                                               1970
                                          26       Po Rm      Grau, 1976
   21 lucida L.                         24 26       Ga Gr      Grau, ined.
   22 myriophylla Boiss. & Heldr.
   23 heterophylla Willd.
      (a) heterophylla                     24       Gr         Grau, ined.
      (b) laciniata (Waldst. & Kit.) Maire
          & Petitmengin                    26
   24 taygetea Boiss.
   25 frutescens L.                        26       Lu         Rodrigues, 1953
   26 ramosissima Loisel.
   27 spinulescens Degen & Hausskn.
   28 canina L.
      (a) canina                        24 26       Ga It      Vaarama & Leikas,
                                                               1970
```

(b) crithmifolia Boiss.	24	Hs	Grau, ined.
(c) hoppii (Koch) P. Fourn.	24 26	He	Vaarama & Hiirschmi, 1967

29 cretacea Fischer ex Sprengel
30 rupestris Bieb. ex Willd.

9 Anarrhinum Desf.

1 bellidifolium (L.) Willd.	18	Co	Contandriopoulos, 1962

2 laxiflorum Boiss.
3 longipedicellatum R. Fernandes
4 duriminium (Brot.) Pers.

5 corsicum Jordan & Fourr.	18	Co	Contandriopoulos, 1962

10 Antirrhinum L.
  1 valentinum Font Quer
  2 sempervirens Lapeyr.
  3 pulverulentum Laz.-Ibiza
  4 pertegasii Rothm.
  5 microphyllum Rothm.
  6 charidemi Lange
  7 molle L.
  8 grosii Font Quer
  9 hispanicum Chav.
    (a) hispanicum
    (b) mollissimum (Roth.) D.A. Webb
  10 meonanthum Hoffmanns. & Link
  11 braun-blanquetii Rothm.
  12 siculum Miller
  13 barrelieri Boreau
  14 graniticum Rothm.
  15 australe Rothm.

16 latifolium Miller	16	It	Pogliani, 1964

  17 majus L.
    (a) majus

(b) linkianum (Boiss. & Reuter) Rothm.	16	Hs	Björkqvist et al., 1969

    (c) cirrhigerum (Ficalho) Franco
    (d) tortuosum (Bosc) Rouy

11 Asarina Miller
  1 procumbens Miller

12 Misopates Rafin.

1 orontium (L.) Rafin.	16	Ga	van Loon et al., 1971

  2 calycinum Rothm.

13 Chaenorhinum (DC.) Reichenb.

1 origanifolium (L.) Fourr.	14	Ga	van Loon et al., 1971

  (a) origanifolium
  (b) cadevallii (O. Bolós & Vigo) Laínz
  (c) crassifolium (Cav.) Rivas Goday & Borja
  (d) segoviense (Willk.) R. Fernandes

2 glareosum (Boiss.) Willk.	14	Hs	Küpfer, 1968

3 villosum (L.) Lange
4 tenellum (Cav.) Lange
5 macropodum (Boiss. & Reuter) Lange
  (a) macropodum
  (b) degenii (Hervier) R. Fernandes

```
  6 serpyllifolium (Lange) Lange
   (a) serpyllifolium
   (b) lusitanicum R. Fernandes
  7 robustum Loscos
  8 grandiflorum (Cosson) Willk.
  9 rubrifolium (Robill. & Cast. ex DC.)
     Fourr.
   (a) rubrifolium
   (b) formenterae (Gand.) R. Fernandes
   (c) raveyi (Boiss.) R. Fernandes
 10 minus (L.) Lange
   (a) minus
   (b) litorale (Willd.) Hayek          14        Ju        Nilsson & Lassen,
                                                            1971

   (c) idaeum (Rech. fil.) R. Fernandes

14 Linaria Miller
  1 reflexa (L.) Desf.
  2 pedunculata (L.) Chaz.
  3 flava (Poiret) Desf.
  4 pseudolaxiflora Lojac.
  5 cretacea Fischer ex Sprengel
  6 sabulosa Czern. ex Klokov.
  7 tonzigii Lona
  8 cavanillesii Chav.
  9 oligantha Lange
 10 huteri Lange
 11 bipunctata (L.) Dum.-Courset
 12 ficalhoana Rouy
 13 hirta (L.) Moench
 14 trinhylla (I ) Millei
 15 genistifolia (L.) Miller             12        Cz        Damboldt, 1968
   (a) genistifolia                      12        Bu        Markova & Ivanova,
                                                            1970

   (b) sofiana (Velen.) Chater & D.A. Webb
   (c) dalmatica (L.) Maire & Petitmengin
 16 peloponnesiaca Boiss. & Heldr.
 17 spartea (L.) Willd.
 18 viscosa (L.) Dum.-Courset            12        Hs        Valdés, 1970
 19 incarnata (Vent.) Sprengel
 20 hellenica Turrill
 21 algarviana Chav.
 22 elegans Cav.
 23 nigricans Lange
 24 clementei Haenseler ex Boiss.
 25 chalepensis (L.) Miller
 26 canadensis (L.) Dum.-Courset
 27 purpurea (L.) Miller                 12        Si        Valdés, 1970
 28 capraria Moris & De Not.
 29 nivea Boiss. & Reuter
 30 repens (L.) Miller
 31 microsepala A. Kerner
 32 triornithophora (L.) Willd.
 33 pelisseriana (L.) Miller             24        Bl        Dahlgren et al.,
                                                            1971

 34 vulgaris Miller                      12        Ho        Gadella & Kliphuis,
                                                            1966

 35 angustissima (Loisel.) Borbás
 36 biebersteinii Besser
 37 macroura (Bieb.) Bieb.
```

```
38 debilis Kuprian.
39 odora (Bieb.) Fischer
40 loeselii Schweigger
41 latifolia Desf.
42 platycalyx Boiss.
43 anticaria Boiss. & Reuter          12        Hs        Valdés, 1970
44 lilacina Lange
45 verticillata Boiss.
46 thymifolia (Vahl) DC.
47 lamarckii Rouy
48 tristis (L.) Miller                12        Hs        Valdés, 1970
49 aeruginea (Gouan) Cav.             12        Hs        Küpfer, 1969
   (a) aeruginea
   (b) pruinosa (Sennen & Pau) Chater &
       Valdés
50 amoi Campo ex Amo                  12        Hs        Valdés, 1970
51 caesia (Pers.) DC. ex Chav.        12        Hs        Valdés, 1970
52 supina (L.) Chaz.                  12        Ga        van Loon, et al.,
                                                          1971
53 oblongifolia (Boiss.) Boiss. & Reuter 12     Hs        Valdés, 1970
   (a) oblongifolia
   (b) haenseleri (Boiss. & Reuter) Valdés
54 badalii Willk.
55 propinqua Boiss. & Reuter
56 glauca (L.) Chaz.
   (a) glauca
   (b) olcadium Valdés & D.A. Webb
   (c) aragonensis (Lange) Valdés
   (d) bubanii (Font Quer) Valdés
57 diffusa Hoffmanns. & Link
58 coutinhoi Valdés
59 satureioides Boiss.
60 depauperata Leresche ex Lange
61 alpina (L.) Miller                 12        Ga        Favarger & Huynh,
                                                          1964
62 faucicola Leresche & Levier
63 glacialis Boiss.                   12        Hs        Küpfer, 1968
64 ricardoi Coutinho
65 amethystea (Lam.) Hoffmanns. & Link
   (a) amethystea
   (b) multipunctata (Brot.) Chater & D.A.
       Webb
66 saxatilis (L.) Chaz.               12        Hs        Valdés, 1970
67 arenaria DC.
68 arvensis (L.) Desf.
69 simplex (Willd.) DC.
70 micrantha (Cav.) Hoffmanns. & Link

15 Cymbalaria Hill.
 1 muralis P. Gaertner, B. Meyer &
     Scherb.
   (a) muralis
   (b) visianii D.A. Webb
   (c) pubescens (C. Presl)
 2 longipes (Boiss. & Heldr.) A. Cheval.
 3 aequitriloba (Viv.) A. Cheval.
   (a) aequitriloba                   56        Bl        Dahlgren et al., 1971
   (b) fragilis (Rodr.) D.A. Webb
 4 muelleri (Moris) A. Cheval.
 5 pallida (Ten.) Wettst.
 6 hepaticifolia (Poiret) Wettst.     56        Co        Contandriopoulos,
                                                          1962
```

```
    7 pilosa (Jacq.) L.H. Bailey
    8 microcalyx (Boiss.) Wettst.
      (a) ebelii (Cuf.) Cuf.
      (b) minor (Cuf.) W. Greuter          28     Gr     Damboldt, 1971
      (c) dodekanesi W. Greuter
      (d) microcalyx

16 Kickxia Dumort.
    1 cirrhosa (L.) Fritsch
    2 commutata (Bernh. ex Reichenb.) Fritsch
      (a) commutata
      (b) graeca (Bory & Chaub.) R. Fernandes
    3 elatine (L.) Dumort.
      (a) elatine                          36     Ge     Wulff, 1939
      (b) crinita (Mabille) W. Greuter
    4 spuria (L.) Dumort
      (a) spuria
      (b) integrifolia (Brot.) R. Fernandes
    5 lanigera (Desf.) Hand.-Mazz.

17 Digitalis L.
    1 obscura L.                           56     Hs     Carpio, 1957
      (a) obscura
      (b) laciniata (Lindley) Maire
    2 purpurea L.
      (a) purpurea                         56     Ga     Lungeanu, 1967
      (b) mariana (Boiss.) Rivas Goday
      (c) heywoodii P. & M. Silva          56     Lu     Lungeanu, 1967
    3 thapsi L.
    4 dubia Rodr.
    5 grandiflora Miller                   56     Po     Skalińska et al.,
                                                          1959
    6 lutea L.
      (a) lutea                            56     Hs     Carpio, 1957
      (b) australis (Ten.) Arcangeli       56     It     Larsen, 1955b
    7 viridiflora Lindley                  56     Bu     Lungeanu, 1967
    8 parviflora Jacq.                     56     Hs     Lungeanu, 1967
    9 laevigata Waldst. & Kit.             56     Ju     Lungeanu, 1967
      (a) laevigata
      (b) graeca (Ivanina) Werner
   10 ferruginea L.                        56     Bu     Markova, 1970
   11 lanata Ehrh.                         56     Bu     Markova, 1970
   12 leucophaea Sibth. & Sm.

18 Erinus L.
    1 alpinus L.                           14     He     Favarger, 1953

19 Lagotis Gaertner
    1 minor (Willd.) Standley
    2 uralensis Schischkin

20 Wulfenia Jacq.
    1 carinthiaca Jacq.                    18     Au     Favarger, 1965
    2 baldaccii Degen

21 Veronica L.
    1 gentianoides Vahl
    2 bellidioides L.
      (a) bellidioides                     36     Au     Fischer, 1969
      (b) lilacina (Townsend) Nyman        18     Ga     Küpfer, 1968
    3 serpyllifolia L.
```

(a) serpyllifolia	14	Is	Löve & Löve, 1956
(b) humifusa (Dickson) Syme	14	Br	Rutland, 1941
4 repens Clarion ex DC.	14	Co	Contandriopoulos, 1962
5 alpina L.	18	Au	Fischer, 1969
	18 36	No	Knaben & Engelskjon, 1967
6 fruticans Jacq.	16	Au	Fischer, 1969
7 fruticulosa L.			
8 nummularia Gouan	16	Ca	Favarger & Kupfer, 1968
9 saturejoides Vis.			
10 erinoides Boiss. & Spruner	16	Gr	Quezel & Contandrio-poulos, 1965
contandriopouli Quezel	32	Gr	Quezel, 1967
11 ponae Gouan			
12 urticifolia Jacq.	18	He	Brandt, 1961
13 austriaca L.			
(a) teucrium (L.) D.A. Webb	64	He	Brandt, 1952
(b) vahlii (Gaudin) D.A. Webb	16	Ga	Küpfer, 1969
(c) dentata (F.W. Schmidt) Watzl	64	He	Brandt, 1952
	48	Cz	Májovský et al., 1970
(d) austriaca	48	Au	Fischer, 1969
14 multifida L.			
15 orientalis Miller			
16 tenuifolia Asso			
17 prostrata L.			
(a) prostrata	16	Ju	Fischer, 1969
(b) scheereri J.P. Brandt	32	Ge He	Brandt, 1952
18 turrilliana Stoj. & Stefanov			
19 rhodopaea (Velen.) Degen ex Stoj. & Stefanov			
20 pectinata L.			
21 rosea Desf.			
22 aragonensis Stroh			
23 thymifolia Sibth. & Sm.			
24 thessalica Bentham	16	Gr	Quezel & Contandrio-poulos, 1965
25 aphylla L.	18	Po	Skalińska et al., 1964
26 baumgartenii Roemer & Schultes			
27 officinalis L.	18	Da Su	Bücher, 1944
	36	Ho	Gadella & Kliphuis, 1968
28 allionii Vill.	18	Ga It	Bocquet et al., 1967
29 debneyi Hochst.			
30 chamaedrys L.			
(a) chamaedrys	32	Ho	Gadella & Kliphuis, 1963
(b) vindobonensis M. Fischer	16	Au	Fischer, 1970
31 melissifolia Poiret			
32 micrantha Hoffmanns. & Link			
33 peduncularis Bieb.			
34 montana L.	18	Ho	Gadella & Kliphuis, 1963

207

35 scutellata L.	18	Ho	Gadella & Kliphuis, 1963
36 beccabunga L.	18	Ho	Gadella & Kliphuis, 1963
	18 36	Po	Skalińska et al., 1964
37 scardica Griseb.	18	Hu Ju Rm	Marchant, ined.
38 anagalloides Guss.	18 + 0-2 B	Hs	Bjorkqvist et al., 1969
39 anagallis-aquatica L.	36	Ho	Gadella & Kliphuis, 1968
40 catenata Pennell	36	Ho	Gadella & Kliphuis, 1966
41 acinifolia L.	14	Europe	Fischer, ined.
	16	Rm	Tarnavschi, 1948
42 praecox All.	18	Au	Fischer, 1969
43 triphyllos L.	14	Au	Fischer, 1969
44 glauca Sibth. & Sm.			
45 aznavourii Dörfler			
46 arvensis L.	16	Au	Fischer, 1969
	?14	Is	Löve & Löve, 1956
47 verna L.	16	Co	Contandriopoulos, 1962
48 dillenii Crantz	16	Ge	Hofelich, 1935
49 peregrina L.			
50 chamaepithyoides Lam.			
51 grisebachii Walters			
52 agrostis L.	28	Ge	Wulff, 1937
53 polita Fries	14	Au	Fischer, 1969
	18	Rs(K)	Borhidi, 1968
54 opaca Fries			
55 persica Poiret	28	It	Gadella & Kliphuis, 1970
56 filiformis Sm.	14	Ge	Beatus, 1936
57 hederifolia L.			
(a) triloba (Opiz) Celak.	18	Au	Fischer, 1967
(b) sibthorpioides (Debeaux, Degen & Hervier) Walters			
(c) lucorum (Klett & Richter) Hartl	36	Au He Po	Fischer, 1967
(d) hederifolia	54	Au Ga Ge Gr It Si Su	Fischer, 1967
58 cymbalaria Bodard	18	Bl	Nilsson & Lassen, 1971
	36	Sa Si	Fischer, 1969
	54	Sa Tu	Fischer, 1969
59 longifolia L.	34	Rm	Borhidi, 1968
	68	Fe	Raitanen, 1967
60 paniculata L.	34	Hu	Graze, 1933
61 bachofenii Heuffel			
62 spicata L.			
(a) spicata	34	He	Graze, 1933
		Fe	Raitanen, 1967
(b) incana (L.) Walters	68	Hu	Graze, 1933

```
    (c) barrelieri (Schott ex Roemer &
        Schultes) Murb.                   34      Ju      Graze, 1933
    (d) orchidea (Crantz) Hayek           34      Au      Fischer, 1969

    (e) crassifolia (Nyman) Hayek

22 Hebe Commerson
    1 salicifolia (G. Forster) Pennell
    2 speciosa (R. Cunn. ex A. Cunn.) Andersen
    3 elliptica (G. Forster) Pennell

23 Paederota L.
    1 bonarota (L.) L.                     36      It      Favarger, 1965b
    2 lutea Scop.                          36      Au      Favarger, 1965b
                                        c.54      Au      Fischer, 1969

24 Sibthorpia L.
    1 europaea L.                          18      Br      I. & O. Hedburg, 1961
    2 africana L.
    3 peregrina L.

25 Lafuentea Lag.
    1 rotundifolia Lag.

26 Castilleja Mutis ex L. fil.
    1 pallida (L.) Sprengel
    2 schrenkii Rebr.
    3 lapponica (Gand.) Rebr.              48      Rs(N)   Löve & Löve, 1961b

27 Melampyrum L.
    1 cristatum L.                         18      Br      Hambler, 1954
    2 ciliatum Boiss. & Heldr.
    3 arvense L.                           18      Au      van Witsch, 1932
    4 barbatum Waldst. & Kit. ex Willd.
     (a) barbatum
     (b) carstiense Ronniger
    5 variegatum Huter, Porta & Rigo
    6 fimbriatum Vandas
    7 nemorosum L.
     (a) nemorosum                         18      Ge      Reese, 1953
     (b) debreceniense (Rapaics) Soó
    8 polonicum (Beauverd) Soó
    9 velebiticum Borbás                   18      He      Favarger, 1969b
   10 vaudense (Ronniger) Soó
   11 catalaunicum Freyn
   12 italicum Soó
   13 bohemicum A. Kerner
   14 subalpinum (Juratzka) A. Kerner
   15 bihariense A. Kerner
   16 hoermannianum K. Maly
   17 doerfleri Ronniger
   18 scardicum Wettst.
   19 trichocalycinum Vandas
   20 heracleoticum Boiss. & Orph.
   21 sylvaticum L.                        18      Is      Löve & Löve, 1956
   22 herbichii Woloszczak
   23 saxosum Baumg.
   24 pratense L.                          18      Fe      Sorsa, 1962

28 Tozzia L.
    1 alpina L.
```

(a) alpina	20	Au	Mattick, 1950
(b) carpathica (Woloszczak) Dostál			

29 Euphrasia L.
 1 grandiflora Hochst.
 2 azorica H.C. Watson
 3 rostkoviana Hayne

(a) montana (Jordan) Wettst.	22	Au	van Witsch, 1932
(b) rostkoviana	22	Au	van Witsch, 1932
(c) campestris (Jordan) P. Fourn.			
4 rivularis Pugsley	22	Br	Yeo, 1954
5 anglica Pugsley	22	Br	Yeo, 1954
6 vigursii Davey	22	Br	Yeo, 1956
7 hirtella Jordan ex Reuter	22	He	Yeo, 1970
8 drosocalyx Freyn			
9 picta Wimmer			
(a) picta	22	Au	Yeo, 1970
(b) kerneri (Wettst.) Yeo			
10 marchesettii Wettst. ex Marchesetti			
11 arctica Lange ex Rostrup			
(a) tenuis (Brenner) Yeo	44	Is	Löve & Löve, 1956
(b) arctica			
(c) minor Yeo			
(d) borealis (Townsend) Yeo	44	Br	Yeo, 1954
(e) slovaca Yeo			
12 tetraquetra (Breb.) Arrondeau	44	Br	Yeo, 1954
13 nemorosa (Pers.) Wallr.	44	Br	Yeo, 1954
14 coerulea Hoppe & Fürnrohr			
15 pseudokerneri Pugsley	44	Br	Yeo, 1954
16 confusa Pugsley	44	Br	Yeo, 1954
17 stricta D. Wolff ex J.F. Lehm.	44	Br	Yeo, 1954
18 hyperborea Joerg.			
19 pectinata Ten.	44	He	Favarger, 1969
20 liburnica Wettst.			
21 alpina Lam.	22	Ga	Favarger, 1969
22 christii Favrat			
23 cisalpina Pugsley			
24 minima Jacq. ex DC.			
(a) minima	44	Au	van Witsch, 1932
(b) tatrae (Wettst.) Hayek			
25 frigida Pugsley	44	Is	Löve & Löve, 1956
26 foulaensis Townsend ex Wettst.	44	Br	Yeo, 1954
27 cambrica Pugsley			
28 ostenfeldii (Pugsley) Yeo			
29 marshallii Pugsley	44	Br	Yeo, 1954
30 rotundifolia Pugsley			
31 dunensis Wiinst.			
32 campbelliae Pugsley			
33 micrantha Reichenb.	44	Br	Yeo, 1954
34 scottica Wettst.	44	Br	Yeo, 1954
35 calida Yeo			
36 saamica Juz.			
37 atropurpurea (Rostrup) Ostenf.			
38 bottnica Kihlman			
39 willkommii Freyn			
40 taurica Ganeschin			
41 salisburgensis Funck	44	Hb	Yeo, 1966
42 portae Wettst.			
43 illyrica Wettst.			
44 dinarica (G. Beck) Murb.			

45 cuspidata Host
46 tricuspidata L.

30 Odontites Ludwig
  1 tenuifolia (Pers.) G. Don fil.
  2 longiflora (Vahl) Webb
  3 glutinosa (Bieb.) Bentham
  4 viscosa (L.) Clairv.
   (a) viscosa
   (b) hispanica (Boiss. & Reuter) Rothm.
  5 granatensis Boiss.
  6 purpurea (Desf.) G. Don fil.
  7 rigidifolia (Biv.) Bentham
  8 lutea (L.) Clairv.   20   Ge   Schneider, 1964
  9 lanceolata (Gaudin) Reichenb.
10 jaubertiana (Boreau) D. Dietr. ex
    Walpers
   (a) jaubertiana
   (b) chrysantha (Boreau) P. Fourn.
   (c) cebennensis (Coste & Soulie)
       P. Fourn.
11 kaliformis (Pourret) Pau
12 bocconei (Guss.) Walpers
13 linkii Heldr. & Sart. ex Boiss.
14 corsica (Loisel.) G. Don fil.
15 verna (Bellardi) Dumort.
   (a) verna   40   Ge   Schneider, 1964
   (b) litoralis (Fries) Nyman   20   Ge   Schneider, 1964
   (c) serotina (Dumort.) Corb. 20   Ge   Rottgardt, 1956

31 Bartsia L.
  1 alpina L.   12 Au Ga He Favarger, 1953
                  36 Ga   Favarger &
                              Küpfer, 1968
                  24 Is   Löve & Löve, 1956
  2 spicata Ramond
  3 aspera (Brot.) Lange

32 Parentucellia Viv.
  1 viscosa (L.) Caruel   48   Br   Hambler, 1955
  2 latifolia (L.) Caruel

33 Bellardia All.
  1 trixago (L.) All.   24   Ga   Delay, 1969

34 Pedicularis L.
  1 acaulis Scop.
  2 sceptrum-carolinum L.   32   No   Knaben & Engelskjon,
                                    1967
  3 foliosa L.   16   Au   Witsch, 1932
  4 hoermanniana K. Maly
  5 hacquetii Graf
  6 exaltata Besser
  7 recutita L.   16   Au   Mattick, 1950
  8 limnogena A. Kerner
  9 rosea Wulfen   16   It   Favarger & Huynh,
                                  1964
   (a) rosea
   (b) allionii (Reichenb. fil.) E. Mayer
10 orthantha Griseb.
11 dasyantha Hadač

12 hirsuta L.	16	No	Knaben & Engelskjon, 1967
13 oederi Vahl	16	No	Knaben, 1950
14 flammea L.	16	Is	Löve & Löve, 1956
15 verticillata L.	12	Ga	Küpfer, 1968
16 amoena Adams ex Steven			
17 arguteserrata Vved.			
18 palustris L.	16	Ge	Reese, 1953
(a) palustris			
(b) opsiantha (E.L. Ekman) Almquist			
(c) borealis (J.W. Zett.) Hyl.			
19 labradorica Wirsing			
20 sylvatica L.			
(a) sylvatica	16	Ho	Gadella & Kliphuis, 1968
(b) lusitanica (Hoffmanns. & Link) Coutinho			
(c) hibernica D.A. Webb			
21 sudetica Willd.	16	Rs(N)	Sokolovskaya & Strelkova, 1960
22 comosa L.			
(a) comosa			
(b) campestris (Griseb.) Soó			
23 kaufmannii Pinzger			
24 sibthorpii Boiss.			
25 uralensis Vved.			
26 asparagoides Lapeyr.			
27 schizocalyx (Lange) Steininger			
28 brachyodonta Schlosser & Vuk.			
(a) brachyodonta			
(b) moesiaca (Stadlm.) Hayek			
(c) montenegrina (Janka ex Nyman) D.A. Webb			
(d) grisebachii (Wettst.) Hayek			
29 heterodonta Pančić			
30 leucodon Griseb.			
(a) leucodon			
(b) occulta (Janka) E. Mayer			
31 physocalyx Bunge			
32 dasystachys Schrenk			
33 graeca Bunge			
34 friderici-augusti Tommasini			
35 petiolaris Ten.			
36 ferdinandi Bornm.			
37 compacta Stephan ex Willd.			
38 lapponica L.	16	Su	Löve & Löve, 1944
39 tuberosa L.	16	Au	Van Witsch, 1932
40 elongata A. Kerner	16	It	Favarger, 1965
41 julica E. Mayer			
42 ascendens Schleicher ex Gaudin	16	He	Favarger, 1953
43 baumgartenii Simonkai			
44 rostratospicata Crantz	16		
45 rostratocapitata Crantz	16	Au	Mattick, 1950
(a) rostratocapitata			
(b) glabra H. Kunz			
46 kerneri Dalla Torre	16	He	Favarger, 1959a
47 pyrenaica Gay	16	Ga	Küpfer & Favarger, 1968
48 mixta Gren.			
49 cenisia Gaudin			
50 gyroflexa Vill.			

```
    (a) gyroflexa                          16      He      Favarger, 1963
    (b) praetutiana (Levier ex
         Steininger) H. Kunz & E. Mayer
 51 elegans Ten.
 52 portenschlagii Sauter ex Reichenb.
 53 asplenifolia Floerke ex Willd.        16      Au      Mattick, 1950
 54 resupinata L.

35 Rhinanthus L.
    1 groenlandicus (Ostenf.) Chab.        22      Is      Löve & Löve, 1956
    2 minor L.                             22      Au      Tschermak-Woess, 1967
    3 asperulus (Murb.) Soó
    4 wettsteinii (Steneck) Soó
    5 pubescens (Sterneck) Boiss. & Heldr.
      ex Soó
    6 pindicus (Sterneck) Soó
    7 antiquus (Sterneck) Schinz & Thell.
    8 dinaricus Murb.
    9 aristatus Čelak.
   10 pampaninii Chab.
   11 alpinus Baumg.
   12 carinthiacus Widder
   13 burnatii (Chab.) Soó
   14 rumelicus Velen.
   15 wagneri Degen
   16 mediterraneus (Sterneck) Adamović    22      Ga      Campion-Bourget, 1967
   17 melampyroides (Borbás & Degen) Soó
   18 songeonii Chab.
   19 ovifugus Chab.
   20 subulatus (Chab.) Soó
   21 borbasii (Dörfler) Soó
      (a) borbasii
      (b) songaricus (Sterneck) Soó
   22 halophilus U. Schneider
   23 angustifolius C.C. Gmelin            22      Au      Tschermak-Woess, 1967
      (a) angustifolius
      (b) cretaceus (Vassilcz.) Soó
      (c) grandiflorus (Wallr.) D.A. Webb
   24 freynii (A. Kerner ex Sterneck) Fiori
   25 alectorolophus (Scop.) Pollich       22      Au      Tschermak-Woess, 1967

36 Rhynchocorys Griseb.
    1 elephas (L.) Griseb.

37 Siphonostegia Bentham
    1 syriaca (Boiss. & Reuter) Boiss.

38 Cymbaria L.
    1 borysthenica Pallas ex Schlecht.

39 Lathraea L.
    1 squamaria L.                         36      Au      van Witsch, 1932
    2 rhodopea Dingler
    3 clandestina L.


CLV GLOBULARIACEAE

1 Globularia L.
   1 incanescens Viv.                      16      It      Schwarz, 1964
   2 trichosantha Fischer & C.A. Meyer
   3 punctata Lapeyr.                      16   Au Ga Hu   Larsen, 1957b
```

```
 4 vulgaris L.                               32      Gs Su    Larsen, 1957b
 5 valentina Willk.
 6 cambessedesii Willk.                      64      Bl       Schwarz, 1964
 7 spinosa L.
 8 alypum L.                                 16      Ga       Schwarz, 1964
 9 cordifolia L.                             32      Ga He    Larsen, 1957
10 meridionalis (Podp.) O. Schwarz           16      Al Bu    Schwarz, 1964
11 repens Lam.                               16      Ga Hs    Schwarz, 1964
12 neapolitana O. Schwarz
13 stygia Orph. ex Boiss.
14 nudicaulis L.                          16 24      Ge       Schwarz, 1964
   x fuxeensis Giraud.                       16      Ga       Contandriopoulos
                                                              & Cauwet, 1968

15 gracilis Rouy & Richter
```

CLVI ACANTHACEAE

```
1 Acanthus L.
  1 mollis L.                               56       Ga       van Loon et al.,
                                                              1971

  2 balcanicus Heywood & I.B.K. Richardson
  3 spinosus L.

2 Justicia L.
  1 adhatoda L.
```

CLVII PEDALIACEAE

```
1 Sesamum L.
  1 indicum L.
```

CLVIII MARTYNIACEAE

```
1 Proboscidea Schmidel
  1 louisianica (Miller) Thell.
```

CLIX GESNERIACEAE

```
1 Haberlea Friv.
  1 rhodopensis Friv.                       38       Bu       Borhidi, 1968
                                            48       Gr       Contandriopoulos,
                                                              1967

2 Ramonda L.C.M. Richard
  1 myconi (L.) Reichenb.                   48       Ga       Contandriopoulos,
                                                              1967
  2 nathaliae Pančić & Petrović             48       Ju Gr    Contandriopoulos,
                                                              1967
  3 serbica Pančić                        c.96       Gr       Contandriopoulos,
                                                              1967

3 Jankaea Boiss.
  1 heldreichii (Boiss.) Boiss.            56        Gr       Contandriopoulos,
                                                              1967
```

CLX OROBANCHACEAE

```
1 Cistanche Hoffmanns. & Link
```

```
  1 phelypaea (L.) Coutinho              40      Lu        Gardé, 1952
  2 salsa (C.A. Meyer) G. Beck

2 Orobanche L.
  1 ramosa L.                            24      Lu        Gardé, 1952
    (a) ramosa
    (b) Reuter) Coutinho
    (c) mutelii (F.W. Schultz) Coutinho
  2 aegyptiaca Pers.                     24      Rs(E)     Zhukov, 1939
  3 lavandulacea Reichenb.
  4 trichocalyx (Webb & Berth.) G. Beck
  5 oxyloba (Reuter) G. Beck
  6 schultzii Mutel
  7 caesia Reichenb.
  8 arenaria Borkh.
  9 purpurea Jacq.                       24      Br        Hambler, 1958
 10 coerulescens Stephan
 11 cernua Loefl.                        24      Rs(E)     Zhukov, 1939
 12 amoena C.A. Meyer
 13 crenata Forskål                      38      Lu        Gardé, 1952
 14 alba Stephan ex Willd.
 15 reticulata Wallr.                    38      Br        Hambler, 1958
 16 haenseleri Reuter
 17 pancicii G. Beck
 18 serbica G. Beck & Petrović
 19 pubescens D'Urv.
 20 densiflora Salzm. ex Reuter
 21 amethystea Thuill.
    (a) amethystea
    (b) castellana (Reuter) Rouy
 22 canescens C. Presl
 23 esulae Pančić
 24 calendulae Pomel
 25 grisebachii Reuter
 26 loricata Reichenb.                  c.38     Br        Hambler, 1958
 27 minor Sm.                            38      Europe    Löve, 1954a
 28 clausonis Pomel
 29 hederae Duby                         38      Br        Hambler, 1958
 30 latisquama (F.W. Schultz) Batt.
 31 caryophyllacea Sm.                   38      Ga        Lévêque &
                                                           Gorenflot, 1969
 32 teucrii Holandre
 33 lutea Baumg.                         38      Rs(E)     Zhukov, 1939
 34 elatior Su-ton                       38      Br        Hambler, 1958
 35 alsatica Kirschleger
 36 chironii Lojac.
 37 laserpitii-sileris Reuter ex Jordan
 38 flava C.F.P. Mart. ex F.W. Schultz   38      Po        Skalińska et al.,
                                                           1966
 39 lucorum A. Braun
 40 salviae F.W. Schultz ex Koch
 41 rapum-genistae Thuill.               38      Br        Hambler, 1958
    (a) rapum-genistae
    (b) benthamii (Timb.-Lagr.) P. Fourn.
    (c) rigens (Loisel.) P. Fourn.
 42 gracilis Sm.
 43 variegata Wallr.
 44 foetida Poiret
 45 sanguinea C. Presl

3 Boschniakia C.A. Meyer
  1 rossica (Cham. & Schlecht.) B. Fedtsch.
```

4 Phelypaea L.
  1 coccinea Poiret
  2 boissieri (Reuter) Stapf

CLXI LENTIBULARIACEAE

1 Pinguicula L.
  1 lusitanica L.   12   Ga   Contandriopoulos, 1962
  2 hirtiflora Ten.   16   It   Honsell, 1959
  3 alpina L.   32   Rs(N)   Sokolovskaya &
                                                                     Strelkova, 1960
  4 villosa L.   16   No   Knaben, 1950
  5 corsica Bernard & Gren.   16   Co   Favarger &
                                                                     Contandriopoulos, 1961
  6 leptoceras Reichenb.   32   Au   Casper, 1962
  7 balcanica Casper   32   Bu   Casper, 1966
  8 nevadensis (Lindb.) Casper
  9 vallisneriifolia Webb
 10 longifolia Ramond ex DC.
    (a) longifolia
    (b) caussensis Casper
    (c) reichenbachiana (Schindler) Casper   32   Ga   Casper, 1962
 11 grandiflora Lam.   32   Ga   Contandriopoulos, 1962
 12 vulgaris L.   64   Da   Larsen, 1965

2 Utricularia L.
  1 subulata L.
  2 gibba L.
  3 minor L.   c.40   Co   Reese, 1952
  4 intermedia Hayne
    ochroleuca R. Hartman   c.40   Ge   Reese, 1952
  5 vulgaris L.   c.40   Ge   Reese, 1952
  6 australis R. Br.   c.40   Ge   Reese, 1952

CLXII MYOPORACEAE

1 Myoporum Solander ex G. Forster
  1 tenuifolium G. Forster
  2 tetrandrum (Labill.) Domin

CLXIII PLANTAGINACEAE

1 Plantago L.
  1 major L.
    (a) major   12   Cz   Májovský et al.,
                                                                     1970
    (b) winteri (Wirtgen.) W. Ludwig
    (c) intermedia (DC.) Arcangeli   12   Cz   Májovský et al.,
                                                                     1970
  2 tenuiflora Waldst. & Kit.   24   Cz   Májovský et al.,
                                                                     1970
  3 cornuti Gouan   12   Rm   Rahn, 1966
  4 coronopus L.
    (a) coronopus   10 + 0 - 1B   Lu   Fernandes &
                                                                     Franca, 1972
    (b) commutata (Guss.) Pilger   20   Co   Gorenflot, 1960
    (c) cupanii (Guss.) Nyman

(d) purpurascens Pilger

5 macrorhiza Poiret	10	Lu	Böcher et al., 1955
6 serraria L.	10 + 0 - 3B	Lu	Fernandes & Franca, 1972
	20 + 2B		
7 crassifolia Forskål	20	Bl	Dahlgren et al., 1971
8 maritima L.	12	Ho	Gadella & Kliphuis, 1967a
	18	Po	Skalińska et al., 1961
	24	Fe	V. Sorsa, 1962
9 subulata L.	12	It	Rahn, 1966
holosteum Scop.	12	Lu	Fernandes & Franca, 1973
insularis (Gren. & Godron) Nyman	24	Co	Contandriopoulos, 1962
10 alpina L.	12	Ga	Cartier & Lennoir, 1968
	24	Ga	Cartier & Lennoir, 1968
11 schwarzenbergiana Schur	12	Rm	Rahn, 1966
12 media L.	12	Ju	Gadella & Kliphuis, 1972
	24	Ho	Gadella & Kliphuis, 1968
13 maxima Juss. ex Jacq.			
14 reniformis G. Beck	12	Ju	Sušnik & Lovka, 1973
15 atrata Hoppe	12	It	Rahn, 1966
	24 + 0 - 2B	Cz	Holub et al., 1971
	36	Ga	Cartier, 1965
16 monosperma Pourret	12	Ga	Cartier, 1965
17 nivalis Boiss.	12	Hs	Rahn, 1966
18 gentianoides Sibth. & Sm.	12	Bu	Kozuharov & Petrova, 1974
19 amplexicaulis Cav.	10	Gr	Runemark, 1967b
20 lanceolata L.	12 + 0 - 1B	Lu	Fernandes & Franca, 1972
21 altissima L.	72	Cz	Májovský et al., 1970
22 argentea Chaix	12	Rm	Rahn, 1966
	12 + 2B		
23 lagopus L.	12 + 0 - 1B	Lu	Fernandes & Franca, 1972
24 albicans L.	20	Hs	Lorenzo-Andreu, 1950
	30	Gr	Runemark, 1967b
25 ovata Forskål	8	Hs	Rahn, 1957
26 minuta Pallas			
27 loeflingii L.			
28 notata Lag.			
29 bellardii All.			
(a) bellardii	10	Lu	Rodriguez, 1953
(b) deflexa (Pilger) Rech. fil.	10	Gr	Runemark, 1967b
30 cretica L.	10	Gr	Runemark, 1967b
31 squarrosa Murray			
32 arenaria Waldst. & Kit.	12	Bu	Kozuharov & Petrova, 1974
33 afra L.	12	Bl	Dahlgren et al., 1971

217

34 sempervirens Crantz   12   Ga   Rahn, 1966
35 asperrima (Gand.) Hervier

2 Littorella Bergius
  1 uniflora (L.) Ascherson   24   Ge   Wulff, 1939a

CLXIV CAPRIFOLIACEAE

1 Sambucus L.
  1 ebulus L.
  2 nigra L.   36   Hu   Pólya, 1950
  3 racemosa L.   36   Fe   V. Sorsa, 1962

2 Viburnum L.
  1 opulus L.   18   Su   Löve, 1954 a
  2 lantana L.   18   Br   Egolf, 1962
  3 tinus L.
    (a) tinus
    (b) subcordatum (Trelease) P. Silva

3 Symphoricarpos Duh.
  1 albus (L.) S.F. Blake

4 Linnaea L.
  1 borealis L.   32   Fe   V. Sorsa, 1962

5 Leycesteria Wall.
  1 formosa Wall.

6 Lonicera L.
  1 caerulea L.
    (a) caerulea   18   Su   Löve, 1954a
    (b) pallasii (Ledeb.) Browicz
  2 pyrenaica L.
    (a) pyrenaica
    (b) majoricensis (Gand.) Browicz
  3 alpigena L.
    (a) alpigena   18   Au   Löve, ined.
    (b) formanekiana (Halácsy) Hayek
  4 glutinosa Vis.
  5 hellenica Orph. ex Boiss.
  6 nigra L.   18   Cz   Májovský et al.,
                                     1970
  7 tatarica L.
  8 xylosteum L.   18   Fe   Sorsa, 1963
  9 arborea Boiss.
  10 nummulariifolia Jaub. & Spach
  11 biflora Desf.
  12 japonica Thunb.   18   Ga   Löve, ined.
  13 implexa Aiton
  14 splendida Boiss.
  15 caprifolium L.   18   It   Löve, ined.
  16 etrusca G. Santi
  17 periclymenum L.
    (a) periclymenum   18   Br   Rutland, 1941
                       36   Br   Janaki-Ammal &
                                   Saunders, 1952
                     54   Ho   Gadella & Kliphuis,
                                   1963
    (b) hispanica (Boiss. & Reuter) Nyman

CLXV ADOXACEAE

1 Adoxa L.
  1 moschatellina L.

CLXVI VALERIANACEAE

1 Patrinia Juss.
  1 sibirica (L.) Juss.

2 Valerianella Miller
  1 coronata (L.) DC.   14   Bl   Dahlgren et al.,
                                                                  1971
  2 divaricata Lange
  3 lasiocarpa (Steven) Betcke
  4 pumila (L.) DC.   14   Ga Rm   Ernet, 1972
  5 hirsutissima Link
  6 kotschyi Boiss.
  7 discoidea (L.) Loisel.   14   Ga Lu   Ernet, 1972
  8 obtusiloba Boiss.
  9 vesicaria (L.) Moench
 10 locusta (L.) Laterrade   16   Ge   Ernet, 1972
 11 martinii Loscos
 12 carinata Loisel.   16   Au   Ernet, 1972
 13 turgida (Steven)Betcke
 14 costata (Steven) Betcke
 15 echinata (L.) DC.   16   It   Ernet, 1972
 16 dentata (L.) Pollich   16   Rm   Ernet, 1972
 17 rimosa Bast.   16   Ga Ge Su   Ernet, 1972
 18 eriocarpa Desv.   16   Si   Ernet, 1972
    muricata (Steven ex Bieb.) J.W.
    Loudon   16   Hs It Lu   Ernet, 1972
 19 microcarpa Loisel.   16   It   Ernet, 1972
 20 puberula (Bertol. ex Guss.) DC.
 21 pontica Lipsky
 22 uncinata (Bieb.) Dufresne

3 Fedia Gaertner
  1 cornucopiae (L.) Gaertner

4 Valeriana L.
  1 officinalis L.   14   Au   Titz, 1964
                                               28   Cz   Májovský et al.,
                                                                  1970
                                               49   Po   Skalinska, 1958
                                               56   Ho   Gadella & Kliphuis,
                                                                  1963
    (a) officinalis   14   Po   Skalińska, 1950b
    (b) collina (Wallr.) Nyman   28   Po   Skalińska, 1950b
    (c) sambucifolia (Mikan fil.) Čelak.   56   Po   Skalińska, 1950b
    salina Pleijel   56   Su   Jonsell in litt.,
                                                                  1973
    repens Host   56   Ge   Walther, 1949
  2 dioscoridis Sibth. & Sm.   16   Gr   Seitz, 1969
  3 tuberosa L.   16   Ga   Guinochet &
                                                                  Logeois, 1962
  4 asarifolia Dufresne
  5 alliariifolia Adams
  6 pyrenaica L.
  7 globulariifolia Ramond ex DC.   16   Ga Hs   Küpfer &
                                                                  Favarger, 1968

8 olenaea Boiss. & Heldr.
9 dioica L.
  (a) dioica

	16	Ju	Lovka et al., 1971
	32	Ge	Engel, 1972

  (b) simplicifolia (Reichenb.) Nyman 16   Po   Skalińska, 1950
10 tripteris L.   16   He   Larsen, 1954
11 montana L.   32   Au   Tischler, 1950
   phitosiana   32   Cr   Quézel & Contand-
                                              riopoulos, 1965

12 bertiscea Pančić
13 crinii Orph. ex Boiss.
14 capitata Link   56   Rs(N) Sokolovskaya &
                                                Strelkova, 1960

15 saxatilis L.
  (a) saxatilis   24   Au   Favarger, 1965
  (b) pancicii (Halácsy & Bald.)
      Ockendon   24   Ju   Engel, 1972
16 celtica L.   c.48 Ga   Favarger, 1965
                                  c.72 Au   Polatschek, 1966
                                  c.96 Au   Favarger, 1965
17 elongata Jacq.   24   It   Favarger, 1965
18 supina Ard.   16   He   Favarger, 1965
19 saliunca All.   16   Ga   Favarger, 1965
20 longiflora Willk.

5 Centranthus DC.
  1 ruber (L.) DC.
    (a) ruber   32   It   Larsen, 1958b
    (b) sibthorpii (Heldr. & Sart. ex
      Boiss.) Hayek
  2 angustifolius (Mill.) DC.
  3 lecoqii Jordan
  4 longiflorus Steven
    (a) junceus (Boiss. & Heldr.) I.B.K.
      Richardson
    (b) kellereri (Stoj. Stefanov &
      Georgiev) I.B.K. Richardson
  5 nevadensis Boiss.
    (a) nevadensis
    (b) sieberi (Heldr.) I.B.K.
      Richardson
  6 trinervis (Viv.) Béguinot   28   Co   Contandriopoulos,
                                              1962
  7 calcitrapae (L.) Dufresne   32   Bl   Dahlgren et al., 1971
    (a) calcitrapae
    (b) trichocarpus I.B.K. Richardson
  8 macrosiphon Boiss.   32   Hs   Murray, 1975

CLXVII DIPSACACEAE

1 Morina L.
  1 persica L.

2 Cephalaria Schrader
  1 squamiflora (Sieber) W. Greuter
    (a) balearica (Willk.) W. Greuter
    (b) squamiflora
  2 leucantha (L.) Roemer & Schultes 18   Ga   Ritter, 1973
  3 radiata Griseb. & Schenk
  4 laevigata (Waldst. & Kit.) Schrader

    5 coriacea (Willd.) Roemer & Schultes ex
      Steudel
    6 uralensis (Murray) Roemer & Schultes   18   Rm   Lungeanu, 1972
    7 syriaca (L.) Roemer & Schultes
    8 transylvanica (L.) Roemer & Schultes   18   Cz   Májovský et al.,
                                                                1970
    9 joppica (Sprengel) Beguinot
   10 ambrosioides (Sibth. & Sm.) Roemer
      & Schultes
   11 flava (Sibth. & Sm.) Szabo
   12 pastricensis Dörfler & Hayek
   13 alpina (L.) Roemer & Schultes
   14 litvinovii Bobrov

3 Dipsacus L.
   1 gmelinii Bieb.
   2 sativus (L.) Honckeny   18   Ga   Poucques, 1948
   3 fullonum L.   18   Po   Skalińska et al.,
                                                                1964
   4 laciniatus L.   16   Bot. Gard. Poucques, 1948
                                              18   Bot. Gard. Kachidze, 1929
   5 comosus Hoffmanns. & Link
   6 ferox Loisel.
   7 pilosus L.   18   Bot. Gard. Kachidze, 1929
   8 strigosus Willd.

4 Succisa Haller
   1 pratensis Moench   20   Fe   Sorsa, 1963
   2 pinnatifida Lange

5 Succisella G. Beck
   1 inflexa (Kluk) G. Beck   20   Hu   Baksay, 1955
   2 petteri (J. Kerner & Murb.) G. Beck
   3 carvalhoana (Mariz) Baksay
   4 microcephala (Willk.) G. Beck

6 Knautia L.
   1 drymeia Heuffel
     (a) nympharum (Boiss. & Heldr.)
         Ehrend.   20   Ju   Ehrendorfer, 1962
     (b) tergestina (G. Beck) Ehrend.   20   It Ju   Ehrendorfer, 1962
     (c) centrifrons (Borbas) Ehrend.   20   It   Ehrendorfer, 1962
     (d) drymeia (Jacq.) Wettst.   40   Au Hu Ju   Ehrendorfer, 1962
                                           38 42-44   Ju   Ehrendorfer, 1962
     (e) intermedia (Pernh. & Wettst.)
         Ehrend.   20   Au   Ehrendorfer, ined.
                                              40   Au   Ehrendorfer, 1962
   2 gussonei Szabo
   3 arvernensis (Briq.) Szabo   40   Ga   Ehrendorfer, 1962
   4 sarajevensis (G. Beck) Szabo   40   Ju   Ehrendorfer, 1962
   5 dipsacifolia Kreutzer
     (a) lancifolia (Heuffel) Ehrend.
     (b) pocutica (Szabo) Ehrend.   40   Hu   Baksay, 1956
     (c) turocensis (Borbas) Jav. ex Kiss   40   Cz   Májovský et al.,
                                                                1978
     (d) gracilis (Szabo) Ehrend.   40   Ge   Breton-Sintes, 1971
     (e) sixtina (Briq.) Ehrend.   60   Ehrendorfer, ined.
     (f) dipsacifolia   60   Au Ga He   Ehrendorfer, 1962
   6 ressmannii (Pacher) Briq.   60   It   Ehrendorfer, 1962
   7 nevadensis (M. Winkler ex Szabo)
     Szabo   64   Hs   Ehrendorfer, 1962

8 basaltica Chassagne & Szabo	20	Ga	Breton-Sintes, 1971
9 foreziensis Chassagne & Szabo	40	Ga	Breton-Sintes, 1971
10 godetii Reuter	20	Ga	Sintes & Cauderon, 1965
11 salvadoris Sennen ex Szabo	20		Ehrendorfer, ined.
12 longifolia (Waldst. & Kit.) Koch	20	Au It	Ehrendorfer, 1962
13 pancicii Szabo			
14 midzorensis Form.	20	Ju	Ehrendorfer, 1962
15 magnifica Boiss. & Orph.			
16 dinarica (Murb.) Borbás			
(a) dinarica	20	Ju	Ehrendorfer, 1962
	40	Ju	Ehrendorfer, 1962
(b) silana (Grande) Ehrend.	40		Ehrendorfer, ined.
17 lucana Lacaita & Szabo			
18 subcanescens Jordan	40	Ga	Ehrendorfer, 1962
19 baldensis A. Kerner ex Borbás	40	It	Ehrendorfer, 1962
20 velutina Briq.	20	It	Ehrendorfer, 1962
21 transalpina (Christ) Briq.	40	It	Ehrendorfer, 1962
22 persicina A. Kerner	40	It	Ehrendorfer, 1962
23 carinthiaca Ehrend.	20	Au	Ehrendorfer, 1962
24 x norica Ehrend.	40	Au	Ehrendorfer, 1962
25 albanica Briq.	20	Ju	Ehrendorfer, 1962
26 velebitica Szabo	20	Ju	Ehrendorfer, 1962
27 mollis Jordan	20	Ga	Ehrendorfer, ined
28 calycina (C. Presl) Guss.	20	It	Ehrendorfer, 1962
29 arvensis (L.) Coulter	20	Au Hu Ju	Ehrendorfer, 1962
	40	Ga He It	Ehrendorfer, 1962
	43	Au	Ehrendorfer, 1962
	46	Au	Ehrendorfer, 1962
30 kitaibelii (Schultes) Borbás			
(a) kitaibelii	40	Au Po	Ehrendorfer, 1962
(b) tomentella (Szabo) Baksay	40	Hu	Baksay, 1956
31 ambigua Boiss. & Orph.	20	Gr Ju	Ehrendorfer, 1962
32 macedonica Griseb.	20	Ju	Ehrendorfer, 1962
33 visianii Szabo	20	Ju	Ehrendorfer, 1962
34 purpurea (Vill.) Borbás	20	He It	Ehrendorfer, 1962
35 rupicola (Willk.) Szabo			
36 subscaposa Boiss. & Reuter	20	Hs	Ehrendorfer, ined.
37 illyrica G. Beck	40	Ju	Ehrendorfer, 1962
38 pectinata Ehrend.	20	Ju	Ehrendorfer, 1962
39 clementii (G. Beck) Ehrend.	40	Ju	Ehrendorfer, 1962
40 adriatica Ehrend.	40	Ju	Ehrendorfer, ined.
41 dalmatica G. Beck	20	Ju	Ehrendorfer, ined.
42 travnicensis (G. Beck) Szabo	60	Ju	Ehrendorfer, 1962
43 fleischmannii (Hladnik ex Reichenb.) Pacher	40	Ju	Ehrendorfer, 1962
44 tatarica (L.) Szabo	20	Rs(C)	Ehrendorfer, ined.
45 byzantina Fritsch			
46 integrifolia (L.) Bertol.	20	Gr Ju	Ehrendorfer, 1962
47 degenii Borbás			
48 orientalis L.	16	Gr	Ehrendorfer, ined.

7 Pterocephalus Adanson
  1 papposus (L.) Coulter
  2 brevis Coulter
  3 diandrus (Lag.) Lag.
  4 intermedius (Lag.) Coutinho
  5 spathulatus (Lag.) Coulter
  6 perennis Coulter
    (a) perennis
    (b) bellidifolius (Boiss.) Vierh.

8 Scabiosa L.
  1 limonifolia Vahl
  2 saxatilis Cav.
   (a) saxatilis
   (b) grosii Font Quer
  3 cretica L.
  4 albocincta W. Greuter
  5 minoana (P.H. Davis) W. Greuter
  6 variifolia Boiss.
  7 hymettia Boiss. & Spruner
  8 epirota Halácsy & Bald.
  9 sphaciotica Roemer & Schultes
 10 pulsatilloides Boiss.
   (a) pulsatilloides
   (b) macropoda (Costa ex Willk.) Nyman
 11 isetensis L.
 12 crenata Cyr.
   (a) crenata
   (b) dallaportae (Heldr. ex Boiss.)
       Hayek
   (c) breviscapa (Boiss. & Heldr.) Hayek
 13 graminifolia L.   16   Ga   Ritter, 1973
                                         18   Ju   Lovka et al., 1971
 14 rhodopensis Stoj. & Stefanov   18   Bu   Frey, 1970
 15 argentea L.   16   Ju   Gadella & Kliphuis,
                                                         1972

 16 stellata L.
   (a) stellata
   (b) simplex (Desf.) Coutinho
 17 monspeliensis Jacq.
 18 rotata Bieb.   18   Bu   Frey, 1970
 19 micrantha Desf.
 20 sicula L.
 21 hispidula Boiss.
 22 cosmoides Boiss.
 23 atropurpurea L.   16   Ga   van Loon et al.,
                                                         1971

 24 semipapposa Salzm. ex DC.
 25 silenifolia Waldst. & Kit.
 26 vestina Facch. ex Koch   16   It   Damboldt, 1966
 27 canescens Waldst. & Kit.   16   Ge Po   Frey, 1969a
 28 parviflora Desf.
 29 tenuis Spruner ex Boiss.
 30 cinerea Lapeyr. ex Lam.   16   Ju   Sušnik & Lovka, 1973
   (a) cinerea
   (b) hladnikiana (Host) Jasiewicz
 31 turolensis Pau ex Willk.
 32 holosericea Bertol.
 33 taygetea Boiss. & Heldr.
 34 triandra L.   16   Co   Contandriopoulos,
                                                         1964a

 35 achaeta Vis. & Pančić
 36 lucida Vill.   16   Hu   Baksay, 1956
 37 nitens Roemer & Schultes
 38 columbaria L.   16   Bu Ge   Frey, 1969a
   (a) columbaria
   (b) pseudobanatica (Schur) Jáv. &
       Csapody
   (c) portae (A. Kerner ex Huter) Hayek
 39 balcanica Velen.
 40 ochroleuca L.   16   Ge Po   Frey, 1969a

41 webbiana D. Don
42 fumarioides Vis. & Pančić
43 triniifolia Friv.   16   Bu   Frey, 1969

9 Tremastelma Rafin.
  1 palaestinum (L.) Janchen   16   Al   Strid, 1971

10 Pycnocomon Hoffmanns. & Link
   1 rutifolium (Vahl) Hoffmanns. & Link   18   Co   Contandriopoulos,
                                                               1962

CLXVIII CAMPANULACEAE

1 Campanula L.
  1 fastigiata Dufour ex A. DC.
  2 uniflora L.   34   No   Knaben & Engelskjon,
                                                               1967
  3 cenisia L.   34   He   Podlech & Damboldt,
                                                               1964
                                          34 + 3B   He   Podlech & Damboldt,
                                                               1964
  4 zoysii Wulfen   34   Ju   Damboldt & Podlech,
                                                               1964
  5 carpatica Jacq.   34   Rm   Borhidi, 1968
  6 arvatica Lag.   28   Hs   Damboldt & Podlech,
                                                               1964
    (a) arvatica
    (b) adsurgens (Leresche & Levier)
        Damboldt
  7 raineri Perpenti   32   It   Damboldt & Podlech,
                                                               1964
  8 aizoon Boiss. & Spruner   16   Gr   Contandriopoulos,
                                                               1964b
    (a) aizoon
    (b) aizoides (Zaffran) Fedorov
  9 primulifolia Brot.   36   Lu   Gadella, 1964
 10 lusitanica L.   18   Hs   Damboldt & Podlech,
                                                               1964
    (a) lusitanica
    (b) transtagana (R. Fernandes) Fedorov
 11 phrygia Jaub. & Spach   16   Ju   Damboldt & Podlech,
                                                               1964
 12 sparsa Friv.
    (a) sparsa
    (b) frivaldskyi (Steudel) Hayek   20   Gr   Damboldt & Podlech,
                                                               1964
    (c) sphaerothrix (Griseb.) Hayek   20   Gr   Contandriopoulos,
                                                               1966
 13 ramosissima Sibth. & Sm.   20   Gr   Damboldt & Podlech,
                                                               1964
 14 spatulata Sibth. & Sm.
    (a) spatulata   20   Gr   Contandriopoulos,
                                                               1966
    (b) sprunerana (Hampe) Hayek   20   Gr   Contandriopoulos,
                                                               1966
    (c) filicaulis (Halácsy) Phitos
 15 patula L.
    (a) patula   20   Ju   Gadella & Kliphuis,
                                                               1972
                                               40   Ju   Gadella & Kliphuis,
                                                               1972

(b) costae (Willk.) Fedorov
(c) epigaea (Janka) Hayek 20 Bu Ancev, 1975
(d) abietina (Griseb.) Simonkai 40 Po Gadella, 1966
16 decumbens A. DC.
17 hemschinica C. Koch
18 stevenii Bieb.
  (a) wolgensis (P. Smirnov) Fedorov
  (b) altaica (Ledeb.) Fedorov
19 rapunculus L. 20 Ga It Larsen, 1956a
20 persicifolia L. 16 Au Bu Ga Gadella, 1966
21 andrewsii A. DC. 34 Gr Phitos, 1965
  (a) andrewsii
  (b) hirsutula Phitos
22 topaliana Beauverd 34 Gr Phitos, 1965
23 lavrensis (Tocl & Rohlena) Phitos 34 Gr Phitos, 1965
24 goulimyi Turrill 34 Gr Phitos, 1965
25 celsii A. DC. 34 Gr Phitos, 1965
  (a) celsii
  (b) parnesia Phitos
  (c) spathulifolia (Turrill) Phitos
  (d) carystea Phitos
26 anchusiflora Sibth. & Sm. 34 Gr Phitos, 1965
27 euboica Phitos 34 Gr Phitos, 1965
28 reiseri Halácsy 34 Gr Phitos, 1963
29 rechingeri Phitos 34 Gr Phitos, 1964
30 merxmuelleri Phitos 34 Gr Phitos, 1965
31 medium L. 34 Ga Gadella, 1964
32 pelviformis Lam. 34 Gr Phitos, 1964
33 tubulosa Lam. 34 Cr Phitos, 1965
34 carpatha Halácsy 34 Cr Phitos, 1965
35 lyrata Lam. 34 Gr Phitos, 1963
36 saxatilis L. 34 Gr Phitos, 1963
3 (a) saxatilis
  (b) cytherea Rech. fil. & Phitos
37 laciniata L. 34 Gr Phitos, 1964
38 rupestris Sibth. & Sm. 34 Gr Phitos, 1964
39 cymaea Phitos 34 Gr Phitos, 1964
40 constantini Beauverd & Top. 34 Gr Phitos, 1964
41 scopelia Phitos 34 Gr Phitos, 1964
42 sciathia Phitos 34 Gr Phitos, 1964
43 thessala Maire 34 Gr Phitos, 1964
44 barbata L. 34 Au Mattick, 1950

45 alpina Jacq. 34 Po Skalińska, 1959
  (a) alpina
  (b) orbelica (Pančić) Urum.
46 speciosa Pourret 34 Ga Gadella, 1964
47 affinis Schultes
  (a) affinis
  (b) bolsoii (Vayr.) Fedorov
48 formanekiana Degen & Dörfler 24 Gr Contandriopoulos, 1966

49 lingulata Waldst. & Kit. 34 Gr Ju Contandriopoulos, 1966

50 sibirica L. 34 Au Cz Hu Rm Gadella, 1964
  (a) sibirica
  (b) taurica (Juz.) Fedorov
  (c) talievii (Juz.) Fedorov
  (d) elatior (Fomin) Fedorov
  (e) divergentiformis (Jáv.) Domin
  (f) charkeviczii (Fedorov) Fedorov

51 incurva Aucher ex A. DC.	32	Gr	Contandriopoulos, 1964
52 grossekii Heuffel			
53 lanata Friv.			
54 dichotoma L.			
55 alpestris All.	34	Ga	Gadella, 1966
56 oreadum Boiss. & Heldr.	34	Gr	Contandriopoulos, 1966
57 calaminthifolia Lam.			
58 hierapetrae Rech. fil.			
59 amorgina Rech. fil.			
60 heterophylla L.			
61 mollis L.	26	Hs	Podlech & Damboldt, 1964
62 papillosa Halácsy			
63 orphanidea Boiss.			
64 rupicola Boiss. & Spruner	32	Gr	Contandriopoulos, 1964
65 petraea L.	34	Ga	Guinochet & Logeois, 1962
66 tymphaea Hausskn.	34	Gr	Contandriopoulos, 1966
67 stenosiphon Boiss. & Heldr.	34	Gr	Contandriopoulos, 1964
68 transsilvanica Schur ex Andrae			
69 moesiaca Velen.			
70 glomerata L.	30	Au Cz Ga Da It Su	Gadella, 1964
(a) glomerata			
(b) serotina (Wettst.) O. Schwarz			
(c) farinosa (Rochel) Kimmelberger	30	Cz	Májovský et al., 1974
(d) elliptica (Kit. ex Schultes) O. Schwarz			
(e) subcapitata (M. Popov) Fedorov			
(f) cervicarioides (Schultes) P. Fourn.			
(g) hispida (Witasek) Hayek			
71 foliosa Ten.	34	Gr	Contandriopoulos, 1966
72 cervicaria L.	34	Bu Rm	Gadella, 1964
73 macrostachya Waldst. & Kit. ex Willd.	18	Hu	Baksay, 1958
74 spicata L.	34	Au Ga It	Gadella, 1964
75 thyrsoides L.	34	Ga He	Gadella, 1964
(a) thyrsoides			
(b) carniolica (Sünd.) Podl.			
76 pyramidalis	34	Ju	Podlech & Damboldt, 1964
77 versicolor Andrews	34	Gr	Damboldt, 1971
78 morettiana Reichenb.	34	It	Podlech & Damboldt, 1964
79 radicosa Bory & Chaub.	34	Gr	Contandriopoulos, 1964
80 secundiflora Vis. & Pančić			
81 hawkinsiana Hausskn. & Heldr.	22	Gr	Contandriopoulos, 1964
82 sartorii Boiss. & Heldr.	34	Gr	Damboldt & Podlech, 1964
83 herminii Hoffmanns. & Link	32	Hs	Damboldt & Podlech, 1964
84 waldsteiniana Schultes	34	Ju	Merxmüller & Damboldt, 1962
85 tommasiniana Koch	34	Ju	Damboldt, 1965
86 isophylla Moretti	32	It	Merxmüller & Damboldt, 1962

87 fragilis Cyr.	32	It	Damboldt, 1965
(a) fragilis			
(b) cavolinii (Ten.) Damboldt			
88 elatinoides Moretti	34	It	Damboldt, 1965
89 elatines L.	34	It	Damboldt, 1965
90 portenschlagiana Schultes	34	Ju	Merxmüller & Damboldt, 1962
91 poscharskyana Degen	34	Ju	Merxmüller & Damboldt, 1962
92 garganica Ten.	34	It	Merxmüller & Damboldt, 1962
(a) garganica			
(b) cephallenica (Feer) Hayek			
(c) acarnanica (Damboldt) Damboldt			
93 fenestrellata Feer			
(a) fenestrellata			
(b) istriaca (Feer) Fedorov			
(c) debarensis (Rech. fil.) Damboldt	34	Ju	Merxmüller & Damboldt, 1962
94 specularioides Cosson			
95 scutellata Griseb.	14	Ju	Contandriopoulos, 1966
96 drabifolia Sibth. & Sm.	28	Gr	Contandriopoulos, 1964
(a) drabifolia			
(b) creutzburgii (W. Greuter) Fedorov			
(c) pinatzii (W. Greuter & Phitos) Fedorov			
97 delicatula Boiss.			
98 erinus L.	28	Lu	Gadella, 1964
99 latifolia L.	34	No	Laane, 1971
	34 + 5B	Cz Su	Gadella, 1964
100 trachelium L.	34	He	Gadella, 1966
(a) trachelium			
(b) athoa (Boiss. & Heldr.) Hayek			
101 rapunculoides L.	68	Rm	Gadella, 1964
	102	Au Cz Ga	Gadella, 1964
102 bononiensis L.	34	Bu Cz Ge Hu	Gadella, 1964
103 aparinoides Pursh			
104 trichocalycina Ten.	32	Gr	Contandriopoulos, 1966
105 macrorhiza Gay ex A. DC.	34	Ga	Podlech, 1962
gracillima Podl.	34	Ga	Podlech, 1965
106 sabatia De Not.	34	It	Podlech, 1962
107 carnica Schiede ex Mert. & Koch	34	It	Gadella, 1964
108 tanfanii Podl.	34	It	Podlech, 1965
109 xylocarpa Kovanda	34	Cz	Kovanda, 1966
110 crassipes Heuffel	34	Ju	Lovka et al., 1972
111 praesignis G. Beck	34	Au	Podlech, 1962
112 forsythii (Arcangeli) Podl.	34	Sa	Podlech, 1962
113 hispanica Willk.			
(a) hispanica	34	Hs	Podlech & Damboldt, 1964
ruscinonensis	34	Ga	Podlech & Damboldt, 1964
(b) catalanica Podl.			
114 justiniana Witasek	34	Ju	Podlech, 1962
115 hercegovina Degen & Fiala	34	Ju	Podlech & Damboldt, 1964
116 albanica Witasek			
(a) albanica	34	Gr	Podlech, 1962

(b) sancta (Hayek) Podl.	34	Gr	Podlech & Damboldt, 1964
117 romanica Savul.			
118 moravica (Spitzner) Kovanda			
(a) moravica	68	Cz	Kovanda, 1966
(b) xylorrhiza (O. Schwarz) Kovanda	102	Cz	Kovanda, 1970
119 apennina (Podl.) Podl.	34	It	Podlech, 1970
120 willkommii Witasek	68	He	Podlech & Damboldt, 1964
121 fritschii Witasek	68	Ga	Podlech, 1962
122 longisepala Podl.	34	Ga	Podlech, 1965
	68	Ga	Podlech, 1965
123 marchesettii Witasek	68	Ju	Podlech, 1962
124 velebitica Borbás			
125 bertolae Colla	102	It	Podlech & Damboldt, 1964
126 pseudostenocodon Lacaita	102	It	Podlech, 1962
127 rhomboidalis L.	34	He	Favarger, 1949a
128 cantabrica Feer	34	Hs	Podlech & Damboldt, 1964
129 serrata (Kit.) Hendrych	34	Cz	Kovanda, 1970a
	34 + 2B	Rm	Podlech & Damboldt, 1964
130 recta Dulac	34	Ga	Podlech & Damboldt, 1964
131 precatoria Timb.-Lagr.			
132 witasekiana Vierh.	34	Ju	Podlech, 1962
133 cochlearifolia Lam.	34	Au Cz Ge	Gadella, 1964
134 cespitosa Scop.	34	Ju	Podlech, 1962
135 jaubertiana Timb.-Lagr.			
(a) jaubertiana			
(b) andorrana (Rm. Pl.) F. Monts.			
136 excisa Schleicher ex Murith	34	He	Podlech, 1962
137 stenocodon Boiss. & Reuter	34	It	Podlech, 1962
138 pulla L.	34	Au	Gadella, 1964
139 scheuchzeri Vill.	68	Au Ge	Kovanda, 1970a
	102	It	Podlech & Damboldt, 1964
bohemica Hruby	68	Ga Hs	Podlech & Damboldt, 1964
gelida Kovanda	68	Cz	Kovanda, 1970a
140 ficarioides Timb.-Lagr.	102	Ga	Podlech & Damboldt, 1963
141 rotundifolia L.	34	Ga	Guinochet, 1942
	68	Ga	Guinochet, 1942
	102	Ga	Kovanda, 1970b
142 giesekiana Vest	34	Sb	Flovik, 1940
143 baumgartenii J. Becker	68	Ge	Podlech, 1962
144 beckiana Hayek	68	Au	Gutermann, 1961
2 Azorina Feer			
1 vidalii (H.C. Watson) Feer	56	Lu	Rodrigues, 1954
3 Symphyandra A. DC.			
1 cretica A. DC.			
(a) cretica	36	Cr	Phitos, 1966
(b) samothracica (Degen) Hayek	36	Gr	Phitos, 1966
(c) sporadum (Halácsy) Hayek	36	Gr	Phitos, 1966
2 wanneri (Rochel) Heuffel			
3 hofmannii Pant.			

4 Adenophora Fischer
 1 lilifolia (L.) Ledeb. ex A. DC.   34   Ju   Lovka et al., 1971
 2 taurica (Suk.) Juz.

5 Legousia Durande
 1 falcata (Ten.) Fritsch
 2 castellana (Lange) Samp.
 3 hybrida (L.) Delarbre   20   Br   Rutland, 1941
 4 speculum-veneris (L.) Chaix   20   Ga   Gadella, 1966
 5 pentagonia (L.) Druce

6 Trachelium L.
 1 caeruleum L.
   (a) caeruleum   32   Lu   Damboldt, ined.
   (b) lanceolatum (Guss.) Arcangeli
 2 jacquinii (Sieber) Boiss.
   (a) jacquinii
   (b) rumelianum (Hampe) Tutin   32 34   Gr   Damboldt, 1968
 3 asperuloides Boiss. & Orph.   34   Gr   Damboldt, 1968

7 Petromarula Vent. ex Hedwig fil.
 1 pinnata (L.) A. DC.   30   Cr   Podlech & Damboldt,
                                                                  1964

8 Asyneuma Griseb. & Schenk
 1 anthericoides (Janka) Bornm.
 2 limonifolium (L.) Janchen   24   Gr   Damboldt, 1968
 3 comosiforme Hayek & Janchen
 4 canescens (Waldst. & Kit.) Griseb. &
   Schenk
   (a) canescens   32   Bu   Ančev, 1975
   (b) cordifolium (Bornm.) Damboldt

9 Phyteuma L.
 1 spicatum L.
   (a) spicatum   22   Au   Polatschek, 1966
                                             22 + 1 - 4B Po   Ochlewska, 1966
   (b) coeruleum R. Schulz   22   Au   Polatschek, 1966
 2 vagneri A. Kerner
 3 pyrenaicum R. Schulz
 4 ovatum Honckeny   22   Au   Polatschek, 1966
                                             26   He   Contandriopoulos,
                                                                  1962
 5 nigrum F.W. Schmidt   22   Cz   Kovanda, 1970
 6 gallicum R. Schulz
 7 tetramerum Schur
 8 michelii All.
 9 scorzonerifolium Vill.
10 zahlbruckneri Vest   24   Au   Polatschek, 1966
11 betonicifolium Vill.   24   He   Favarger, 1953
12 cordatum Balbis   24   Ga   Favarger, 1953
13 orbiculare L.   22 + 0 - 1B Au   Polatschek, 1966
14 pseudorbiculare Pant.
15 sieberi Sprengel   20   Au   Polatschek, 1966
16 scheuchzeri All.   26   He It   Contandriopoulos,
                                                                  1962

   (a) scheuchzeri
   (b) columnae (Gaudin) Becherer
17 charmelii Vill.   26   Ga   Contandriopoulos,
                                                                  1962

    villarsii R. Schulz   26   Ga   Contandriopoulos,
                                                                  1962

```
18 serratum Viv.                            28      Co      Contandriopoulos,
                                                            1962
19 hemisphaericum L.                        28      Ga      Contandriopoulos,
                                                            1962
20 hedraianthifolium R. Schulz              28      He      Contandriopoulos,
                                                            1962
21 humile Schleicher ex Gaudin              28      He      Contandriopoulos,
                                                            1962
22 globulariifolium Sternb. & Hoppe
   (a) globulariifolium                     28      Au      Polatschek, 1966
   (b) pedemontanum (R. Schulz) Becherer    28      Ga      Contandriopoulos,
                                                            1962
23 rupicola Br.-Bl.
24 confusum A. Kerner                       28      Au      Polatschek, 1966
                                       28 + 1B


10 Physoplexis (Endl.) Schur
   1 comosa (L.) Schur                      34      It      Damboldt, 1965


11 Wahlenbergia Schrader ex Roth
   1 hederacea (L.) Reichenb.
   2 nutabunda (Guss.) A. DC.


12 Edraianthus A. DC.
   1 parnassicus (Boiss. & Spruner) Halácsy 32     Gr      Contandriopoulos,
                                                            1964b
   2 graminifolius (L.) A. DC.              32      Gr      Contandriopoulos,
                                                            1964b
   (a) graminifolius
   (b) niveus (G. Beck) Janchen
   3 tenuifolius (Waldst. & Kit.) A. DC.    00      Gi      Contandriopoulos,
                                                            1966
   4 dalmaticus (A. DC.) A. DC.
   5 serbicus Petrović                      32      Bu      Kuzmanov, 1973
   6 serpyllifolius (Vis.) A. DC.
   7 pumilio (Portenschl.) A. DC.
   8 dinaricus (A. Kerner) Wettst.
   9 wettsteinii Halácsy & Bald.


13 Jasione L.
   1 montana L.                             12      Au Cz Lu   Kovanda,
     blepharodon Boiss. & Reuter            12      Hs         Björkqvist et al., 1969
   2 penicillata Boiss.
   3 corymbosa Poiret ex Schultes
   4 lusitanica A. DC.
   5 crispa (Pourret) Samp.
     (a) amethystina (Lag. & Rodr.) Tutin   36      Hs      Küpfer & Favarger,
                                                            1967
     (b) centralis (Rivas Martínez)  Rivas
         Martínez
     (c) arvernensis Tutin
     (d) crispa                             36      Ga Hs   Küpfer & Favarger,
                                                            1967
     (e) serpentinica P. Silva
     (f) mariana (Willk.) Rivas Martínez
     (g) maritima (Duby) Tutin
     (h) sessiliflora (Boiss. & Reuter)
         Rivas Martínez
     (i) tomentosa (A. DC.) Rivas Martínez
     (j) cavanillesii (C. Vicioso)
```

```
  6 laevis Lam.                        12    Ju      Gadella & Kliphuis,
                                                       1972
                                       24    Hs      Küpfer, 1971
    (a) laevis
    (b) carpetana (Boiss. & Reuter) Rivas
        Martínez
    (c) orbiculata (Griseb. ex Velen.)
        Tutin
    (d) rosularis (Boiss. & Reuter) Tutin
  7 bulgarica Stoj. & Stefanov
  8 heldreichii Boiss. & Orph.          12    Gr      Contandriopoulos,
                                                       1966
  9 foliosa Cav.
    (a) foliosa
    (b) minuta (Agardh ex Roemer & Schultes)
        Font Quer


14 Lobelia L.
   1 urens L.
   2 dortmanna L.                       14    Fe Ge Po   Reese, 1952

15 Laurentia Adanson
   1 gasparrinii (Tineo) Strobl


CLXIX COMPOSITAE

Eupatorieae Cass.

1 Eupatorium L.
   1 cannabinum L.
     (a) cannabinum                     20    Hu      Pólya, 1949
                                        40    Po      Skalińska et al.,
                                                       1971
     (b) corsicum (Req. ex Loisel.) P. Fourn.
   2 adenophorum Sprengel               51    Lu      Fernandes &
                                                       Queirós, 1971
Astereae Cass.

2 Grindelia Willd.
   1 squarrosa (Pursh) Dunal

3 Solidago L.
   1 virgaurea L.                       18    It      Garbari &
                                                       Tornadore, 1972
     ssp. minuta (L.) Arcangeli         18    Ga      Ritter, 1973
   2 sempervirens L.
   3 canadensis L.
   4 gigantea Aiton                     36    Ga      Beaudry, 1970
   5 graminifolia (L.) Salisb.

4 Dichrocephala L'Hér. ex DC.
   1 integrifolia (L. fil.) O. Kuntze   18    non-Europe

5 Bellis L.
   1 annua L.
     (a) annua                          18    Lu      Fernandes &
                                                       Queirós, 1971
     (b) vandasii (Velen.) D.A. Webb
   2 perennis L.                        18    Hu      Felföldy, 1947
   3 bernardii Boiss. & Reuter          18    Co      Contandriopoulos,
                                                       1957
```

```
  4 azorica Hochst.
  5 longifolia Boiss. & Heldr.
  6 sylvestris Cyr.                       36      Lu      Fernandes &
                                                         Queirós, 1971

  7 rotundifolia (Desf.) Boiss. & Reuter


6 Bellium L.
  1 bellidioides L.                       18      Bl      Dahlgren et al.,
                                                          1971

  2 minutum (L.) L.
  3 crassifolium Moris


7 Aster L.
  1 macrophyllus L.
  2 schreberi Nees
  3 divaricatus L.
  4 novae-angliae L.
  5 puniceus L.
  6 laevis L.
  7 novi-belgii L.
  8 x salignus Willd.
  9 x versicolor Willd.
 10 lanceolatus Willd.
 11 pilosus Willd
 12 squamatus (Sprengel) Hieron.          20      Lu      Fernandes &
                                                         Queirós, 1971

 13 sibiricus L.
 14 pyrenaeus Desf. ex DC.                18      Ga      Delay, 1969
 15 amellus L.                            18      Cz      Holub et al., 1970
                                          36
                                          54      Cm      Májovský et al.,
                                                          1970

 16 willkommii Schultz Bip.
 17 alpinus L.                            18      Ga      Favarger &
                                                         Küpfer, 1968

                                          36      Cz      Favarger, 1959a
 18 bellidiastrum (L.) Scop.              18      He      Larsen, 1954
 19 tripolium L.
    (a) tripolium                         18      Su      I. & O. Hedberg, 1964
    (b) pannonicus (Jacq.) Soó            18      Cz      Májovský et al.,
                                                          1970

 20 sedifolius L.
    (a) sedifolius                        36      It      Garbari &
                                                         Tornadore, 1972

    (b) dracunculoides (Lam.) Merxm.
    (c) canus (Waldst. & Kit.) Merxm.
    (d) illyricus (Murb.) Merxm.
    (e) trinervis (Pers.) Thell.
    (f) angustissimus (Tausch) Merxm.
 21 albanicus Degen
 22 kirghisorum (Fischer ex Bieb.) Korsh.
 23 aragonensis Asso                      20      Lu      Fernandes &
                                                         Queirós, 1971

 24 creticus (Gand.) Rech. fil.
 25 linosyris (L.) Bernh.                 18      Cz      Májovský et al.,
                                                          1970
                                          36      Hu      Baksay, 1958

 26 oleifolius (Lam.) Wagenitz
 27 tarbagatensis (C. Koch) Merxm.

8 Erigeron L.
```

1 annuus (L.) Pers.	26	Po	Bijok et al., 1972
	27 36	N. Amer.	Montgomery & Yang, 1960

(a) annuus
(b) septentrionalis (Fernald & Wieg.)
    Wagenitz
(c) strigosus (Muhl. ex Willd.)
    Wagenitz

2 karvinskianus DC.	36	Lu	Fernandes & Queirós, 1971

3 acer L.
(a) acer	18	Hs	Gadella & Kliphuis, 1968

(b) angulosus (Gaudin) Vacc.
(c) droebachiensis (O.F. Mueller)
    Arcangeli
(d) macrophyllus (Herbich) Guterm.
(e) politus (Fries) H. Lindb. fil.	18	No	Knaben & Engelskjon, 1967
4 orientalis Boiss.			
5 atticus Vill.	18	He	Favarger, 1954
6 gaudinii Brügger	18	He	Favarger, 1954
7 alpinus L.	18	Po	Skalińska et al., 1966
8 epiroticus (Vierh.) Halácsy			
9 neglectus A. Kerner			
10 nanus Schur	18	Po	Skalińska et al., 1966
11 borealis (Vierh.) Simmons	18	Is	Löve & Löve, 1956
12 glabratus Hoppe & Hornsch. ex Bluff & Fingerh.	18	Au	Reese, 1953
13 major (Boiss.) Vierh.			
14 uniflorus L.	18	Is	Löve & Löve, 1956
subsp. eriocephalus (J. Vahl) Cronq.	18	Is	Löve & Löve, 1956
aragonensis Vierh.	18	Hs Ga	Küpfer & Favarger, 1967
15 humilis R.C. Graham	36	Is	Löve & Löve, 1956
16 frigidus Boiss. ex DC.	18	Hs	Küpfer & Favarger, 1967
17 silenifolius (Turcz. ex DC.) Botsch.			

9 Conyza Less.
1 canadensis (L.) Cronq.	18	Po	Frey, 1971
2 bonariensis (L.) Cronq.	36	Lu	Fernandes & Queirós, 1971
	54	Al	Strid, 1971

10 Nolletia Cass.
  1 chrysocomoides (Desf.) Cass. ex Less.

11 Baccharis L.
  1 halimifolia L.

Inuleae Cass.

12 Karelinia Less.
  1 caspia (Pallas) Less.

13 Filago L.
  1 vulgaris Lam.
2 eriocephala Guss.	28	Gr	Runemark, 1970

3 aegaea Wagenitz
  (a) aegaea   28   Gr   Runemark, 1970
  (b) aristata Wagenitz
4 cretensis Gand.
  (a) cretensis   28   Gr   Runemark, 1970
  (b) cycladum Wagenitz   28   Gr   Runemark, 1970
5 lutescens Jordan   28   Lu   Fernandes &
                                            Queirós, 1971

  (a) lutescens
  (b) atlantica Wagenitz
6 fuscescens Pomel
7 pyramidata L.   28   Bl   Dahlgren et al.,
                                            1971

8 desertorum Pomel
9 ramosissima Lange
10 congesta Guss. ex DC.
11 duriaei Cosson ex Lange
12 micropodioides Lange
13 mareotica Delile
14 filaginoides (Kar. & Kir.) Wagenitz
15 eriosphaera (Boiss. & Heldr.) Chrtek
   & J. Holub
16 hispanica (Degen & Hervier) Chrtek &
   J. Holub

14 Ifloga Cass.
1 spicata (Forskål) Schultz Bip.

15 Logfia Cass.
1 heterantha (Rafin.) J. Holub
2 arvensis (L.) J. Holub   28   Gr   Holub et al., ined.
3 minima (Sm.) Dumort.   28   Ge   Reese, 1952
4 clementei (Willk.) J. Holub
5 gallica (L.) Cosson & Germ. 28   Lu   Fernandes &
                                            Queirós, 1971

6 neglecta (Soyer-Willemet) J. Holub

16 Evax Gaertner
1 pygmaea (L.) Brot.
  (a) pygmaea
  (b) ramosissima (Mariz) R. Fernandes &
     Nogueira
2 contracta Boiss.
3 asterisciflora (Lam.) Pers.
4 carpetana Lange
5 lusitanica Samp.
6 rotundata Moris   26   Co   Contandriopoulos,
                                          1962

7 perpusilla Boiss. & Heldr.
8 nevadensis Boiss.

17 Bombycilaena (DC.) Smolj.
1 erecta (L.) Smolj.
2 discolor (Pers.) Lainz

18 Micropus L.
1 supinus L.

19 Evacidium Pomel
1 discolor (DC.) Maire

20 Omalotheca Cass.
  1 sylvatica (L.) Schultz Bip. & F.W.
    Schultz
  2 norvegica (Gunn.) Schultz Bip. &
    F.W. Schultz   56   Is   Löve & Löve, 1956
  3 hoppeana (Koch) Schultz Bip. & F.W.
    Schultz
  4 roeseri (Boiss. & Heldr.) J. Holub
  5 pichleri (Murb.) J. Holub
  6 supina (L.) DC.   28   Cz   Májovský et al.,
                                            1970

21 Gamochaeta Weddell
  1 subfalcata (Cabrera) Cabrera
  2 purpurea (L.) Cabrera   28   Lu   Fernandes &
                                            Queirós, 1971

22 Filaginella Opiz
  1 uliginosa (L.) Opiz   14   Ho   Gadella & Kliphuis,
                                          1966
    (a) uliginosa
    (b) sibirica (Kirp.) J. Holub
    (c) kasachstanica (Kirp.) J. Holub
    (d) rossica (Kirp.) J. Holub

23 Gnaphalium L.
  1 luteo-album L.   14   Lu   Fernandes &
                                            Queirós, 1971
  2 undulatum L.

24 Helichrysum Miller
  1 amorginum Boiss. & Orph.
  2 sibthorpii Rouy
  3 doerfleri Rech. fil.
  4 frigidum (Labill.) Willd.   28   Co   Contandriopoulos,
                                          1962
  5 stoechas (L.) Moench   28   Lu   Fernandes &
                                          Queirós, 1971
    (a) stoechas
    (b) barrelieri (Ten.) Nyman
  6 rupestre (Rafin.) DC.
  7 heldreichii Boiss.
  8 ambiguum (Pers.) C. Presl   28   Bl   Guinochet & Le-
                                          franc, 1972
  9 saxatile Moris
    (a) saxatile
    (b) errerae (Tineo) Nyman
  10 italicum (Roth) G. Don fil.   28   It   Mori, 1957
    (a) italicum
    (b) microphyllum (Willd.) Nyman
    (c) serotinum (Boiss.) P. Fourn. 28   Lu   Rodrigues, 1953
  11 orientale (L.) Gaertner
  12 plicatum DC.
  13 arenarium (L.) Moench   14
                              28   Po   Skalińska et al.,
                                          1968
    (a) arenarium
    (b) ponticum (Velen.) Clapham
  14 graveolens (Bieb.) Sweet
  15 foetidum (L.) Cass.   14   Lu   Fernandes &
                                            Queirós, 1971

16 bracteatum (Vent.) Andrews

25 Lasiopogon Cass.
  1 muscoides (Desf.) DC.

26 Antennaria Gaertner

1 dioica (L.) Gaertner	28	Ga He It	Urbanska-Worytkiewicz, 1968
2 nordhageniana Rune & Rønning	28	No	Rune & Rønning, 1956
3 alpina (L.) Gaertner	70	No	Urbanska-Worytkiewicz, 1967
	84	Is	Löve & Löve, 1956
	85	No	Urbanska-Worytkiewicz, 1967
boecherana A.E. Porsild	56	Is	Löve, ined.
4 porsildii Elis. Ekman	63	No	Urbanska-Worytkiewicz, 1967
	70	No	Urbanska-Worytkiewicz, 1967
5 carpatica (Wahlenb.) Bluff & Fingerh.	56	Au He Po	Urbanska-Worytkiewicz, 1967
6 villifera Boriss.	28	No	Urbanska-Worytkiewicz, 1970
	42	No	Urbanska-Worytkiewicz, 1970

27 Leontopodium (Pers.) R. Br.

1 alpinum Cass.	52	He It	Urbanska-Worytkiewicz, 1968b

  (a) alpinum
  (b) nivale (Ten.) Tutin

28 Anaphalis DC.

1 margaritacea (L.) Bentham	28	Br	Maude, 1940

29 Phagnalon Cass.
  1 sordidum (L.) Reichenb.

2 rupestre (L.) DC.	18	Hs	Gadella et al., 1966

  3 metlesicsii Pignatti
  4 graecum Boiss. & Heldr.

5 saxatile (L.) Cass.	18	Lu	Fernandes & Queirós, 1971

30 Leysera L.
  1 leyseroides (Desf.) Maire

31 Inula L.

1 helenium L.	20	Br	Rutland, 1941
2 helvetica Weber			
3 germanica L.	16	Hu	Baksay, 1958
4 salicina L.			
(a) salicina	16	Hu	Baksay, 1958
(b) aspera (Poiret) Hayek			
5 spiraeifolia L.	16	Ju	Nilsson & Lassen, 1971
6 hirta L.	16	Cz	Májovský et al., 1970
7 ensifolia L.	16	Cz	Májovský et al., 1970
8 britannica L.	32	Po	Skalińska, 1959

9 caspica Blume
10 oculus-christi L.   32   Hu   Baksay, 1958
11 helenioides DC.
12 montana L.   16   Ga   Guinochet &
                                          Logeois, 1962
13 candida (L.) Cass.
   (a) candida
   (b) decalvans (Halácsy) P.W. Ball ex
       Tutin
   (c) limonella (Heldr.) Rech. fil.
14 verbascifolia (Willd.) Hausskn.
   (a) verbascifolia   16   Ju   Larsen, 1956a
   (b) aschersoniana (Janka) Tutin
   (c) parnassica (Boiss. & Heldr.) Tutin
   (d) methanea (Hausskn.) Tutin
   (e) heterolepis (Boiss.) Tutin
15 subfloccosa Rech. fil.
16 conyza DC.   32   Cz   Májovský et al.,
                                          1970
17 thapsoides Sprengel
18 bifrons (L.) L.
19 crithmoides L.   18   Ga   Delay, 1968

32 Dittrichia W. Greuter
  1 viscosa (L.) W. Greuter   18   Lu   Fernandes &
                                          Queirós, 1971
                        34   Ga   Gadella et al.,
   (a) viscosa   1966
   (b) revoluta (Hoffmanns. & Link) P.
       Silva & Tutin
  2 graveolens (L.) W. Greuter   18   Lu   Fernandes &
                                          Queirós, 1971
              20 + 0 - 2B   Bl   Nilsson & Lassen,
                                           1971
33 Pulicaria Gaertner
  1 odora (L.) Reichenb.   18   Lu   Fernandes &
                                          Queirós, 1971
             18 + 0 - 6B   Lu   Queirós, 1973
  2 dysenterica (L.) Bernh.   18   Al   Strid, 1971
                          20   Cz   Májovský et al.,
                                          1970
  3 vulgaris Gaertner   18   Ge   Wulff, 1937b
  4 paludosa Link   18   Bu   Kuzmanov & Kozuharov,
                                          1970
                15 + 1B   Lu   Fernandes &
                                          Queirós, 1971
  5 sicula (L.) Moris

34 Carpesium L.
  1 cernuum L.
  2 abrotanoides L.   40   Hu   Baksay, 1956

35 Jasonia Cass.
  1 glutinosa (L.) DC.
  2 tuberosa (L.) DC.   18   Lu   Fernandes &
                                          Queirós, 1971

36 Buphthalmum L.
  1 salicifolium L.   20   Cz   Májovský et al.,
                                          1970
  2 inuloides Moris

37 Telekia Baumg.
  1 speciosa (Schreber) Baumg.   20   Cz   Májovský et al.,
                                                1970

  2 speciosissima (L.) Less.   20   It   Favarger, 1959a

38 Pallenis (Cass.) Cass.
  1 spinosa (L.) Cass.
    (a) spinosa   10   Lu   Fernandes &
                                              Queirós, 1971

    (b) microcephala (Halácsy) Rech. fil.

39 Asteriscus Miller
  1 aquaticus (L.) Less.   14   Bl   Dahlgren et al.,
                                              1971

  2 maritimus (L.) Less.   12   Lu   Fernandes &
                                              Queirós, 1971

 Heliantheae Cass.

40 Guizotia Cass.
  1 abyssinica (L. fil.) Cass.

41 Bidens L.
  1 tripartita L.   48   Hu   Pólya, 1950
  2 connata Muhl. ex Willd.
  3 radiata Thuill.   48   Da   Hagerup, 1944
  4 aurea (Aiton) Sherff   72   Lu   Fernandes &
                                              Queirós, 1971
  5 cernua L.   24   Po   Skalińska et al.,
                                              1959
  6 frondosa L.   48   Lu   Fernandes &
                                              Queirós, 1971
  7 pilosa L.   72   Lu   Fernandes &
                                              Queirós, 1971

  8 bipinnata L.
  9 subalternans DC.

42 Sigesbeckia L.
  1 orientalis L.   60   Rm   Solbrig et al.,
                                              1972

  2 jorullensis Kunth

43 Eclipta L.
  1 prostrata (L.) L.   22   Lu   Fernandes &
                                              Queirós, 1971

44 Rudbeckia L.
  1 hirta L.
  2 laciniata L.   76   Cz   Holub et al.,
                                              1972

45 Helianthus L.
  1 annuus L.
  2 tuberosus L.
  3 x laetiflorus Pers.

46 Verbesina L.
  1 encelioides (Cav.) Bentham & Hooker
    (a) encelioides
    (b) exauriculata (Robinson & Greenman)
        J.R. Coleman

47 Silphium L.
  1 perfoliatum L.

48 Iva L.
  1 xanthifolia Nutt.   36   Cz   Feráková, 1968

49 Ambrosia L.
  1 maritima L.
  2 artemisiifolia L.   36   non-Europe
  3 tenuifolia Sprengel
  4 coronopifolia Torrey & A. Gray   72   non-Europe
  5 trifida L.   24   non-Europe

50 Xanthium L.
  1 strumarium L.
    (a) strumarium   36   Cz   Májovský et al.,
                                                             1970
    (b) italicum (Moretti) D. Love   36   It   Fagidi & Fabbri,
                                                             1971
  2 spinosum L.   36   Cz   Májovský et al.,
                                                             1970

51 Heliopsis Pers.
  1 helianthoides (L.) Sweet
    (a) scabra (Dunal) Fisher
    (b) occidentalis Fisher

52 Galinsoga Ruiz & Pavón
  1 parviflora Cav.   16   Lu   Fernandes & Queirós,
                                                             1971
  2 ciliata (Rafin.) S.F. Blake   32   Po   Skalińska et al.,
                                                             1964
                                           36   Ge   Reese, 1952

53 Schkuhria Roth
  1 pinnata (Lam.) O. Kuntze

54 Gaillardia Foug.
  1 aristata Pursh

55 Tagetes L.
  1 minuta L.

Anthemideae Cass.

56 Santolina L.
  1 oblongifolia Boiss.
  2 elegans Boiss. ex DC.
  3 viscosa Lag.
  4 rosmarinifolia L.   18   Lu   Fernandes & Queirós,
                                                             1971
                                           36   Lu   Fernandes & Queirós,
                                                             1971
    (a) rosmarinifolia
    (b) canescens (Lag.) Nyman
  5 chamaecyparissus L.
    (a) chamaecyparissus
    (b) squarrosa (DC.) Nyman
    (c) tomentosa (Pers.) Arcangeli
    (d) insularis (Genn. ex Fiori) Yeo

57 Anthemis L.
  1 trotzkiana Claus ex Bunge
  2 carpatica Willd.
    (a) carpatica   36   Hs   Küpfer &
                                                     Favarger, 1967
                                         54   Hs   Küpfer &
                                                     Favarger, 1967
    (b) pyrethriformis (Schur) Beldie
    (c) petraea (Ten.) R. Fernandes
  3 sibthorpii Griseb.
  4 aetnensis Schouw   36   Si   Larsen &
                                                     Laegaard, 1971
  5 punctata Vahl
  6 cretica L.
    (a) cretica
    (b) calabrica (Arcangeli) R. Fernandes
    (c) saxatilis (DC.) R. Fernandes
    (d) alpina (L.) R. Fernandes
  7 sterilis Steven
  8 abrotanifolia (Willd.) Guss.
  9 panachaica Halácsy
 10 argyrophylla (Halácsy & Georgiev)
    Velen.
 11 gerardiana Jordan
 12 alpestris (Hoffmanns. & Link) R.
    Fernandes
 13 maritima L.   36   Lu   Fernandes & Queirós,
                                                     1971
 14 virescens Velen.
 15 orbelica Pančić
 16 hydruntina Grover
 17 tuberculata Boiss.
    (a) tuberculata
    (b) turolensis (Pau ex A. Caballero)
        R. Fernandes & Borja
 18 ismelia Lojac.
 19 tenuiloba (DC.) R. Fernandes
    (a) tenuiloba
    (b) cronia (Boiss. & Heldr.) R. Fernandes
 20 anatolica Boiss.
 21 spruneri Boiss. & Heldr.
 22 orientalis (L.) Degen
 23 pindicola Heldr. ex Halácsy
 24 meteorica Hausskn.
 25 rumelica (Velen.) Stoj. & Acht.
 26 stribrnyi Velen.
 27 arvensis L.
    (a) arvensis   18   Lu   Rodriguez, 1953
    (b) incrassata (Loisel.) Nyman
    (c) cyllenea (Halácsy) R. Fernandes
    (d) sphacelata (C. Presl) R. Fernandes
 28 ruthenica Bieb.   18   Cz   Májovský et al.,
                                                     1970
 29 auriculata Boiss.
 30 werneri Stoj. & Acht.
 31 flexicaulis Rech. fil.
 32 scopulorum Rech. fil
 33 tomentosa L.
    (a) tomentosa   18   It   Carpineri, 1971
    (b) heracleotica (Boiss. & Heldr.)
        R. Fernandes

34 rigida (Sibth. & Sm.) Boiss. & Heldr.
35 chrysantha Gay
36 secundiramea Biv.
  (a) secundiramea
  (b) intermedia (Guss.) R. Fernandes

37 muricata (DC.) Guss.	18	Si	Carpineri, 1971
38 macedonica Boiss. & Orph.			
39 cotula L.	18	Cz	Májovský et al., 1970

40 lithuanica (DC.) Besser ex Trautv.
41 chia L.
42 tinctoria L.

(a) tinctoria	18	Cz	Májovský et al., 1970

  (b) subtinctoria (Dobrocz.) Soó
  (c) australis R. Fernandes
  (d) fussii (Griseb.) Beldie
43 gaudium-solis Velen.
44 sancti-johannis Turrill
45 cretacea Zefirov

46 monantha Willd.	18	Rs(K)	Chouksanova et al., 1968
47 parnassica (Boiss. & Heldr.) R. Fernandes			

48 dubia Steven
49 triumfetti (L.) DC.
50 dumetorum D. Sossn.
51 macrantha Heuffel

52 jailensis Zefirov	18	Rs(K)	Chouksanova et al., 1968
53 altissima L.	18	Ga	Delay, 1971
54 coelopoda Boiss.			
55 syriaca Bornm.			
56 segetalis Ten.	18	It	Carpineri, 1971
57 austriaca Jacq.	18	Bu	Kuzmanov & Kozuharov, 1970

58 brachmannii Boiss. & Heldr.
59 filicaulis (Boiss. & Heldr.) W. Greuter
60 tomentella W. Greuter
61 ammanthus W. Greuter
  (a) ammanthus
  (b) paleacea W. Greuter
62 glaberrima (Rech. fil.) W. Greuter

58 Achillea L.

1 ageratifolia (Sibth. & Sm.) Boiss.	18	Gr	Contandriopoulos & Martin, 1967
2 servica Nyman	18	Ju	Lovka et al., 1971
3 oxyloba (DC.) Schulz Bip.	18	It	Favarger & Huynh, 1964
(a) oxyloba			

  (b) mucronulata (Bertol.) I.B.K. Richardson
  (c) schurii (Schultz Bip.) Heimerl
4 barrelieri (Ten.) Schultz Bip.

5 erba-rotta All.	18	He	Favarger & Huynh, 1964

  (a) olympica (Heimerl) I.B.K. Richardson
  (b) moschata (Wulfen) I.B.K. Richardson
  (c) ambigua (Heimerl) I.B.K. Richardson

(d) erba-rotta	18	Ga	Favarger & Huynh, 1964

  (e) calcarea (Porta) I.B.K. Richardson
  (f) rupestris (Porta) I.B.K. Richardson

241

6 nana L.	18	He	Favarger, 1953
7 umbellata Sibth. & Sm.	18	Gr	Contandriopoulos & Martin, 1967
	18 + 1 - 3B	Gr	Contandriopoulos & Martin, 1967
9 atrata L.	18	Au	Reese, 1952
10 clusiana Tausch	18	Au	Favarger & Huynh, 1964
11 clavennae L.	18	Gr	Contandriopoulos & Martin, 1967
12 ambrosiaca (Boiss. & Heldr.) Boiss.	18	Gr	Contandriopoulos & Martin, 1967
13 abrotanoides (Vis.) Vis.	18	Gr	Contandriopoulos & Martin, 1967
14 fraasii Schultz Bip.	18	Gr	Contandriopoulos & Martin, 1967
15 macrophylla L.			
16 lingulata Waldst. & Kit.			
17 ptarmica L.	18	Ho	Gadella & Kliphuis, 1968
18 cartilaginea Ledeb. ex Reichenb.	18	Po	Dabrowska, 1971
19 pyrenaica Sibth. ex Godron	18	Hs	Küpfer, 1968
20 impatiens L.			
21 grandifolia Friv.	18	Gr	Contandriopoulos & Martin, 1967
22 distans			
(a) distans	54	Cz	Májovský et al., 1970
(b) tanacetifolia Janchen			
23 stricta (Koch) Schleicher ex Gremli	54	Europe	Ehrendorfer, 1952
24 millefolium L.			
(a) millefolium	54	Ho	Gadella & Kliphuis, 1966
(b) sudetica (Opiz) Weiss	54	Cz	Májovský et al., 1970
monticola	72	Hs	Ehrendorfer, 1959
25 pannonica Scheele	72	Cz	Májovský et al., 1970
26 setacea Waldst. & Kit.	18	Cz	Májovský et al., 1970
27 asplenifolia Vent.	18	Hu	Pólya, 1949
28 roseo-alba Ehrend.	18	It	Ehrendorfer, 1959
	36	Au	Ehrendorfer, 1959
29 collina J. Becker ex Reichenb.	36	Hu	Pólya, 1948
30 crithmifolia Waldst. & Kit.	18	Hu	Baksay, 1958
31 odorata L.	18	Au	Ehrendorfer, 1959
32 nobilis L.			
(a) nobilis	18	Po	Skalińska et al., 1970
(b) neilreichii (A. Kerner) Velen.	45	Cz	Májovský et al., 1970
33 virescens (Fenzl) Heimerl	36		
34 ligustica All.	18	It	Caldo, 1971
	c.54	Si	Larsen & Laeggaard, 1971
35 chamaemelifolia Pourret			
36 tomentosa L.			
37 chrysocoma Friv.			
38 holosericea Sibth. & Sm.	18	Gr	Contandriopoulos & Martin, 1967
39 absinthoides Halácsy			

40 leptophylla Bieb.			
41 glaberrima Klokov			
42 ageratum L.	18	Lu	Fernandes & Queirós, 1971
43 ochroleuca Ehrh.	18	Hu	Baksay, 1958
44 depressa Janka			
45 aegyptiaca L.	18	Gr	Contandriopoulos & Martin, 1967
46 clypeolata Sibth. & Sm.	18	Bu	Kuzmanov & Kozuharov, 1970
	18 + 1B	Gr	Contandriopoulos & Martin, 1967
47 thracica Velen.			
48 coarctata Poiret	18	Rm	Lungeanu, 1973
	36	Bu	Kuzmanov & Kozuharov, 1970
49 micrantha Willd.			
50 biebersteinii C. Afan.			
51 santolinoides Lag.			
52 cretica L.	18	Cr	Contandriopoulos & Martin, 1967

59 Chamaemelum Miller
1 nobile (L.) All.
2 mixtum (L.) All.
3 fuscatum (Brot.) Vasc.

60 Matricaria L.
1 trichophylla (Boiss.) Boiss.
2 conoclinia (Boiss. & Balansa) Nyman
3 caucasica (Willd.) Poiret
4 maritima L.

(a) maritima	18	Da	Larsen, 1965
	18 + 1 - 4B	Br	Kay, 1969
(b) subpolaris (Pobed.) Rauschert			
(c) phaeocephala (Rupr.) Rauschert	18	Fe	Hamet-Ahti
5 perforata Merat	18	Ge	Rottgardt, 1956
	18 + 1B	Br	Kay, 1969
	36	Ge	Fritsch, 1973

6 rosella (Boiss. & Orph.) Nyman
7 tempskyana (Freyn & Sint.) Rauschert
8 parviflora (Willd.) Poiret

61 Chamomilla S.F. Gray

1 recutita (L.) Rauschert	18	Lu	Fernandes & Queirós, 1971
2 tzvelevii (Pobed.) Rauschert			
3 suaveolens (Pursh) Rybd.	18	Lu	Fernandes & Queirós, 1971

4 aurea (Loefl.) Gay ex Cosson & Kralik

62 Cladanthus Cass.
1 arabicus (L.) Cass

63 Anacyclus L.

1 clavatus (Desf.) Pers.	18	Ga	Van Loon et al., 1971
2 radiatus Loisel.	18	Lu	Fernandes & Queirós, 1971

3 valentinus L.
4 pyrethrum (L.) Link

243

64 Lonas Adanson
  1 annua (L.) Vines & Druce

65 Otanthus Hoffmanns. & Link
  1 maritimus (L.) Hoffmanns. & Link   18   Lu   Fernandes & Queirós,
                                                    1971
66 Chrysanthemum L.
  1 segetum L.   18   Ho   Gadella & Kliphuis,
                                                    1968
  2 coronarium L.   18   Lu   Fernandes & Queirós,
                                                    1971

67 Heteranthemis Schott
  1 viscidehirta Schott

68 Dendranthema (DC.) Desmoulins
  1 zawadskii (Herbich) Tzvelev   54   Po   Skalińska et al., 1959
  2 arcticum (L.) Tzvelev

69 Tanacetum L.
  1 vulgare L.   18   Cz   Májovský et al., 1970
  2 microphyllum DC.
  3 annuum L.   18   Lu   Queirós, 1975
  4 achilleifolium (Bieb.) Schultz Bip.
  5 paczoskii (Zefirov) Tzvelev
  6 millefolium (L.) Tzvelev
  7 bipinnatum (L.) Schultz Bip.   54   Rs(N) Sokolovskaya &
                                                    Strelkova, 1960
  8 santolina Winkler
  9 corymbosum (L.) Schultz Bip.
   (a) corymbosum   36   Lu   Fernandes & Queirós,
                                                    1971
   (b) clusii (Fischer ex Reichenb.)
       Heywood   18   Cz   Májovský et al.,
                                                    1970
 10 mucronulatum (Hoffmanns. & Link)
    Heywood
 11 parthenium (L.) Schultz Bip.   18   Ju   Lovka et al., 1972
 12 parthenifolium (Willd.) Schultz Bip.
 13 macrophyllum (Waldst. & Kit.) Schultz
    Bip.
 14 cinerariifolium (Trev.) Schultz Bip.

70 Leucanthemella Tzvelev
  1 serotina (L.) Tzvelev   18   Hu   Baksay, 1958

71 Balsamita Miller
  1 major Desf.

72 Phalacrocarpum Willk.
  1 oppositifolium (Brot.) Willk.   18   Lu   Fernandes & Queirós,
                                                    1971
  2 hoffmannseggii (Samp.) Lainz   18   Lu   Fernandes & Queirós,
                                                    1971

73 Otospermum Willk.
  1 glabrum (Lag.) Willk.   18   Lu   Fernandes & Queirós,
                                                    1971

74 Leucanthemopsis (Giroux) Heywood
  1 alpina (L.) Heywood

244

(a) alpina	18	Cz	Májovský et al., 1970
	36	Ga	Tombal, 1969
	54	Ga Hs	Küpfer & Favarger, 1967
(b) tomentosa (Loisel.) Heywood	18	Co	Contandriopoulos, 1962
(c) cuneata (Pau) Heywood	54	Hs	Küpfer, 1971
2 pallida (Miller) Heywood	36	Hs	Küpfer, 1971
(a) pallida			
(b) virescens (Pau) Heywood			
(c) spathulifolia (Gay) Heywood			
3 flaveola (Hoffmanns. & Link	36	Lu	Küpfer, 1971
4 radicans (Cav.) Heywood	18	Hs	Favarger & Küpfer, 1967
5 pulverulenta (Lag.) Heywood			
(a) pulverulenta			
(b) pseudopulverulenta (Heywood) Heywood			

75 Prolongoa Boiss.
1 pectinata (L.) Boiss.

76 Lepidophorum Cass.

1 repandum (L.) DC.	18	Lu	Fernandes & Queirós, 1971

77 Daveaua Willk. ex Mariz
1 anthemoides Mariz

78 Glossopappus G. Kunze
1 macrotus (Durieu) Briq.

79 Hymenostemma (G. Kunze) Willk.
1 pseudanthemis (G. Kunze) Willk.

80 Coleostephus Cass.

1 myconis (L.) Reichenb. fil.	18	Lu	Fernandes & Queirós, 1971
2 clausonis Pomel			

81 Leucanthemum Miller

1 vulgare Lam.	36 54	Po	Skalińska et al., 1961
praecox (Horvatić) Horvatić	18	Po	Frey, 1969b
leucolepis (Briq. & Cavillier) Horvatic	18	Hs	Favarger, 1975
	36	Ju	Mirković, 1966
	54	Bu Ju	Polatschek, 1966
adustum (Koch) Gremli	54	Po	Przywara, 1970
heterophyllum (Willd.) DC.	45	Ju	Papeš, 1975
	54	Au	Polatschek, 1966
	72	It	Favarger & Villard, 1965
cuneifolium Le Grand ex Coste	72	Ga	Villard, 1970
maximum (Ramond) DC.	90	It	Favarger & Villard, 1965
	108	Ga	Favarger & Villard, 1965
pallens (Gay) DC.	54	Ga	Favarger & Villard, 1965
subglaucum De Laramb.	54	Ge	Favarger, 1975b

245

	90	Ga It	Favarger & Villard, 1965
	72 + 1B	Ga	Favarger, 1975
sylvaticum (Hoffmanns. & Link) Nyman	36	Lu	Fernandes & Queirós, 1971
	54	Lu	Fernandes & Queirós, 1971
2 graminifolium (L.) Lam.	18	Ga	Favarger & Villard, 1965
3 gracilicaule (Dufour) Alavi & Heywood			
4 burnatii Briq. & Cavillier	18	Ga	Favarger & Villard, 1965
5 monspeliense (L.) Coste	36	Ga	Favarger & Villard, 1965
6 atratum (Jacq.) DC.			
(a) atratum	54	It	Favarger & Villard, 1965
(b) halleri (Suter) Heywood	18	Au	Polatschek, 1966
(c) coronopifolium (Vill.) Horvatić	54	It	Marchi, 1971
(d) ceratophylloides (All.) Horvatić	54	He	Favarger & Villard, 1965
(e) tridactylites (A. Kerner & Huter ex Rigo) Heywood			
(f) platylepis (Borbás) Heywood	27	Ju	Papeš, 1972
	36	Ju	Polatschek, 1966
	45	Ju	Papes, 1975
	54	Ju	Favarger & Villard, 1965
	108	Ju	Mirković, 1966
(g) lithopolitanicum (F. Mayer) Horvatić	18	Ju	Lovka et al., 1972
	72	Ju	Mirković, 1966
7 chloroticum A. Kerner & Murb.			
8 corsicum (Less.) DC.	36	Co	Contandriopoulos, 1964a
9 waldsteinii (Schultz Bip.) Pouzar	18	Cz	Villard, 1970
10 discoideum (All.) Coste	18	Ga	Forissier, 1975
11 arundanum (Boiss.) Cuatrec.			
12 paludosum (Poiret) Bonnet & Barratte			
82 Plagius L'Hér. ex DC.			
1 flosculus (L.) Alavi & Heywood	18	Co	Contandriopoulos, 1962
83 Cotula L.			
1 coronopifolia L.	20	Lu	Fernandes & Queirós, 1971
2 australis (Sieber ex Sprengel) Hooker fil.			
84 Chlamydophora Ehrenb. ex Less.			
1 tridentata (Delile) Ehrenb. ex Less.			
85 Nananthea DC.			
1 perpusilla (Loisel.) DC.	18	Co	Contandriopoulos, 1962
86 Soliva Ruiz & Pavón			
1 pterosperma (Juss.) Less.	c.110	Lu	Fernandes & Queirós, 1971

87 Gymnostyles Juss.
  1 stolonifera (Brot.) Tutin

88 Artemisia L.

1 vulgaris L.	16	Ho	Gadella & Kliphuis, 1963
2 verlotiorum Lamotte	54	It	Vignoli, 1945
3 tilesii Ledeb.			
4 absinthium L.	18	Hu	Pólya, 1950
5 siversiana Ehrh. ex Willd.			
6 arborescens L.	18	It	Martinoli, 1943
7 stellerana Besser			
8 alba Turra	36	It	Cesca, 1972
9 maritima L.			
(a) maritima	36	Ge	Löve & Löve, in litt. 1972
	54	Ho	Gadella & Kliphuis, 1968
	(50-56)	Br Da Ge Su	Persson, 1974
(b) humifusa (Fries ex Hartman) K. Persson			
10 caerulescens L.			
(a) caerulescens	18	It	Fagioli & Fabbri, 1971
(b) gallica (Willd.) K. Persson	18	It	Löve & Löve, unpub.
11 vallesiaca All.	36	It	Löve & Löve, unpub.
12 lerchiana Weber			
13 nitrosa Weber			
14 taurica Willd.			
15 santonicum L.			
(a) santonicum			
(b) patens (Neilr.) K. Persson	18	Hu	Pólya, 1945
16 pauciflora Weber			
17 gracilescens Krasch. & Iljin			
18 lessingiana Besser			
19 laciniata Willd.	18	Au	Ehrendorfer, 1964
20 armeniaca Lam.			
21 latifolia Ledeb.			
22 pancicii (Janka) Ronniger	54	Au Ju	Ehrendorfer, 1964
23 insipida Vill.			
24 oelandica (Besser) Komarov.	54	Su	Ehrendorfer, 1964
25 atrata Lam.	18	Ga	Ehrendorfer, 1964
26 norvegica Fries	18	Ho	Ehrendorfer, 1964
27 abrotanum L.	18	Po	Skalińska et al., 1959
28 molinieri Quezel, Barbero & R. Loisel			
29 santolinifolia Turcz. ex Krasch.			
30 pontica L.	18	Hu	Pólya, 1948
31 austriaca Jacq.	16	Po	Skalińska et al., 1959
32 chamaemelifolia Vill			
(a) chamaemelifolia	18	It	Löve & Löve, in litt. 1972
(b) cantabrica Lainz			
33 macrantha Ledeb.			
34 umbelliformis Lam.	34	He	Favarger & Huynh, 1964
gabriellae Br.-Bl.	(not 34) 36	Ga	Favarger & Küpfer, 1968

35 nitida Bertol.	54	It	Löve & Löve, in litt. 1972
36 eriantha Ten.	18	Po	Skalińska et al., 1959
37 genipi Weber	18	He	Favarger, 1953
38 glacialis L.	16	It	Favarger & Huynh, 1964
39 granatensis Boiss.			
40 pedemontana Balbis	16	It	Marchetti, 1962
41 frigida Willd.			
42 sericea Weber			
43 reptans C. Sm. ex Link			
44 barrelieri Besser			
45 herba-alba Asso			
46 hololeuca Bieb. ex Besser			
47 rupestris L.	18	Su	Ehrendorfer, 1964
48 annua L.	18	Hu	Pólya, 1948
49 dracunculus L.	18	Su	Rousi, 1969
	36	Ga	Rousi, 1969
	90	Da Fe	Rousi, 1969
50 glauca Pallas ex Willd.			
51 trautvetterana Besser			
52 salsoloides Willd.			
53 tschernieviana Besser			
54 commutata Besser			
55 bargusinensis Sprengel			
56 campestris L.			
(a) campestris	36	Cz	Májovský et al., 1970
(b) glutinosa (Gay ex Besser) Rouy			
(c) maritima Arcangeli	54	Lu	Fernandes & Queirós, 1971
(d) alpina (DC.) Arcangeli	36		
(e) bottnica A.N. Lundstrom ex Kinb.			
(f) borealis (Pallas) H.M. Hall & Clements	18	He	Damboldt, 1964
	36	Non-Europe	
57 scoparia Waldst. & Kit.			

Senecioneae Cass.

89 Tussilago L.			
1 farfara L.	60	Ho	Gadella & Kliphuis, 1966
90 Petasites Miller			
1 albus (L.) Gaertner	60	Ge	Scheerer, 1939
2 hybridus (L.) P. Gaertner, B. Meyer & Scherb.	60	Ho	Gadella & Kliphuis, 1966
(a) hybridus			
(b) ochroleucus (Boiss. & Huet) Sourek			
3 paradoxus (Retz.) Baumg.	60	Au	Sørensen & Christiansen, 1964
4 kablikianus Tausch ex Berchtold	60	Cz	Löve & Löve, 1961
5 radiatus (J.F. Gmelin) J. Toman			
6 spurius (Retz) Reichenb.	60	Da	Sørensen & Christiansen, 1964
7 frigidus (L.) Fries	60	Sb	Flovik, 1940
8 fragrans (Vill.) C. Presl	58 59 60 61	Br It	Sørensen & Christiansen, 1964

```
    9 japonicus (Siebold & Zucc.) Maxim.    84-87    Da    Sørensen &
                                                            Christiansen, 1964
   10 doerfleri Hayek
   11 sibiricus (J.F. Gmelin) Dingwall

91 Homogyne Cass.
    1 alpina (L.) Cass.                       120     Au    Favarger, 1971
                                              160     Cz    Májovský et al.,
                                                            1970
    2 discolor (Jacq.) Cass.                   60     Au    Favarger, 1971
    3 sylvestris Cass.                         58     Au    Favarger, 1969

92 Adenostyles Cass.
    1 alliariae (Gouan) A. Kerner
      (a) alliariae
      (b) hybrida (Vill.) DC.
    2 alpina (L.) Bluff & Fingerh.             38     He    Contandriopoulos,
                                                            1964a
      (a) alpina
      (b) briquetii (Gamisans) Tutin
    3 leucophylla (Willd.) Reichenb.           38     Ga    Delay, 1970

93 Arnica L.
    1 angustifolia Vahl
      (a) alpina (L.) I.K. Ferguson            76     Su    Löve, 1954a
      (b) iljinii (Maguire) I.K. Ferguson      56     Sb    Flovik, 1940
    2 montana L.                               38     Lu    Fernandes & Queirós,
                                                            1971
      (a) montana
      (b) atlantica A. Bolos

94 Doronicum L.
    1 corsicum (Loisel.) Poiret                60     Co    Contandriopoulos,
                                                            1962
    2 austriacum Jacq.                         60     Hu    Baksay, 1956
    3 columnae Ten.
    4 orientale Hoffm.                         60     Hu    Baksay, 1956
    5 carpetanum Boiss. & Reuter ex Willk.     60     Hs    Küpfer, 1969
                                              120     Lu    Fernandes & Queirós,
                                                            1971
    6 plantagineum L.                         120     Lu    Fernandes & Queirós,
                                                            1971
    7 hungaricum Reichenb. fil.                60     Hu    Baksay, 1956
    8 pardalianches L.                         60     Ga    Delay, 1970
    9 cataractarum Widder
   10 grandiflorum Lam.                        60     Ga    Küpfer &
                                                            Favarger, 1968
   11 glaciale (Wulfen) Nyman
      (a) glaciale
      (b) calcarium (Vierh.) Hegi             60     Au    Polatschek, 1966
   12 clusii (All.) Tausch                    60     Ju    Lovka et al.,
                                                            1972
                                             120     Po    Skalińska, 1950

95 Erechtites Rafin.
    1 hieracifolia (L.) Rafin. ex DC.
```

96 Senecio L.

1 mikanioides Otto ex Walpers	20	Lu	Fernandes & Queirós, 1971

2 gnaphalodes Sieber
3 ambiguus (Biv.) DC.
  (a) ambiguus
  (b) gibbosus (Guss.) Chater
4 bicolor (Willd.) Tod.
  (a) bicolor
  (b) nebrodensis (Guss.) Chater

(c) cineraria (DC.) Chater	40	Lu	Fernandes & Queirós, 1971

5 thapsoides DC.
  (a) thapsoides
  (b) visianianus (Papaf. ex Vis.) Vandas
6 incanus L.

(a) incanus	40	He	Favarger, in litt. 1970
(b) carniolicus (Willd.) Br.-Bl.	c.120	It	Favarger, in litt. 1974
	c.140	He	Favarger, in litt. 1974
7 persoonii De Not.	40	It	Küpfer, 1971
8 leucophyllus DC.	40	Ga	Küpfer & Favarger, 1967

9 halleri Dandy

10 boissieri DC.	40	Hs	Küpfer, 1968
11 linifolius L.	40	Bl	Dahlgren et al., 1971

12 inaequidens DC.
13 nevadensis Boiss. & Reuter
14 quinqueradiatus Boiss. ex DC.
15 othonnae Bieb.

16 fluviatilis Wallr.	40	Cz	Májovský et al., 1970

17 nemorensis L.

(a) nemorensis	40	Cz	Májovský et al., 1970
(b) fuchsii (C.C. Gmelin) Čelak.	40	Cz	Májovský et al., 1970

18 cacaliaster Lam.
19 doria L.
  (a) doria

(b) umbrosus (Waldst. & Kit.) Soó	40	Cz	Májovský et al., 1970

  (c) kirghisicus (DC.) Chater
  (d) legionensis (Lange) Chater
20 lopezii Boiss.

21 paludosus L.	40	Ga	Zickler, 1968

22 eubaeus Boiss. & Heldr.
23 macedonicus Griseb.
24 auricula Bourgeau ex Cosson

25 pyrenaicus L.	40	Hs	Küpfer & Favarger, 1967

26 doronicum (L.) L.

(a) doronicum	40	Lu	Fernandes & Queirós, 1971
	80	Ga	Guinochet, 1967
(b) ruthenensis (Mazuc & Timb.-Lagr.)			
(c) gerardii (Godron & Gren.) Nyman	40	Ga	Guinochet, 1967

27 scopolii Hoppe & Hornsch. ex Bluff & Fingerh.

250

28 lagascanus DC.
29 eriopus Willk.
30 smithii DC.
31 malvifolius (L'Hér.) DC.
32 petasitis (Sims) DC.
33 integrifolius (L.) Clairv.
  (a) integrifolius   48 Br   Rutland, 1941
              c.90 Rs(N) Sokolovskaya &
                          Strelkova, 1960
  (b) maritimus (Syme) Chater
  (c) serpentini (Gayer) Jáv.
  (d) capitatus (Wahlenb.) Cuf. 64 Cz   Májovský,et al.,
                          1970
              96 Ga   Zickler, 1968
         96 + 1 - 4B Ga   Favarger, 1965
  (e) aurantiacus (Hoppe ex Willd.)
   Briq. & Cavillier
  (f) czernjaevii (Minder) Chater
  (g) tundricola (Tolm.) Chater
  (h) atropurpureus (Ledeb.) Cuf.
34 balbisianus DC.
35 elodes Boiss. ex DC.
36 helenitis (L.) Schinz & Thell.
  (a) helenitis   48 Ga   Delay, 1971
  (b) candidus (Corb.) Brunerye
  (c) macrochaetus (Willk.) Brunerye
  (d) salisburgensis Cuf.
37 rivularis (Waldst. & Kit.) DC.
  (a) rivularis
  (b) pseudocrispus (Fiori) C. Mayer
38 ovirensis (Koch) DC.
  (a) ovirensis
  (b) gaudinii (Gremli) Cuf.
39 papposus (Reichenb.) Less. 40 Po   Frey, 1969b
  (a) papposus
  (b) wagneri (Degen) Cuf.
  (c) kitaibelii (Jáv.) Cuf.
40 congestus (R. Br.) DC.
41 cordatus Koch   40 It   Zickler, 1968
42 subalpinus Koch   40 Po   Skalińska et al.,
                          1966
43 pancicii Degen   100 Bu   Kozuharov &
                          Kuzmanov, 1964
44 jacobaea L.   40 Ho   Gadella & Kliphuis,
                          1973
              80 Cz   Májovský et al.,
                          1970
45 aquaticus Hill
  (a) aquaticus
  (b) barbareifolius (Wimmer & Grab.)
   Walters
46 erucifolius L.
47 carpetanus Boiss. & Reuter
48 squalidus L.
              20 Ga   Zickler, 1968
49 siculus All.   20 Si   Larsen &
                          Laegaard, 1971
50 cambrensis Rosser   60 Br   Rosser, 1955
51 aethnensis Jan ex DC.
52 nebrodensis L.
53 abrotanifolius L.

(a) abrotanifolius L.	c.40	He	Favarger, 1965
(b) carpathicus (Herbich) Nyman	40	Cz	Májovský et al., 1970
54 adonidifolius Loisel.	40	Ga	Delay, 1970
55 resedifolius Less.			
56 delphinifolius Vahl			
57 minutus (Cav.) DC.			
58 gallicus Chaix	20	Lu	Fernandes & Queirós, 1971
59 leucanthemifolius Poiret	20	Lu	Fernandes & Queirós, 1971
rodriguezii Willk. ex Rodr.	20	Bl	Dahlgren et al., 1971
60 vernalis Waldst. & Kit.	40	Cz	Májovský et al., 1970
61 petraeus Boiss. & Reuter			
62 sylvaticus L.	40	Lu	Fernandes & Queirós, 1971
63 lividus L.	40	Lu	Fernandes & Queirós, 1971
64 viscosus L.	40	Ga	Zickler, 1968
65 vulgaris L.	40	Ho	Gadella & Kliphuis, 1963
66 elegans L.			
67 flavus (Decne) Schultz Bip.			

97 Ligularia Cass.

1 sibirica (L.) Cass.	60	Cz	Májovský et al., 1970
2 glauca (L.) O. Hoffm.	60	Ru	Tarnavschi, 1948
3 Jamucta (A. Gray) Hara			

98 Kleinia Miller
  1 mandraliscae Tineo
  2 repens (L.) Haw.

99 Cacalia L.
  1 hastata L.

Calenduleae Cass.

100 Calendula L.
  1 suffruticosa Vahl

(a) suffruticosa			
(b) lusitanica (Boiss.) Ohle	32	Lu	Fernandes & Queirós, 1971
(c) fulgida (Rafin.) Ohle	32	Si	Meusel & Ohle, 1966
(d) tomentosa Murb.			
(e) maritima (Guss.) Meikle			
(f) algarbiensis (Boiss.) Nyman	32	Lu	Rodriguez, 1953
	32 + 2B	Lu	Fernandes & Queirós, 1971
2 officinalis L.			
3 stellata Cav.			
4 arvensis L.	44	Lu	Fernandes & Queirós, 1971
5 tripterocarpa Rupr.			

101 Chrysanthemoides Fabr.
  1 monilifera (L.) T. Norlindh

Arctotideae Cass.

102 Arctotis L.
   1 stoechadifolia Bergius

103 Arctotheca Wendl.
   1 calendula (L.) Levyns   18   Lu   Fernandes & Queirós,
                                                   1971

104 Gazania Gaertner
   1 rigens (L.) Gaertner

Cardueae Cass.

105 Amphoricarpos Vis.
   1 neumayeri Vis
     (a) neumayeri   24   Ju   Sušnik & Lovka
                                                 (Love, in litt.)
     (b) murbeckii Bošnjak

106 Carlina L.
   1 diae (Rech. fil.) Meusel &
     Kästner
   2 tragacanthifolia Klatt
   3 corymbosa L.   18   Lu   Fernandes & Queirós,
                                                 1971
                          20   Si   Arata, 1944
     (a) corymbosa
     (b) graeca (Boiss.) Nyman
     (c) curetum (Heldr. ex Halácsy) Rech. fil.
   4 sicula Ten.
   5 barnebiana B.L. Burtt & P.H. Davis
   6 vulgaris L.   20   Cz   Májovský et al.,
     (a) vulgaris   1970
     (b) intermedia (Schur) Hayek
     (c) longifolia Nyman
   7 frigida Boiss. & Heldr.
   8 fiumensis Simonkai
   9 macrocephala Moris
     (a) macrocephala   20   Co   Contandriopoulos,
                                                     1962
     (b) nebrodensis (Guss. ex DC.) D.A. Webb
  10 acaulis L.
     (a) acaulis
     (b) simplex (Waldst. & Kit.) Nyman   20   It   Arata, 1944
  11 acanthifolia All.   20   Ga   Delay, 1970
     (a) acanthifolia
     (b) cynara (Pourret ex Duby) Rouy
  12 lanata L.   20   Si   Arata, 1944
  13 racemosa L.   20   Lu   Fernandes & Queirós,
                                                    1971

107 Atractylis L.
   1 gummifera L.   20   Lu   Fernandes & Queirós,
                                                 1971
   2 tutinii Franco
   3 humilis L.
   4 cancellata L.

108 Xeranthemum L.
   1 annuum L.   12   Ju   Gadella & Kliphuis,
                                               1972

```
    2 inapertum (L.) Miller
    3 cylindraceum Sibth. & Sm.              20      Bu      Kuzmanov &
                                                             Kozuharov, 1970


109 Cardopatum Pers.
    1 corymbosum (L.) Pers.

110 Echinops L.
    1 spinosus L.
    2 spinosissimus Turra
     (a) spinosissimus
     (b) bithynicus (Boiss.) Kozuharov
     (c) neumayeri (Vis.) Kozuharov
    3 orientalis Trautv.
    4 strigosus L.                           32      Lu      Fernandes & Queirós,
                                                             1971
    5 graecus Miller
    6 fontqueri Pau
    7 sphaerocephalus L.                      32      Cz      Májovský et al.,
                                                             1970
     (a) sphaerocephalus
     (b) albidus (Boiss. & Spruner) Kozuharov
     (c) taygeteus (Boiss. & Heldr.)
         Kozuharov
    8 exaltatus Schrader
    9 bannaticus Rochel ex Schrader
   10 ritro L.                               32      Bu      Kuzmanov &
                                                             Kozuharov, 1970
     (a) ritro
     (b) thracicus (Velen.) Kozuharov
     (c) sartorianus (Boiss & Heldr.)
         Kozuharov
     (d) meyeri (DC.) Kozuharov
     (e) ruthenicus (Bieb.) Nyman
   11 microcephalus Sibth. & Sm.
   12 oxyodontus Bornm. & Diels.


111 Berardia Vill.
    1 subacaulis Vill.                       36      He      Favarger, 1959a

112 Arctium L.
    1 tomentosum Miller                      36      Cz      Májovský et al.,
                                                             1970
    2 lappa L.                               36      Cz      Májovský et al.,
                                                             1970
    3 pubens Bab.
    4 minus Bernh.                           36      Cz      Májovský et al.,
                                                             1970
    5 nemorosum Lej.                         36      Cz      Májovský et al.,
                                                             1970

113 Cousinia Cass.
    1 astracanica (Sprengel) Tamamsch.

114 Saussurea DC.
    1 amara (L.) DC.
    2 salsa (Pallas) Sprengel
    3 turgaiensis B. Fedtsch.
    4 pygmaea (Jacq.) Sprengel              52      Ju      Lovka et al., 1972
    5 parviflora (Poiret) DC.
    6 porcii Degen
```
254

7 alpina (L.) DC.
　(a) alpina   52   No   Knaben &
　 Engelskjon, 1967
　 54   Su   Löve & Löve, 1944
　(b) macrophylla (Sauter) Nyman
　(c) depressa (Gren.) Nyman   50-54   Ga   Favarger, 1965
　(d) esthonica (Baer ex Rupr.) Kupffer
8 discolor (Willd.) DC.   26   Ju   Lovka et al.,
　 1972
9 controversa DC.

115 Staehelina L.
　1 fruticosa (L.) L.
　2 uniflosculosa Sibth. & Sm.
　3 arborea Schreber
　4 baetica DC.
　5 dubia L.   30   Lu   Fernandes & Queirós,
　 1971

116 Jurinea Cass.
　1 linearifolia DC.
　2 stoechadifolia (Bieb.) DC.
　3 tzar-ferdinandii Davidov
　4 pinnata (Lag.) DC.
　5 tanaitica Klokov
　6 albicaulis Bunge
　(a) kilaea (Aznav.) Kozuharov   30   Bu   Kuzmanov & Ancev, 1973
　(b) laxa (Fischer ex Iljin)
　　　Kozuharov
　7 kirghisorum Janisch.
　8 cyanoides (L.) Reichenb.
　(a) cyanoides
　(b) tenuiloba (Bunge) Nyman
　9 ewersmanii Bunge
　10 mollis (L.) Reichenb.   30   Bu   Kuzmanov &
　 Kozuharov, 1967
　 34   Cz   Holub et al., 1971
　 35   Cz   Holub et al., 1971
　 36   Rm   Lungeanu, 1973
　(a) mollis
　(b) moschata (DC.) Nyman
　(c) transylvanica (Sprengel) Hayek
　(d) anatolica (Boiss.) Stoj. &
　　　Stefanov
　11 polyclonos (L.) DC.
　12 ledebourii Bunge   36   Rm   Lungeanu, 1973
　13 consanguinea DC.
　(a) consanguinea
　(b) neicevii Kozuharov
　(c) arachnoidea (Bunge) Kozuharov   36   Asia   Griff, 1965
　(d) bulgarica (Velen.) Kozuharov
　14 glycacantha (Sibth. & Sm.) DC.   30   Hu   Baksay, 1956
　15 humilis (Desf.) DC.   34   Hs   Küpfer, 1969
　16 taygetea Halácsy
　17 fontqueri Cuatrec.

117 Carduus L.
　1 macrocephalus Desf.
　(a) macrocephalus
　(b) sporadum (Halácsy) Franco
　(c) siculus Franco

255

```
 2 granatensis Willk.
 3 taygeteus Boiss. & Heldr
   (a) taygeteus
   (b) insularis Franco
 4 thoermeri Weinm.
 5 nutans L.                              16      Cz      Májovský et al.,
                                                          1970

   (a) natans
   (b) alpicola (Gillot) Chassagne &
       J. Arenes
   (c) platylepis (Reichenb. & Sauter)
       Nyman
 6 micropterus (Borbás) Teyber
   (a) micropterus
   (b) perspinosus (Fiori) Kazmi
 7 broteroi Welw. ex Coutinho            20      Lu      Fernandes & Queirós,
                                                          1971

 8 sandwithii Kazmi
 9 platypus Lange
10 chrysacanthus Ten.
   (a) chrysacanthus
   (b) hispanicus Franco
11 aurosicus Vill.
12 acanthoides L.                        16      Ju      Podlech, 1964
                                         20      Ge      Fritsch, 1973
                                         22      Hu      Baksay, 1958

13 ramosissimus Pančić
14 tmoleus Boiss.
   (a) tmoleus
   (b) armatus (Boiss. & Heldr.) Franco
15 thessalus Boiss. & Heldr
16 cronius Boiss. & Heldr.
   (a) cronius
   (b) baldaccii Kazmi
17 personata (L.) Jacq.                  18      He      Larsen, 1954
   (a) personata
   (b) albidus (Adamović) Kazmi
18 crispus L.
   (a) crispus                           16      Fe      Hamet-Ahti &
                                                          Virrankoski, 1970

   (b) multiflorus (Gaudin) Franco
19 litigiosus Nocca & Balbis
   (a) litigiosus
   (b) horridissimus (Briq. & Cavillier)
       Franco
20 euboicus Franco
21 vivariensis Jordan
   (a) vivariensis
   (b) australis Nyman
   (c) assoi (Willk.) Kazmi
22 nigrescens Vill.
23 hamulosus Ehrh.
   (a) hamulosus
   (b) hystrix (C.A. Meyer) Kazmi
24 uncinatus Bieb.
25 defloratus L.
   (a) defloratus                        18      Ga It   Favarger &
                                                          Küpfer, 1970
   (b) glaucus Nyman                      22      Hu      Baksay, 1958
26 argemone Pourret ex Lam.
   (a) argemone                          22      Hs      Favarger &
                                                          Küpfer, 1970
```

(b) obtusisquamus Franco	22	He	Favarger & Küpfer, 1970
27 carlinifolius Lam.	22	Ga Hs	Favarger & Küpfer, 1970

28 candicans Waldst. & Kit.
  (a) candicans
  (b) globifer (Velen.) Kazmi

29 collinus Waldst. & Kit.	16	Hu	Baksay, 1958
	32	Ga	Kliphuis & Wieffering, 1972

  (a) collinus
  (b) cylindricus (Borbás) Soó
  (c) glabrescens (Sagorski) Kazmi
30 carduelis (L.) Gren.
31 kerneri Simonkai
  (a) lobulatiformis (Csuros & E.I.
     Nyarady) Soó
  (b) kerneri
  (c) scardicus (Griseb.) Kazmi
  (d) austro-orientalis Franco
32 affinis Guss.
  (a) affinis
  (b) brutius (Porta) Kazmi
33 adpressus C.A. Meyer

34 argyroa Biv.	26	Si	Podlech, 1964

35 bourgeanus Boiss. & Reuter
  (a) bourgeanus
  (b) valentinus (Boiss. & Reuter) Franco
36 myriacanthus Salzm. ex DC.
37 asturicus Franco

38 carpetanus Boiss. & Reuter	16	Lu	Fernandes & Queirós, 1971
39 carlinoides Gouan	18	Hs	Küpfer, 1968

  (a) carlinoides
  (b) hispanicus (Kazmi) Franco
40 corymbosus Ten.
41 acicularis Bertol.
42 argentatus L.

43 meonanthus Hoffmanns. & Link	16	Lu	Queirós, 1973
44 tenuiflorus Curtis	54	Ga	van Loon et al., 1971
45 pycnocephalus L.	62-64	Bl	Dahlgren et al., 1971

  (a) pycnocephalus
  (b) albidus (Bieb.) Kazmi
46 australis L. fil.

47 cephalanthus Viv.	22	Co	Contandriopoulos, 1962
48 fasciculiflorus Viv.	22	Co	Contandriopoulos, 1962

118 Cirsium Miller
  1 ferox (L.) DC.
  2 heldreichii Halácsy
    (a) heldreichii
    (b) euboicum Petrak
  3 bulgaricum DC.
  4 polycephalum DC.
  5 morinifolium Boiss. & Heldr.
  6 hypopsilum Boiss. & Heldr.
  7 vallis-demonis Lojac.

```
 8 lacaitae Petrak
 9 tenoreanum Petrak
10 lobelii Ten.
11 morisianum Reichenb. fil.
12 richteranum Gillot
13 costae (Sennen & Pau) Petrak
14 giraudiasii Sennen & Pau
15 eriophorum (L.) Scop.              34      Cz      Májovský et al.,
                                                     1970
16 odontolepis Boiss. ex DC.
17 spathulatum (Moretti) Gaudin
18 ligulare Boiss.
19 grecescui Rouy
20 decussatum Janka
21 boujartii (Piller & Mitterp.) Schultz Bip.
   (a) boujartii
   (b) wettsteinii Petrak
22 furiens Griseb. & Schenk
23 ciliatum Moench
24 serrulatum (Bieb.) Fischer
25 laniflorum (Bieb.) Fischer
26 scabrum (Poiret) Bonnet & Barratte
27 echinatum (Desf.) DC.
28 vulgare (Savi) Ten.               68      Lu      Fernandes & Queirós,
                                                     1971
                                    102      Lu      Fernandes & Queirós,
                                                     1971
29 italicum (Savi) DC.
30 dissectum (L.) Hill
31 filipendulum Lange                34      Lu      Fernandes & Queirós,
                                                     1971
32 tuberosum (L.) All.               34      He      Favarger, 1969
33 rivulare (Jacq.) All.             34      Ga      Delay, 1970
34 montanum (Waldst. & Kit. ex Willd.)
      Sprengel
35 erisithales (Jacq.) Scop.         34      Hu      Baksay, 1956
36 appendiculatum Griseb.            68      Bu      Kuzmanov & Ancev,
                                                     1973
37 waldsteinii Rouy                  68      Po      Mizianty & Frey,
                                                     1973
38 hypoleucum DC.
39 carniolicum Scop.
   (a) carniolicum                   16      Ju      Lovka & Sušnik, 1973
   (b) rufescens (Ramond ex DC.) P. Fourn.
40 oleraceum (L.) Scop.              34      Cz      Májovský et al.,
                                                     1970
41 spinosissimum (L.) Scop.
   (a) spinosissimum                 34      He      Larsen, 1954
   (b) bertolonii (Sprengel) Werner  34      It      Miceli & Garbari,
                                                     1976
42 glabrum DC.                       34      Ga      Küpfer, 1971
43 candelabrum Griseb.
44 acaule Scop.
   (a) acaule                        34      Po      Skalińska et al.,
                                                     1959
   (b) gregarium (Boiss. ex DC.) Werner 34 + 2B  Hs  Küpfer, 1968
   (c) esculentum (Sievers) Werner
45 mairei Halácsy
46 valentinum Porta & Rigo
47 helenioides (L.) Hill           c.34      Fe      V. Sorsa, 1962
```

48 pannonicum (L. fil.) Link	34	Hu	Baksay, 1956
49 heterotrichum Pančić			
50 canum (L.) All.	34	Ju	Nilsson & Lassen, 1971
51 tymphaeum Hausskn.			
52 monspessulanum (L.) Hill			
53 welwitschii Cosson	34	Lu	Fernandes & Queirós, 1971
54 alatum (S.G. Gmelin) Bobrov			
55 brachycephalum Juratzka			
56 bourgaeanum Willk.			
57 palustre (L.) Scop.	34	Ho	Gadella & Kliphuis, 1966
58 flavispina Boiss. ex DC.			
59 creticum (Lam.) D'Urv.			
(a) creticum			
(b) triumfetti (Lacaita) Werner			
60 arvense (L.) Scop.	34	Cz	Májovský et al., 1970

119 Picnomon Adanson
1 acarna (L.) Cass.	32	Ga	Moore, 1967
	34	Lu	Fernandes & Queirós, 1971

120 Notobasis Cass.
1 syriaca (L.) Cass.	34	Bl	Dahlgren et al., 1971

121 Ptilostemon Cass.
1 strictus (Ten.) W. Greuter			
2 niveus (C. Presl) W. Greuter			
3 afer (Jacq.) W. Greuter			
4 chamaepeuce (L.) Less.			
5 gnaphaloides (Cyr.) Sojak			
(a) gnaphaloides			
(b) pseudofruticosus (Pamp.) W. Greuter			
6 echinocephalus (Willd.) W. Greuter			
7 hispanicus (Lam.) W. Greuter			
8 casabonae (L.) W. Greuter	32	Sa	Renzoni-Cela, 1963
9 stellatus (L.) W. Greuter	24	It	Burdet & Greuter, 1973

122 Lamyropsis (Charadze) Dittrich
1 cynaroides (Lam.) Dittrich
2 microcephala (Moris) Dittrich &
  W. Greuter

123 Galactites Moench
1 tomentosa Moench	22	Lu	Fernandes & Queirós, 1971
2 duriaei Spach ex Durieu	22	Hs	Gadella & Kliphuis, 1968

124 Tyrimnus (Cass.) Cass.
1 leucographus (L.) Cass.

125 Onopordum L.
1 acaulon L.
 (a) acaulon
 (b) uniflorum (Cav.) Franco
2 acanthium L.

(a) acanthium	34	Po	Skalińska et al., 1959

(b) gautieri (Rouy) Franco
(c) parnassicum (Boiss. & Heldr.) Nyman
3 laconicum Heldr. & Sart. ex Rouy

4 nervosum Boiss.	34	Lu	Fernandes & Queirós, 1971

5 corymbosum Willk.
  (a) corymbosum
  (b) visegradense Franco
6 tauricum Willd.
7 argolicum Boiss.
8 caulescens D'Urv.
  (a) caulescens
  (b) atticum Franco
9 messeniacum Halácsy
10 macracanthum Schousboe
11 bracteatum Boiss. & Heldr.
  (a) bracteatum
  (b) ilex (Janka) Franco
  (c) creticum Franco
  (d) myriacanthum (Boiss.) Franco

12 illyricum L.	34	Hs	Valdés, 1970b

  (a) illyricum
  (b) cardunculus (Boiss.) Franco
  (c) horridum (Viv.) Franco
13 majorii Beauverd

126 Cynara L.

1 humilis L.	34	Lu	Fernandes & Queirós, 1971

2 alba Boiss. ex DC.
3 algarbiensis Cosson ex Mariz
4 cornigera Lindley

5 cardunculus L.	34	Lu	Fernandes & Queirós, 1971

6 scolymus L.
7 tournefortii Boiss. & Reuter

127 Silybum Adanson

1 marianum (L.) Gaertner	34	It	Larsen, 1956a

2 eburneum Cosson & Durieu

128 Palaeocyanus Dostál

1 crassifolius (Bertol.) Dostál	30	Si	Damboldt et al., 1973

129 Cheirolophus Cass.

1 sempervirens (L.) Pomel	30	Lu	Guinochet & Foissac, 1962
2 uliginosus (Brot.) Dostál	24	Lu	Guinochet & Foissac, 1962
	32	Lu	Fernandes & Queirós, 1971
3 intybaceus (Lam.) Dostál	32	Ga	Gardou, 1972

130 Serratula L.

1 tinctoria L.	22	Cz	Májovský et al., 1970

2 wolffii Andrae

    3 pinnatofida (Cav.) Poiret
    4 abulensis Pau   90   Lu   Fernandes & Queirós,
                                                          1971
    5 baetica Boiss. ex DC.
    6 alcalae Cosson
    7 pauana Iljin
    8 nudicaulis (L.) DC.   30   Ga   Guinochet &
                                                          Logeois, 1962
    9 flavescens (L.) Poiret
   10 leucantha (Cav.) DC.
   11 cichoracea (L.) DC.
    (a) cichoracea
    (b) mucronata (Desf.) Jahandiez &
        Maire
    (c) cretica Turrill
   12 lycopifolia (Vill.) A. Kerner   60   Hu   Baksay, 1957
   13 bulgarica Acht. & Stoj.
   14 radiata (Waldst. & Kit.) Bieb.   30   Hu   Baksay, 1956
    (a) radiata
    (b) cetingensis (Rohlena) Hayek
   15 gmelinii Tausch
   16 cardunculus (Pallas) Schischkin
   17 erucifolia (L.) Boriss.

131 Leuzea DC.
    1 centauroides (L.) J. Holub   26
    2 rhapontica (L.) J. Holub   26   Ga   Guinochet, 1957
    (a) rhapontica
    (b) heleniifolia (Gren. & Godron)
        J. Holub
    (c) bicknellii (Briq.) J. Holub
    3 altaica (Fischer ex Sprengel) Link
    4 conifera (L.) DC.   26   Hs   Lorenzo-Andreu &
                                                          Garcia, 1950
    5 rhaponticoides Graells   26   Hs   Dittrich, 1968
    6 longifolia Hoffmanns. & Link   26   Lu   Fernandes & Queirós,
                                                          1971

132 Amberboa (Pers.) Less.
    1 turanica Iljin

133 Volutaria Cass.
    1 lippii (L.) Maire

134 Cyanopsis Cass.
    1 muricata (L.) Dostál

135 Mantisalca Cass.
    1 salmantica (L.) Briq. & Cavillier   18   It   Chiappini, 1954
                                             20   Ga   Guinochet, 1957
                                             22   Lu   Guinochet & Foissac,
                                                          1962

136 Acroptilon Cass.
    1 repens (L.) DC.

137 Phalacrachena Iljin
    1 inuloides (Fischer ex Janka) Iljin

138 Centaurea L.
    1 centaurium L.

261

2 amplifolia Boiss. & Heldr.			
3 fraylensis Schultz Bip. ex Nyman			
4 africana Lam.	30	Lu	Fernandes & Queirós, 1971
5 ruthenica Lam.			
6 linaresii Laz.-Ibiza			
7 kasakorum Iljin			
8 taliewii Kleopow			
9 alpina L.			
10 jankae Brandza			
11 collina L.	20	Lu	Fernandes & Queirós, 1971
12 salonitana Vis.	20	Rm	Lungeanu, 1975
	40	Gr	Phitos, 1970
(a) salonitana			
(b) ognianoffii (Urum.) Dostál			
13 centauroides L.			
14 rumelica Boiss.			
15 nicolai Bald.			
16 ornata Willd.			
(a) ornata	20	Lu	Queirós, 1973
	40	Lu	Fernandes & Queirós, 1971
(b) saxicola (Lag.) Dostál	60	Hs	Gardou, 1972
17 granatensis Boiss. ex DC.			
18 tuntasia Heldr. ex Halácsy			
19 macedonica Boiss.			
(a) macedonica			
(b) parnonia (Halácsy) Dostál			
20 mannagettae Podp.			
21 dichroantha A. Kerner	20	Ju	Nilsson & Lassen, 1971
22 rupestris L.	20	Ju	Nilsson & Lassen, 1971
(a) rupestris			
(b) ceratophylla (Ten.) Gugler			
(c) finazzeri (Adamovič) Hayek			
(d) athoa (DC.) Gugler			
23 kosaninii Hayek			
24 clementei Boiss. ex DC.			
25 orientalis L.	20	Bu	Kuzmanov & Kozuharov, 1970
26 neiceffii Degen & H. Wagner			
27 tauromenitana Guss.			
28 chrysolepis Vis.			
29 prolongi Boiss. ex DC.			
30 polymorpha Lag.	40	Hs	Gardou, 1972
31 atropurpurea Waldst. & Kit.	18	Bu	Kozuharov et al., 1968
(a) atropurpurea			
(b) soskae (Stoj. & Acht.) Dostál			
32 kotschyana Heuffel ex Koch			
33 murbeckii Hayek			
34 candelabrum Hayek & Kosanin			
35 immanuelis-loewii Degen			
36 grbavacensis (Rohlena) Stoj. & Acht.			
37 oliverana DC.	22	Gr	Runemark, 1967c
38 ragusina L.			
(a) ragusina			
(b) lungensis (Ginzberger) Hayek			
39 rechingeri Phitos	22	Gr	Runemark, 1967c

40 graeca Griseb.   20   Gr   Phitos, 1970
   (a) graeca
   (b) ceccariniana (Boiss. & Heldr.) Dostál
41 laconica Boiss.
42 achaia Boiss. & Heldr.
43 sibthorpii Halácsy
44 psilacantha Boiss. & Heldr.
45 redempta Heldr.   20   Cr   Runemark, 1967c
46 cytherea Rech. fil.
47 ebenoides Heldr. ex S. Moore
48 spruneri Boiss. & Heldr.
   (a) spruneri   110   Gr   Phitos, 1970
   (b) minoa (Heldr. ex Boiss.) Dostál
   (c) guicciardii (Boiss.) Hayek   20   Gr   Runemark, 1967c
                                           100   Gr   Phitos, 1970
                                           110   Gr   Phitos, 1970
   (d) lineariloba (Halácsy & Dörfler)
       Dostál   20   Gr   Phitos, 1970
49 scabiosa L.   20   Cz   Májovský et al.,
                                                          1970
                                  20 + 0 - 2B   Ga   Lévêque &
                                                          Gorenflot, 1969
50 cephalariifolia Willk.
51 alpestris Hegetschw.   20   Po   Skalińska et al.,
                                                          1959
52 sadlerana Janka   20   Hu   Baksay, 1956
53 badensis Tratt.
54 grinensis Reuter
   (a) grinensis
   (b) fritschii (Hayek) Dostál   20   Ju   Lovka et al., 1971
55 apiculata Ledeb.
   (a) apiculata
   (b) spinulosa (Rochel ex Sprengel) Dostál
   (c) adpressa (Ledeb.) Dostál
56 stereophylla Besser
57 raphanina Sibth. & Sm.
   (a) raphanina   20   Cr   Runemark, 1967
   (b) mixta (DC.) Runemark   20   Gr   Runemark, 1967
58 aegialophila Wagenitz   22   Cr   Runemark, 1967
59 pumilio L.
60 macrorrhiza Willk.   20   Hs   Gardou, 1972
61 toletana Boiss. & Reuter
62 haenseleri (Boiss.) Boiss.
63 argecillensis Gredilla
64 amblensis Graells
65 lagascana Graells
   (a) lagascana
   (b) podospermifolia (Loscos & Pardo) Dostál
66 acaulis L.
67 loscosii Willk.
68 thracica (Janka) Hayek
69 charrelii Halácsy & Dörfler
70 cineraria L.   18   It   Gori, 1954
   (a) cineraria   18   It   Viegi et al., 1972
   (b) busambarensis (Guss.) Dostál   18   Si   Viegi et al., 1972
   (c) cinerea (Lam.) Dostál
   (d) veneris (Sommier) Dostál   18   It   Viegi et al., 1972
71 cuspidata Vis.
72 niederi Heldr.
73 kilaea Boiss.
74 wettsteinii Degen & Dörfler

76 pannosa DC.
77 nicopolitana Bornm.
78 cuneifolia Sibth. & Sm.
  (a) cuneifolia
  (b) pallida (Friv.) Hayek
  (c) sublanata (DC.) Hayek
79 ipsaria Stoj. & Kitanov
80 rutifolia Sibth. & Sm.
  (a) rutifolia
  (b) jurineifolia (Boiss.) Nyman
  (c) pseudobovina (Hayek) Dostál
81 varnensis Velen.
82 crithmifolia Vis.
83 friderici Vis.
  (a) friderici
  (b) jabukensis (Ginzberger & Teyber)
      Dostál
84 affinis Friv.
  (a) affinis
  (b) balcanica (Urum. & H. Wagner)
      Dostál
  (c) peloponnesiaca (Halacsy) Dostál
  (d) candida (Velen.) Dostál
  (e) lacerata (Hausskn.) Dostál
85 pallidior Halácsy
  (a) pallidior
  (b) denudata (Halácsy) Dostál
  (c) vatevii (Degen, Urum. & H. Wagner)
      Dostál
86 pontabulis Heldr
  (a) tenorei (Guss. ex Lacaita) Dostál
  (b) nigra (Fiori) Dostál
  (c) parlatoris
87 horrida Badaro
88 balearica Rodr.
89 arenaria Bieb. ex Willd.
  (a) arenaria
  (b) sophiae (Klokov) Dostál
  (c) majorovii (Dumbadze) Dostál
  (d) odessana (Prodan) Dostál
  (e) borysthenica (Gruner) Dostál
90 ovina Pallas ex Willd.
  (a) besserana (DC.) Dostál
  (b) lavrenkoana (Klokov) Dostál
  (c) steveniana (Klokov) Dostál
  (d) koktebelica (Klokov) Dostál
91 tenuiflora DC.
92 spinosociliata Seenus
  (a) cristata (Bartl.) Dostál
  (b) spinosociliata
  (c) tommasinii (A. Kerner) Dostál
93 incompta Vis.
  (a) incompta
  (b) derventana (Vis. & Pančić) Dostál
94 chalcidicaea Hayek
95 grisebachii (Nyman) Form.
  (a) grisebachii
 · (b) confusa (Halácsy) Dostál
  (c) paucijuga (Halácsy) Dostál
96 tauscheri A. Kerner

```
 97 biokovensis Teyber
 98 gracilenta Velen.
 99 kalambakensis Freyn & Sint.
100 transiens Halácsy
101 subsericans Halácsy
102 attica Nyman
    (a) ossaea (Halácsy) Dostál
    (b) asperula (Halácsy) Dostál
    (c) attica
    (d) megarensis (Halácsy & Hayek) Dostál
    (e) drakiensis (Freyn & Sint.) Dostál
    (f) pentelica (Hausskn.) Dostál
103 soskae Hayek ex Kosanin
104 kartschiana Scop.
105 dalmatica A. Kerner
106 brachtii Reichenb. fil.
107 schousboei Lange
    (a) schousboei
    (b) septentrionalis (J. Arenes) Dostál
108 spinabadia Bubani ex Timb.-Lagr.
    (a) shuttleworthii (Rouy) Dostál
    (b) spinabadia
    (c) isernii (Willk.) Dostál
    (d) hanryi (Jordan) Dostál
109 limbata Hoffmanns. & Link        18      Lu      Queirós, 1973
110 urgellensis Sennen
111 rothmalerana (J. Arenes) Dostál
112 aristata Hoffmanns. & Link
    (a) exilis (J. Arenes) Dostál
    (b) langeana (Willk.) Dostál
    (c) geresensis (J. Arenes) Dostál
    (d) aristata
113 paniculata L.                    18      Ga It   Gardou, 1972
                                     19      Ga      Guinochet, 1957
    (a) paniculata
    (b) rigidula (Jordan) Dostál
    (c) polycephala (Jordan) Nyman
    (d) esterellensis (Burnat) Dostál
    (e) castellana (Boiss. & Reuter) Dostál
    (f) cossoniana (J. Arenes) Dostál
114 micrantha Hoffmanns. & Link
    (a) micrantha
    (b) herminii (Rouy) Dostál
    (c) melanosticta (Lange) Dostál
115 leucophaea Jordan                18      Ga      Delay, 1969
    (a) brunnescens (Briq.) Dostál
    (b) reuteri (Reichenb. fil.) Dostál
    (c) leucophaea                   18      Ga      Guinochet, 1957
    (d) pseudocoerulescens (Briq.) Dostál  18  Ga   Guinochet, 1957
    (e) biformis (Timb.-Lagr.) Dostál
    (f) ochrolopha (Costa) Dostál
116 filiformis Viv.
117 corymbosa Pourret
118 subtilis Bertol.
119 exarata Boiss. ex Cosson
120 maculosa Lam.                    18      Ga      Guinochet, 1957
    (a) maculosa
    (b) chaubardii (Reichenb. fil.) Dostál
    (c) albida (Lecoq & Lamotte) Dostál
    (d) subalbida (Jordan) Dostál
121 vallesiaca (DC.) Jordan
```

265

```
122 rhenana Boreau
    (a) rhenana                          18      Cz      Májovský et al.,
                                                         1970
                                    18 + 0 - 2B   Po     Skalińska et al.,
                                                         1959
    (b) tartarea (Velen.) Dostál
    (c) pseudomaculosa (Dobrocz.) Dostál
    (d) savranica (Klokov) Dostál
123 glaberrima Tausch
124 triniifolia Heuffel
125 reichenbachii DC.
126 calvescens Pančić
127 peucedanifolia Boiss. & Orph.
128 biebersteinii DC.                     36      Hu      Baksay, 1958
    (a) biebersteinii
    (b) australis (Pančić) Dostál
    (c) rhodopaea (Hayek & Wagner) Dostál
    (d) radoslavoffii (Urum.) Dostál
    (e) cylindrocephala (Bornm.) Dostál
129 boissieri DC.
    (a) dufourii Dostál
    (b) resupinata (Cosson) Dostál
    (c) prostrata (Cosson) Dostál
    (d) boissieri
    (e) willkommii (Schultz Bip.) Dostál
    (f) pomeliana (Batt. & Trabut) Dostál
    (g) mariolensis (Rouy) Dostál
    (h) pinae (Pau) Dostál
    (i) paui (Loscos ex Willk.) Dostál
    (j) jaennensis (Degen & Debeaux) Dostál
    (k) spachii (Schultz Bip. ex Willk.) Dostál
130 lagascae Nyman
131 bombycina Boiss. ex DC.
    (a) bombycina
    (b) funkii (Schultz Bip.) Dostál
132 monticola Boiss. ex DC.
133 carratracensis Lange
134 aplolepa Moretti
    (a) aplolepa                          18      It      Viegi et al., 1972
    (b) aeolica (Guss. ex Lojac.) Dostál  18      Si      Viegi et al., 1972
    (c) pandataria (Fiori & Béguinot) Dostál
    (d) carueliana (Micheletti) Dostál
    (e) maremmana (Fiori) Dostál
    (f) subciliata (DC.) Arcangeli
    (g) ligustica (Gremli ex Briq.) Dostál
    (h) aetaliae (Sommier) Dostál
    (i) lunensis (Fiori) Dostál
    (j) cosana (Fiori) Dostál
    (k) gallinariae (Briq. & Cavillier) Dostál
135 lactiflora Halácsy
136 laureotica Heldr. ex Halácsy
137 pelia DC.
138 rufidula Bornm.
139 tymphaea Hausskn.
    (a) tymphaea
    (b) brevispina (Hausskn.) Dostál
140 orphanidea Heldr. & Sart. ex Boiss.
    (a) orphanidea
    (b) thessala (Hausskn.) Dostál
141 diffusa Lam.
142 bovina Velen.
```

```
143 acmulans Klokov
144 zuccariniana DC.
145 spinosa L.                                   36      Gr      Runemark, 1967 c
    (a) spinosa
    (b) cycladum (Heldr.) Hayek
146 hyalolepis Boiss.
147 iberica Trev. ex Sprengel
    (a) iberica
    (b) holzmanniana (Boiss.) Dostál
148 calcitrapa L.                                20      Lu      Fernandes & Queirós,
                                                                 1971
149 pontica Prodan & E.I. Nyárády
150 sonchifolia L.
151 seridis L.
    (a) seridis
    (b) cruenta (Willd.) Dostál
    (c) maritima (Dufour) Dostál
152 sphaerocephala L.
    (a) sphaerocephala
    (b) lusitanica (Boiss. & Reuter) Nyman  20     Lu      Guinochet & Foissac,
                                                                 1962
                                                 22      Lu      Fernandes & Queirós,
                                                                 1971
    (c) malacitana (Boiss.) Dostál
    (d) polyacantha (Willd.) Dostál              22      Lu      Guinochet & Foissac,
                                                                 1962
153 aspera L.                                    22      Lu      Fernandes & Queirós,
                                                                 1971
    (a) aspera
    (b) stenophylla (Dufour) Nyman
    (c) scorpiurifolia (Dufour) Nyman
    (d) pseudosphaerocephala (R.J. Shuttlew.
        ex Rouy) Kugler
154 napifolia L.
155 micracantha Dufour
156 hermannii F. Hermann
157 solstitialis L.                              16      Ju      Gadella & Kliphuis,
                                                                 1972
    (a) solstitialis
    (b) adamii (Willd.) Nyman
    (c) schouwii (DC.) Dostál
    (d) erythracantha (Halácsy) Dostál
158 idaea Boiss. & Heldr.
159 melitensis L.                                24      Ga      Guinochet, 1957
160 sulphurea Willd.
161 nicaeensis All.
162 eriophora L.                                 24      Hs      Björkqvist et al.,
                                                                 1969
163 diluta Aiton
164 margaritacea Ten.
    (a) pseudoleucolepis (Kleopow) Dostál
    (b) breviceps (Iljin) Dostál
    (c) margaritacea
    (d) margaritalba (Klokov) Dostál
    (e) protomargaritacea (Klokov) Dostál
    (f) appendicata (Klokov) Dostál
    (g) konkae (Klokov) Dostál
    (h) protogerberi (Klokov) Dostál
    (i) donetzica (Klokov) Dostál
    (j) pineticola (Iljin) Dostál
    (k) dubjanskyi (Iljin) Dostál
```

    (l) gerberi (Steven) Dostál
    (m) paczoskii (Kotov ex Klokov) Dostál
165 transcaucasica D. Sossn. ex Grossh.
166 sterilis Steven
    (a) sterilis
    (b) semijusta (Juz.) Dostál
    (c) vankovii (Klokov) Dostál
167 alba L.   18   Hs   Gardou, 1972
    (a) costae (Willk.) Dostál
    (b) latronum (Pau) Dostál
    (c) alba
    (d) splendens (L.) Arcangeli   20   Ju   Nilsson & Lassen,
                                                        1971
    (e) tenoreana (Willk.) Dostál
    (f) pestalottii (De Not.) Arcangeli
    (g) diomedea (Gasparr.) Dostál
    (h) albanica (Halácsy) Dostál
    (i) ipecensis (Rech. fil.) Dostál
    (j) deusta (Ten.) Nyman   18   Ju   Damboldt et al.,
                                                        1971
    (k) brunnea (Halácsy) Dostál
    (l) epapposa (Velen.) Dostál
    (m) caliacrae (Prodan) Dostál
    (n) leucomalla (Bornm.) Dostál
    (o) formanekii (Halácsy) Dostál
    (p) vandasii (Velen.) Dostál
    (q) euxina (Velen.) Dostál
    (r) heldreichii (Halácsy) Dostál
    (s) princeps (Boiss. & Heldr.) Gugler
    (t) subciliaris (Boiss. & Heldr.) Dostál
168 deustiformis Adamović
    (a) deustiformis
    (b) ptarmicifolia (Halácsy ex Hayek) Dostál
    (c) pseudocadmea (Wagenitz) Dostál
169 ferulacea U. Martelli   18   Sa   Arrigoni & Mori,
                                                        1971
170 musarum Boiss. & Orph.
171 haynaldii Borbás ex Vuk
172 bracteata Scop.
173 weldeniana Reichenb.
174 rocheliana (Heuffel) Dostál
175 pannonica (Heuffel) Simonkai   22   Ga   Guinochet, 1957
                                          44   Rm   Tarnavschi, 1948
    (a) pannonica
    (b) substituta (Czerep.) Dostál
176 vinyalsii Sennen
    (a) vinyalsii
    (b) approximata (Rouy) Dostál
177 dracuncifolia Dufour   22   Hs   Gardou, 1972
178 jacea L.   22   Ga It Ju Gardou, 1972
                                          44   Au Ga Ju Gardou, 1972
179 decipiens Thuill.
    (a) decipiens
    (b) ruscinonensis (Boiss.) Dostál
180 subjacea (G. Beck) Hayek
181 macroptilon Borbas
    (a) oxylepis (Wimmer & Grab.) Soó
    (b) macroptilon
182 microptilon
    (a) microptilon
    (b) emporitana (Vayr. ex Hayek) Dostál

                          268

```
183 transalpina Schleicher ex DC.          44    Ga       Guinochet, 1957
184 nigrescens Willd.                       44    Ga       Guinochet & Foissac,
                                                            1962
    (a) nigrescens
    (b) ramosa Gugler
    (c) neapolitana (Boiss.) Dostál
    (d) pinnatifida (Fiori) Dostál
    (e) smolinensis (Hayek) Dostál
185 carniolica Host
186 debeauxii Gren. & Godron                22    Ga It    Gardou, 1972
                                            33    Ga       Gardou, 1972
                                            44    Ga       Gardou, 1972
    (a) endressii (Hochst. & Steudel ex
        Lamotte) Dostál
    (b) nevadensis (Boiss. & Reuter) Dostál
    (c) thuillieri Dostál
    (d) nemoralis (Jordan) Dostál
    (e) debeauxii
187 nigra L.                                22    Ga Hs Lu Gardou, 1972
                                            44    Ga       Gardou, 1972
    (a) nigra                               22    Ga       Guinochet, 1957
    (b) carpetana (Boiss. & Reuter) Nyman
    (c) rivularis (Brot.) Coutinho          22    Lu       Fernandes & Queirós,
                                                            1971
188 phrygia L.
    (a) phrygia                             22    Rm       Gardou, 1972
    (b) melanocalathia (Borbás) Dostál
    (c) carpatica (Porc.) Dostál
    (d) moeoinca (Urum. & H. Wagner) Hayek
    (e) nigriceps (Dobrocz.) Dostál
    (f) pseudophrygia (C.A. Meyer) Gugler   22    Au       Favarger & Huynh,
                                                            1964
    (g) abbreviata (C. Koch) Dostál
    (h) ratezatensis (Prodan) Dostál
    (i) rarauensis (Prodan) Dostál
189 stenolepis A. Kerner
    (a) stenolepis                          22    Cz       Májovský et al.,
                                                            1970
    (b) razgradensis (Velen.) Stoj. & Acht.
    (c) bosniaca (Murb.) Dostál
190 indurata Janka
191 uniflora Turra                          22    Ga       Guinochet, 1957
    (a) uniflora
    (b) nervosa (Willd.) Bonnier & Layens   22    Ga It    Gardou, 1972
    (c) ferdinandi (Gren.) Bonnier
    (d) davidovii (Urum.) Dostál
192 kernerana Janka
    (a) kernerana
    (b) gheorghieffii (Halácsy) Dostál
193 pectinata L.                            22    Ga       Guinochet, 1957
    (a) pectinata
    (b) acutifolia (Jordan) Dostál
    (c) supina (Jordan) Dostál
194 antennata Dufour
195 trichocephala Bieb. ex Willd.
    (a) trichocephala
    (b) simonkaiana (Hayek) Dostál
196 janeri Graells
197 linifolia L.                            44    Hs       Gardou, 1972
198 hyssopifolia Vahl
199 parilica Stoj. & Stefanov
```

```
200 procumbens Balbis
    (a) procumbens
    (b) jordaniana (Gren. & Godron) Rouy
    (c) aemilii (Briq.) Dostál
    (d) verguinii (Briq. & Cavillier) Dostál
201 rhaetica                              22    It    Favarger & Huynh,
                                                      1964
202 leucophylla Bieb.
203 declinata Bieb.
204 dealbata Willd.
205 sibirica L.
206 carbonata Klokov
207 marschalliana Sprengel
208 sumensis Kalenicz
209 trinervia Stephan ex Willd.
210 montana L.                            44    Ga    Guinochet, 1957
211 mollis Waldst. & Kit.                 44    Po    Skalińska et al.,
                                                      1961
212 maramarosiensis (Jav.) Czerep.
213 pinnatifida Schur
214 baldaccii Degen ex Bald.
215 pindicola Griseb.
216 triumfetti All.                       22    Ga    Guinochet, 1957
                                          44    Ga    Guinochet, 1957
    (a) aligera (Gugler) Dostál
    (b) tanaitica (Klokov) Dostál
    (c) pirinensis (Degen, Urum. & H.
         Wagner) Dostál
    (d) semidecurrens (Jordan) Dostál
    (e) dominii Dostál
    (f) novakii (Dostál) Dostál
    (g) stricta (Waldst. & Kit.) Dostál
    (h) angelescui (G. Grint.) Dostál
    (i) adscendens (Bartl.) Dostál
    (j) lugdunensis (Jordan) Dostál
    (k) triumfetti                        22    Hu    Baksay, 1956
                                          44    Ga    Guinochet, 1957
    (l) cana (Sibth. & Sm.) Dostál
    (m) lingulata (Lag.) Dostál
217 napulifera Rochel
    (a) pseudaxillaris (Stefanov & Georgiev) Dostál
    (b) tuberosa (Vis.) Dostál
    (c) napulifera
    (d) nyssana (Petrović) Dostál
    (e) thirkei (Schultz Bip.) Dostál
218 depressa Bieb.
219 pinardii Boiss.
220 cyanus L.                             24    Ga    Guinochet, 1957
221 pullata L.                            22    Ga    Guinochet, 1957

139 Crupina (Pers.) Cass.
    1 vulgaris Cass.                       30    Ju    Larsen, 1956a
    2 crupinastrum (Moris) Vis.           28    It    Larsen, 1956a

140 Chartolepis Cass.
    1 glastifolia (L.) Cass.

141 Wagenitzia Dostal
    1 lancifolia (Sieber ex Sprengel) Dostál

142 Cnicus L.
```

1 benedictus L.	22	Lu	Fernandes & Queirós, 1971

143 Carthamus L.

1 arborescens L.	24	Hs	Harvey & Knowles, 1965
2 tinctorius L.	24	Europe	Ashri & Knowles, 1960
3 dentatus (Forskål) Vahl	20	Bu	Hanelt, 1963
(a) dentatus			
(b) ruber (Link) Hanelt			
4 glaucus Bieb.			
5 boissieri Halácsy			
6 leucocaulos Sibth. & Sm.	20	Gr	Hanelt, 1963
7 lanatus L.			
(a) lanatus	44	Lu	Hanelt, 1963
(b) baeticus (Boiss. & Reuter) Nyman	64	Lu	Khidir & Knowles, 1970

144 Carduncellus Adanson

1 monspelliensium All.	48	Ga	Delay, 1969
2 mitissimus (L.) DC.	24	Ga	Delay, 1969
3 pinnatus (Desf.) DC.			
4 dianius Webb			
5 araneosus Boiss. & Reuter			
6 caeruleus (L.) C. Presl	24	Lu	Fernandes & Queirós, 1971

CICHORIOIDEAE

145 Scolymus L.

1 maculatus L.	20	Lu	Fernandes & Queirós, 1971
2 hispanicus L.	20	Lu	Fernandes & Queirós, 1971
3 grandiflorus Desf.			

146 Cichorium L.

1 intybus L.	18	Po	Skalińska et al., 1964
2 endivia L.			
(a) endivia	18	Lu	Fernandes & Queirós, 1971
(b) divaricatum (Schousboe) P.D. Sell			
3 spinosum L.			

147 Catananche L.

1 caerulea L.	18	Ga	Delay, 1968
2 lutea L.			
(a) lutea			
(b) carpholepis (Schultz Bip.) Nyman			

148 Rothmaleria Font Quer
    1 granatensis (Boiss. ex DC.) Font Quer

149 Hymenonema Cass.
    1 laconicum Boiss. & Heldr.
    2 graecum (L.) DC.

150 Tolpis Adanson

1 barbata (L.) Gaertner	18	Lu	Fernandes & Queirós, 1971

271

2 virgata Bertol.
3 fruticosa Schrank
4 azorica (Nutt.) P. Silva
5 staticifolia (All.) Schultz Bip. 18 He Favarger, 1949a

151 Arnoseris Gaertner
1 minima (L.) Schweigger & Koerte 18 Br Stebbins et al., 1953

152 Koelpinia Pallas
1 linearis Pallas

153 Hyoseris L.
1 scabra L. 16 Sa Martinoli, 1953
2 radiata L.
  (a) radiata 16 Sa Martinoli, 1953
  (b) graeca Halácsy 16 Sa Martinoli, 1953
3 taurina (Pamp.) G. Martinoli 16 Sa Martinoli, 1953

154 Hedypnois Miller
1 cretica (L.) Dum.-Courset 8 Lu Fernandes & Queirós, 1971
    11 Lu Fernandes & Queirós, 1971
    12 Ga Fahmy, 1955
    13 Lu Fernandes & Queirós, 1971
    14 Lu Fernandes & Queirós, 1971
    13 Lu Fernandes & Queirós, 1971
    16 Ga Fahmy, 1955
    18 Lu Fernandes & Queirós, 1971
2 arenaria (Schousboe) DC. 6 Lu Fernandes & Queirós, 1971

155 Rhagadiolus Scop.
1 stellatus (L.) Gaertner

156 Aposeris Cass.
1 foetida (L.) Less. 16 Cz Májovský et al., 1970

157 Urospermum Scop.
1 dalechampii (L.) Scop. ex F.W. Schmidt 14 Bl Dahlgren et al., 1971
2 picroides (L.) Scop. ex F.W. Schmidt 10 Ga Delay, 1970

158 Hypochoeris L.
1 robertia Fiori 8 Sa Martinoli, 1953
2 laevigata (L.) Cesati, Passer. & Gibelli
3 achyrophorus L. 12 Ga Kliphuis & Wieffering, 1972
4 cretensis (L.) Bory & Chaub. 6 Gr Stebbins et al., 1953
5 maculata L. 10 Ga Delay, 1968
6 uniflora Vill. 10 Po Skalińska, 1950
7 tenuiflora (Boiss.) Boiss.
8 glabra L. 10 Lu Fernandes & Queirós, 1971

```
 9 radicata L.                                    8      Lu      Fernandes & Queirós,
                                                                  1971

159 Leontodon L.
    1 pyrenaicus Gouan
      (a) pyrenaicus
      (b) cantabricus (Widder) Finch & P.D. Sell
      (c) helveticus (Merat) Finch & P.D.       12      He      Favarger, 1953
          Sell
    2 croceus Haenke
      (a) croceus                                24      Au      Favarger, 1959a
      (b) rilaensis (Hayek) Finch & P.D.         14      Bu      Kozuharov &
          Sell                                                    Kuzmanov, 1964b
    3 montanus Lam.
      (a) montanus                               12      Ga      Delay, 1970
      (b) montaniformis (Widder) Finch &
          P.D. Sell
      (c) pseudotaraxaci (Schur) Finch &         12      Po      Skalińska et al.,
          P.D. Sell                                               1959
    4 autumnalis L.                              12      No      Knaben & Engelskjon,
                                                                  1967
                                                 24      Scand.  Vaarama, 1948

      (a) autumnalis
      (b) pratensis (Koch) Arcangeli
    5 keretinus F. Nyl.
    6 carpetanus Lange
      (a) carpetanus
      (b) nevadensis (Lange) Finch & P.D.
          Sell
    7 duboisii Sennen ex Widder
    8 microcephalus (Boiss. ex DC.) Boiss.
    9 cichoraceus (Ten.) Sanguinetti
   10 muelleri (Schultz Bip.) Fiori
   11 salzmannii (Schultz Bip.) Ball
   12 hispidulus (Delile) Boiss.
   13 repens Schur
   14 schischkinii V. Vassil.
   15 hispidus L.
      (a) hispidus                               14      Cz      Májovský et al.,
                                                                  1970
      (b) alpinus (Jacq.) Finch & P.D. Sell
      (c) pseudocrispus (Schultz Bip. ex
          Bischoff) J. Murr
      (d) danubialis (Jacq.) Simonkai            14      Po      Skalińska et al.,
                                                                  1964
      (e) opimus (Koch) Finch & P.D. Sell
      (f) hyoseroides (Welw. ex Reichenb.)
          J. Murr.
   16 siculus (Guss.) Finch & P.D. Sell
   17 boryi Boiss. ex DC.                        14      Hs      Küpfer, 1968
   18 hirtus L.                                   8      It      Larsen, 1956a
   19 crispus Vill.
      (a) rossianus (Degen & Lengyel) Hayek
      (b) crispus                                 8      Al      Strid, 1971
      (c) asperrimus (Willd.) Finch & P.D.
          Sell
      (d) bourgaeanus (Willk.) Finch & P.D.
          Sell
      (e) graecus (Boiss. & Heldr.) Hayek         8      Gr      Phitos & Damboldt,
                                                                  1971

   20 hellenicus Phitos
```

21 incanus (L.) Schrank
  (a) incanus   8   Hu   Baksay, 1956
  (b) tenuiflorus (Gaudin) Hegi
22 berinii (Bartl.) Roth
23 tuberosus L.   8   Lu   Fernandes & Queirós,
                                                               1971

24 maroccanus (Pers.) Ball   8
25 taraxacoides (Vill.) Mérat
  (a) taraxacoides   8   Ga   Delay, 1969
  (b) longirostris Finch & P.D. Sell   8   Lu   Fernandes & Queirós,
                                                               1971

26 filii (Hochst.) Paiva & Ormonde
27 rigens (Aiton) Paiva & Ormonde

160 Picris L.
  1 aculeata Vahl
  2 echioides L.   10   Lu   Fernandes & Queirós,
                                                               1971

  3 comosa (Boiss.) B.D. Jackson
  4 algarbiensis Franco
  5 spinifera Franco
  6 hispanica (Willd.) P.D. Sell
  7 hispidissima (Bartl.) Koch
  8 scaberrima Guss.
  9 hieracioides L.   10   Lu   Fernandes & Queirós,
                                                               1971

    (a) longifolia (Boiss. & Reuter) P.D.
        Sell
    (b) hieracioides
    (c) spinulosa (Bertol. ex Guss.)
        Arcangeli
    (d) villarsii (Jordan) Nyman
    (e) grandiflora (Ten.) Arcangeli
 10 pauciflora Willd.
 11 sprengerana (L.) Poiret
 12 willkommii (Schultz Bip.) Nyman

161 Scorzonera L.
  1 cana (C.A. Meyer) O. Hoffm.   14   Hu   Pólya, 1948
  2 laciniata L.   14   Lu   Fernandes & Queirós,
                                                               1971

  3 mollis Bieb.
  4 idaea (Gand.) Lipsch.
  5 undulata Vahl
  6 purpurea L.   14   Po   Skalińska et al.,
                                                               1971

    (a) purpurea
    (b) rosea (Waldst. & Kit.) Nyman   14   It   Favarger, 1965b
    (c) peristerica Form.
  7 graminifolia L.
  8 austriaca Willd.   14   Cz   Májovský et al.,
                                                               1970

    (a) austriaca
    (b) crispa (Bieb.) Nyman
    (c) bupleurifolia (Pouzolz) Bonnier
  9 humilis L.   14   Lu   Fernandes & Queirós,
                                                               1971

                                               15   Lu   Fernandes & Queirós,
                                                               1971

 10 parviflora Jacq.   14   Cz   Májovský et al.,
                                                               1970

274

```
  11 baetica (Boiss.) Boiss.
  12 transtagana Coutinho
  13 crocifolia Sibth. & Sm.              14    Gr    Damboldt, 1968
  14 aristata Ramond ex DC.               14    It    Favarger & Huynh,
                                                      1964
  15 hispanica L.                         14    Po    Skalińska et al.,
                                                      1964
  16 crispatula (Boiss.) Boiss.
  17 brevicaulis Vahl
  18 scyria M. Gustafsson & Snogerup      14    Gr    Gustaffson &
                                                      Snogerup, 1972
  19 fistulosa Brot.
  20 pusilla Pallas
  21 villosa Scop.
     (a) villosa
     (b) columnae (Guss.) Nyman
     (c) callosa (Moris) Chater
  22 ensifolia Bieb.
  23 hirsuta L.                           12    Ga    Delay, 1968
                                          14    Ga    Delay, 1968
  24 doria Degen & Bald.
  25 cretica Willd.
  26 albicans Cosson
  27 lanata (L.) Hoffm.
  28 tuberosa Pallas

162 Tragopogon L.
   1 ruber S.G. Gmelin
   2 marginifolius Pawl.
   3 cretaceus S. Nikitin
   4 porrifolius L.
     (a) porrifolius                      12    Br    Edmonds et al., 1974
     (b) australis (Jordan) Nyman         12    Gr    Damboldt, 1968
     (c) cupani (Guss. ex DC.) I.B.K.
         Richardson
   5 pterodes Pančić                      12    Bu    Kozuharov &
                                                      Kuzmanov, 1968
   6 balcanicus Velen.                    12    Bu    Kozuharov &
                                                      Kuzmanov, 1968
   7 longirostris Bischoff ex Schultz Bip.
   8 crocifolius L.
     (a) crocifolius
     (b) samaritani (Heldr. & Sart. ex Boiss.)
         I.B.K. Richardson
   9 kindingeri Adamović
  10 dubius Scop.                         12    Hu    Pólya, 1950
  11 pratensis L.
     (a) pratensis                        12    Bu    Kozuharov &
                                                      Kuzmanov, 1968
     (b) minor (Miller) Wahlenb.          12    Ga    Delay, 1969
     (c) orientalis (L.) Čelak.           12    Po    Skalińska et al.,
                                                      1961
  12 lassithicus Rech. fil.
  13 hayekii (Soó) I.B.K. Richardson
  14 tommasinii Schultz Bip.
  15 brevirostris DC.
     (a) brevirostris
     (b) volgensis (S. Nikitin) C. Regel
     (c) podolicus (DC.) C. Regel
     (d) bjelorussicus (Artemczuk) C. Regel
```

(e) longifolius (Heldr. & Sart. ex Boiss.)
    I.B.K. Richardson
16 dasyrhynchus Artemczuk
   (a) dasyrhynchus
   (b) daghestanicus Artemczuk
17 elatior Steven
18 floccosus Waldst. & Kit.
   (a) floccosus   12   Cz   Májovský et al.,
                                                          1970
   (b) heterospermus (Schweigger) C. Regel
19 ruthenicus Besser ex Krasch. & S.
   Nikitin
   (a) ruthenicus
   (b) donetzicus (Artemczuk) I.B.K.
       Richardson
20 hybridus L.   14   Lu   Queirós, 1973

163 Reichardia Roth
    1 tingitana (L.) Roth   16   Bl   Dahlgren et al.,
                                                          1971
    2 gaditana (Willk.) Coutinho   16   Lu   Fernandes & Queirós,
                                                          1971
    3 picroides (L.) Roth   14   It   Larsen, 1955b
    4 intermedia (Schultz Bip.) Coutinho 14   Lu   Fernandes & Queirós,
                                                          1971

164 Launaea Cass.
    1 resedifolia (L.) O. Kuntze
    2 nudicaulis (L.) Hooker fil.
    3 pumila (Cav.) O. Kuntze   16   Hr   Lorenzo-Andreu &
                                                          Garcia, 1950
    4 cervicornis (Boiss.) Font Quer &   18   Bl   Dahlgren et al.,
      Rothm.   1971
    5 lanifera Pau   16   Hs   Casas, 1973
    6 arborescens (Batt.) Murb.

165 Aetheorhiza Cass.
    1 bulbosa (L.) Cass.   18   Lu   Fernandes & Queirós,
                                                          1971
    (a) bulbosa
    (b) microcephala Rech. fil.
    (c) willkommii (Burnat & W. Barbey)
        Rech. fil.

166 Sonchus L.
    1 asper (L.) Hill
      (a) asper   18   Ho   Gadella & Kliphuis, 1966
      (b) glaucescens (Jordan) Ball
    2 tenerrimus L.   14   It   Larsen, 1956a
    3 oleraceus L.   32   Lu   Rodriguez, 1953
    4 maritimus L.   18   Ga   Delay, 1968
      (a) maritimus
      (b) aquatilis (Pourret) Nyman
    5 palustris L.   18   Da   Henin, 1960
    6 crassifolius Pourret ex Willd.
    7 arvensis L.   36   Fe   V. Sorsa, 1962
                                          54   Ho   Gadella & Kliphuis,
                                                          1968
      (a) arvensis
      (b) uliginosus (Bieb.) Nyman
    8 pustulatus Willk.

167 Cephalorrhynchus Boiss.
  1 tuberosus (Steven) Schchian

168 Steptorhamphus Bunge
  1 tuberosus (Jacq.) Grossh.

169 Lactuca L.
  1 viminea (L.) J. & C. Presl
    (a) alpestris (Gand.) Ferakova

(b) viminea	18	Cz	Feráková, 1968

    (c) chondrilliflora (Boreau) Bonnier
    (d) ramosissima (All.) Bonnier

2 acanthifolia (Willd.) Boiss.	18	Cr	Phitos & Kamari, 1974

  3 longidentata Moris ex DC.

4 tatarica (L.) C.A. Meyer	18	Rm	Tarnavschi, 1935

  5 sibirica (L.) Maxim.
  6 quercina L.

(a) quercina	18	Cz	Feráková, 1968b

    (b) wilhelmsiana (Fischer & C.A. Meyer
      ex DC.) Feráková
  7 watsoniana Trelease

8 aurea (Schultz Bip. ex Pančić) Stebbins	16	Bu	Babcock et al., 1937
9 serriola L.	18	Hu	Pólya, 1950
10 sativa L.	18	Su	Lindqvist, 1960
11 saligna L.	18	Au	Lindqvist, 1960

  12 altaica Fischer & C.A. Meyer

13 virosa L.	18	Lu	Fernandes & Queirós, 1971

  14 livida Boiss. & Reuter

15 perennis L.	18	Ge	Babcock et al., 1937
16 tenerrima Pourret	16	Bl	Babcock et al., 1937

  17 graeca Boiss.

170 Cicerbita Wallr.

1 alpina (L.) Wallr.	18	Bu No	Stebbins et al., 1953

  2 macrophylla (Willd.) Wallr.
  3 plumieri (L.) Kirschleger
  4 pancicii (Vis.) Beauverd

171 Prenanthes L.

1 purpurea L.	18	Bu	Kozuharov & Kuzmanov, 1967

172 Mycelis Cass.

1 muralis (L.) Dumort	18	Su	Löve & Löve, 1942

173 Taraxacum Weber
  1 glaciale Huet ex Hand.-Mazz.

2 bessarabicum (Hornem.) Hand.-Mazz.	16	Hu	Gustaffson, 1932
3 serotinum (Waldst. & Kit.) Poiret	16	Cz	Májovský et al., 1970

  4 phymatocarpum J. Vahl
  5 glabrum DC.

6 pacheri Schultz Bip.	32	Au	Fürnkranz, 1965
7 schroeteranum Hand.-Mazz.	24	He	Hou-Liu, 1963

  8 bithynicum DC.

minimum (Briganti ex Guss.) N. Terracc.	16	Ju	Richards, 1969
9 obovatum (Willd.) DC.	32	Hs	Bjorkqvist et al., 1969
10 glaucanthum (Ledeb.) DC.			
11 ceratophorum (Ledeb.) DC.			
tornense T.C.E. Fries	32	Fe Su	Gustaffson, 1935
12 spectabile Dahlst.			
eximium Dahlst.	40	No	Fagerlind, 1968
faeroense (Dahlst.) Dahlst.	40	Br	Richards, 1969
spectabile Dahlst.	40	Br	Richards, 1969
13 praestans H. Lindb. fil.			
euryphyllum (Dahlst.) M.P. Christiansen	32	Br	Richards, 1969
lainzii Van Soest	24	Br	Richards, 1969
landmarkii Dahlst.	32	No	Fagerlind, 1968
maculigerum H. Lindb. fil.	32	Br	Richards, 1969
naevosiforme Dahlst.	32	Br	Richards, 1972
naevosum Dahlst.	32	No	Fagerlind, 1968
praestans H. Lindb. fil.	32	Br	Richards, 1969
stoctophyllum Dahlst.	32	No	Fagerlind, 1968
14 unguilobum Dahlst.			
fulvicarpum Dahlst.	32	Br	Richards, 1969
unguilobum Dahlst.	32	Br	Richards, 1969
15 croceum Dahlst.			
craspedotum Dahlst.	32	Br	Richards, 1972
croceum Dahlst.	32	Br	Richards, 1972
cymbifolium H. Lindb. fil. ex Dahlst.	32	Sb	Engelskjon, 1968
pycnostictum M.P. Christiansen	32	Br	Richards, 1969
repletum (Dahlst.) Dahlst.	40	No	Fagerlind, 1968
16 adamii Claire	24	Br	Richards, 1969
litorale Raunk.	24	Da	Richards, 1969
holmstedtii Dahlst.	48	Br	Richards, 1969
17 palustre (Lyons) Symons			
austrinum G. Hagl.	24	Cz	Malecka, 1970
	32	Br	Richards, 1969
balticiforme Dahlst.	24	Po	Malecka, 1970
balticum Dahlst.	24		
	31	Su	Richards, 1972
	32	Po	Malecka, 1970
crocodes Dahlst.	40	Su	Richards, 1969
illyricum Dahlst. ex Van Soest	29	Hu	Richards, 1972
palustre (Lyons) Symons	40	Br	Richards, ined.
turfosum (Schultz Bip.) Van Soest	24	Po	Malecka, 1970
18 apenninum (Ten.) Ten.			
helveticum Van Soest	32	Po	Richards, 1972
venustum Dahlst.	32	Au	Fürnkranz, 1965
19 nigricans (Kit.) Reichenb.			
aestivum Van Soest	32	Po	Richards, 1972
nigricans (Kit.) Reichenb.	24	Au Po	Richards, 1972
	32	Po	Malecka, 1962
rhaeticum Van Soest	24	He	Hou-Liu, 1963
20 fontanum Hand.-Mazz.			
fontaneosquameum Van Soest	25	Au	Richards, 1972
pohlii Van Soest	32	Au	Richards, 1972
21 cucullatum Dahlst.			
concullatum A.J. Richards	24	Au	Richards, 1972
22 dissectum (Ledeb.) Ledeb.			
23 obliquum (Fries) Dahlst.	24	Ho	Hou-Liu, 1963
24 erythrospermum Andrz. ex Besser			
austriaeum Van Soest	16	Cz	Richards, 1970
	24	Cz	Richards, 1970

278

brachyglossum (Dahlst.) Dahlst.	16	Br	Richards, 1969
	24	Br	Richards, 1969
disseminatum G. Hagl.	24	Ge	Richards, 1969
dunense Van Soest	24	Br	Richards, 1969
isophyllum G. Hagl.	24	Su	Richards, 1969
lacistophyllum (Dahlst.) Raunk.	24	Br	Richards, 1969
	25	Su	Richards, 1972
laetum (Dahlst.) Dahlst.	24	Su	Richards, 1972
polyschistum Dahlst.	24	Su	Richards, ined.
proximum (Dahlst.) Dahlst.	24	Br Cz	Richards, 1969
rubicundum (Dahlst.) Dahlst.	24	Br	Richards, 1969
scanicum Dahlst.	25	Po	Malecka, 1967
silesiacum Dahlst. ex G. Hagl.	24	Cz	Richards, 1970
taeniatum G. Hagl. ex Holmgren	24	Ho	Hou-Liu, 1963
tenuilobum (Dahlst.) Dahlst.	24	Rs(B)	Gustaffson, 1935
	25	Po	Malecka, 1969
25 simile Raunk.			
dissimile Dahlst.	24	Su	Gustaffson, 1935
placidum A.J. Richards	24	Ga	Richards, 1969
proximiforme Van Soest	24	Br	Richards, 1969
pseudolacistophyllum Van Soest	24	Br	Richards, 1969
purpureomarginatum Van Soest	24	Cz	Richards, 1969
simile Raunk.	32	Br	Richards, 1969
tortilobum Florström	24	Ho	Hou-Liu, 1963
26 fulvum Raunk.			
fulviforme Dahlst.	32	Br	Richards, 1969
fulvum Raunk.	32	Da	Gustaffson, 1935
oxoniense Dahlst.	32	Br	Richards, 1969
27 gasparrinii Tineo ex Lojac.			
28 hoppeanum Griseb.			
amborum G. Hagl.	24	Gr	Gustaffson, 1935
aquilonare Hand.-Mazz.	24	Au	Furnkranz, 1965
capricum Van Soest	24	Ga	Richards, 1969
pieninicum Pawl.	16	Po	Malecka, 1961
29 crassipes H. Lindb.			
cochleatum Dahlst. & H. Lindb. fil.	24	Fe	Richards, 1969
30 officinale Weber			
aequilobum Dahlst.	24	Su	Richards, 1969
alatum H. Lindb. fil.	24	Br	Richards, 1969
cordatum Palmgren	24	Su	Gustaffson, 1932
cyanolepis Dahlst.	24	Su	Richards, 1969
dahlstedtii H. Lindb. fil.	24	Br	Richards, 1969
duplidens H. Lindb. fil.	24	Su	Gustaffson, 1935
	26	Br	Richards, ined.
ekmanii Dahlst.	24	Su	Gustaffson, 1935
fasciatum Dahlst.	24	Su	Gustaffson, 1935
haematicum G. Hagl.	24	Da	Richards, 1969
haematopus H. Lindb. fil.	24	Da	Gustaffson, 1935
hamatiforme Dahlst.	24	Br	Richards, 1969
hamatum Raunk.	24	Br Ga	Richards, 1969
insigne E.L. Ekman ex Wiinst. & K.	25	Br	Richards, 1972
involucratum Dahlst.	24	Su	Gustaffson, 1932
laciniosifrons Wiinst.	19	Da	Sörensen &
	20	Da	Gudjonsson,
	21	Da	1946
	22	Da	
	23	Da	Sörensen &
	24	Da	Gudjónsson,
	48	Da	1946
laeticolor Dahlst.	24	Su	Gustaffson, 1935
longisquameum H. Lindb. fil.	24	Su	Gustaffson, 1932

marklundii Palmgren.	24	Br	Richards, 1969
melanthoides Dahlst.	24	Su	Gustaffson, 1935
mimulum Dahlst.	24	Rs(N)	Gustaffson, 1932
oblongatum Dahlst.	24	Br	Richards, ined.
parvuliceps H. Lindb. fil.	24	Rs(N)	Gustaffson, 1932
polyodon Dahlst.	21	Da	Sørensen &
	22	Da	Gudjónsson,
	23	Da	1946
	24	Br Da	Richards, 1969
	44 45	Da	Sørensen &
	46 47	Da	Gudjónsson,
	48	Da	1946
raunkiaeri Wiinst.	24	Br	Richards, 1969
retroflexum H. Lindb. fil.	24	Su	Gustaffson, 1935
sellandii Dahlst.	24	Br	Richards, ined.
speciosum Raunk.	24	Da	Richards, 1969
stenoschistum Dahlst.	24	Su	Gustaffson, 1932
subcyanolepis M.P. Christiansen	24	Br	Richards, 1969
sublaeticolor Dahlst.	24	Su	Gustaffson, 1935
tenebricans (Dahlst.) Dahlst.	24	No	Fagerlind, 1968

174 Chondrilla L.

1 juncea L.	14 + 1B	Ga	Delay, 1970
	30	Ga	Delay, 1947

2 ramosissima Sibth. & Sm.
3 pauciflora Ledeb.
4 chondrilloides (Ard.) Karsten

175 Calycocorsus F.W. Schmidt

1 stipitatus (Jacq.) Rauschert	10	Ga	Delay, 1970

176 Heteracia Fischer & C.A. Meyer
1 szovitsii Fischer & C.A. Meyer

177 Lapsana L.
1 communis L.

(a) communis	14	Br	Edmonds et al., 1974
	16	Fe	V. Sorsa, 1962
(b) adenophora (Boiss.) Rech. fil.			
(c) intermedia (Bieb.) Hayek	14		
(d) alpina (Boiss. & Balansa) P.D. Sell			

178 Crepis L.

1 sibirica L.	10	?Europe	Babcock, 1947
2 geracioides Hausskn.	12	?Europe	Babcock, 1947
3 viscidula Froelich	12	Bu	Kuzmanov et al., 1981
4 paludosa (L.) Moench	12	Cz	Májovský et al., 1970
5 pygmaea L.			
(a) pygmaea	12	Ga	Delay, 1970
(b) anachoretica Babcock			
6 terglouensis (Hacq.) A. Kerner	12	?Europe	Babcock, 1947
7 jacquinii Tausch			
(a) jacquinii			
(b) kerneri (Rech. fil.) Merxm.			
8 rhaetica Hegetschw.			
9 aurea (L.) Cass.			
(a) aurea	10	Ga	Delay, 1970

(b) glabrescens (Caruel) Arcangeli	10	?Europe	Babcock, 1947
10 chrysantha (Ledeb.) Turcz.	8	?Europe	Babcock, 1947
11 lampsanoides (Gouan) Tausch	12	Ga	Delay, 1969
12 smyrnaea DC. ex Froelich			
13 mollis (Jacq.) Ascherson	12	Ga	Delay, 1970
14 fraasii Schultz Bip.			
(a) fraasii	12	?Europe	Babcock, 1947
(b) mungieri (Boiss. & Heldr.) P.D. Sell	12	?Europe	Babcock, 1947
15 bocconi P.D. Sell	10	Ga	Delay, 1970
16 conyzifolia (Gouan) A. Kerner	8	Ga	Delay, 1970
17 pyrenaica (L.) W. Greuter	8	Ga	Delay, 1970
18 alpestris (Jacq.) Tausch	8	?Europe	Babcock, 1947
19 albida Vill.			
(a) albida	10	Ga	Delay, 1970
(b) asturica (Lacaita & Pau) Babcock	10	?Europe	Babcock, 1947
(c) grosii (Pau) Babcock			
(d) scorzoneroides (Rouy) Babcock			
(e) macrocephala (Willk.) Babcock	10	?Europe	Babcock, 1947
(f) longicaulis Babcock			
20 tingitana Ball	10	?Europe	Babcock, 1947
21 leontodontoides All.	10	?Europe	Babcock, 1947
22 biennis L.	31 36	Br	Edmonds et al., 1974
	38 39 40	Br	Edmonds et al., 1974
23 pannonica (Jacq.) C. Koch	8	Cz	Májovský et al., 1970
24 lacera Ten.	18	?Europe	Babcock, 1947
25 bertiscea Jav.			
26 chondrilloides Jacq.	8	?Europe	Babcock, 1947
27 auriculifolia Sieber ex Sprengel	10		
28 baldaccii Halácsy	10	?Europe	Babcock, 1947
29 turcica Degen & Bald.			
30 pantocsekii (Vis.) Latzel			
31 triasii (Camb.) Nyman	8	?Europe	Babcock, 1947
32 albanica (Jav.) Babcock			
33 macedonica Kitanov			
34 oporinoides Boiss. ex Froelich	8		
35 sibthorpiana Boiss. & Heldr.			
36 incana Sibth. & Sm.	16	?Europe	Babcock, 1947
37 taygetica Babcock	c.40	?Europe	Babcock, 1947
38 guioliana Babcock			
39 crocifolia Boiss. & Heldr.			
40 athoa Boiss.			
41 schachtii Babcock	10	?Europe	Babcock, 1947
42 bithynica Boiss.	10	Bu	Kuzmanov & Kozuharov, 1970
43 praemorsa (L.) Tausch			
(a) praemorsa	8	Ga	Zickler & Lambert, 1967
(b) corymbosa (Gaudin) P.D. Sell			
(c) dinarica (G. Beck) P.D. Sell			
44 tectorum L.	8	Fe	Sorsa, 1963
(a) tectorum			
(b) nigrescens (Pohle) P.D. Sell			
(c) pumila (Liljeblad) Sterner			
45 purpurea (Willd.) Bieb.			
46 reuterana Boiss.			
47 pulchra L.	8	Cz	Májovský et al., 1970

48 stojanovii Georgiev	8		
49 alpina L.	10	Rs(K)	Babcock, 1947
50 rubra L.	10	?Europe	Babcock, 1947
51 foetida L.			
(a) foetida	10	Lu	Fernandes & Queirós, 1971
(b) rhoeadifolia (Bieb.) Čelak.	10	Cz	Májovský et al., 1970
(c) commutata (Sprengel) Babcock	10	Lu	Fernandes & Queirós, 1971
52 multicaulis Ledeb.	10	?Europe	Babcock, 1947
53 sancta (L.) Babcock	10	Bu	Kuzmanov & Juakova, 1977
54 dioscoridis L.	8	?Europe	Babcock, 1947
55 multiflora Sibth. & Sm.	8	?Europe	Babcock, 1947
56 zacintha (L.) Babcock	6	?Europe	Babcock, 1947
57 pusilla (Sommier) Merxm.	10		Merxmüller, 1968
58 nicaeensis Balbis	8	Ga	Delay, 1970
59 foliosa Babcock			
60 capillaris (L.) Wallr.	6	Ge	Gross, 1966
61 micrantha Czerep.	8	?Europe	Babcock, 1947
62 neglecta L.			
(a) neglecta	8	?Europe	Babcock, 1947
(b) corymbosa (Ten.) Nyman	8	?Europe	Babcock, 1947
(c) fuliginosa (Sibth. & Sm.) Vierh.	6	?Europe	Babcock, 1947
(d) cretica (Boiss.) Vierh.	8	?Europe	Babcock, 1947
63 suffreniana (DC.) Lloyd			
(a) suffreniana	8	?Europe	Babcock, 1947
(b) apula (Fiori) P.D. Sell	8		
64 spathulata Guss.			
65 bourgaeui Boiss. ex Maire	8	?Europe	Babcock, 1947
66 vesicaria L.			
(a) vesicaria	8	It	Marchi, 1971
	16	Si	Pavone et al., 1981
(b) hyemalis (Biv.) Babcock	8	Si	
(c) haenseleri (Boiss. ex DC.) P.D.Sell	8	Br	Edmonds et al., 1974
	16		
(d) congenita Babcock			
67 tybakiensis Vierh.			
68 setosa Haller fil.	6	Br	Edmonds & Walters, ined.
	8	Al	Strid, 1971
69 bellidifolia Loisel.	8	?Europe	Babcock, 1947
70 bursifolia L.	8	Ga	Van Loon et al., 1971
179 Hispidella Barnades ex Lam.			
1 hispanica Barnades ex Lam.	18	Lu	Fernandes & Queirós, 1971
180 Andryala L.			
1 integrifolia L.	18	Ga	Kliphuis & Wieffering, 1972
2 laxiflora DC.	18	Lu	Fernandes & Queirós, 1971
3 ragusina L.			
4 agardhii Haenseler ex DC.			
5 levitomentosa (E.I. Nyárády) P.D. Sell			
181 Hieracium L.			
1 castellanum Boiss. & Reuter	18	Hs	Merxmüller, 1975
2 hoppeanum Schultes	18	He Ju	Gadella, 1972
	45		

```
    (a) hoppeanum
    (b) pilisquamum Naegeli & Peter
    (c) testimoniale Naegeli ex Peter
    (d) troicum Zahn
    (e) cilicicum Naegeli & Peter
  3 x viridifolium Peter
  4 x hypeuryum Peter
  5 x ruprechtii Boiss.
  6 x tephrocephalum Vuk.
  7 peleteranum Mérat             18      Br     Richards, 1972
                                  27 36
                                  45
    (a) peleteranum               18      Br     Finch, ined.
    (b) subpeleteranum Naegeli & Peter  18  Ho   Gadella & Kliphuis,
                                                   1968
    (c) sabulosorum Dahlst.       18      Su     G. & B. Turesson,
                                                   1960
    (d) tenuiscapum (Pugsley) P.D. Sell
    (e) ligericum Zahn
  8 x longisquamum Peter          27      Su     G. & B. Turesson,
                                                   1960
  9 argyrocomum (Fries) Zahn      18      Hs     Merxmüller, 1975
 10 pilosella L.                  18      It     Gadella & Kliphuis,
                                                   1970
                                  36      Ho     Gadella & Kliphuis,
                                                   1966
                                  45      Ga     Gadella, 1972
                                  54      It     Scannerini, 1971
                                  63      Su     G. & B. Turesson,
    (a) micradenium Naegeli & Peter  18          1960
                                  36 45
    (b) euronotum Naegeli & Peter  45     Br     Finch, ined.
    (c) pilosella                 36      Su     G. & B. Turesson,
    (d) trichosoma Peter          36             1960
                                  45
                                  54
                                  63
    (e) tricholepium Naegeli & Peter  36  Br     Finch, ined.
    (f) melanops Peter            36
                                  45
    (g) trichoscapum Naegeli & Peter  45  Su     B. Turesson, 1972
    (h) velutinum Fries           45      Ga     Gadella, 1972
                                  54
 11 x florentoides Arvet-Touvet
 12 x brachiatum Bertol. ex Lam.
 13 pseudopilosella Ten.
    (a) pseudopilosella           18      Hs It  Merxmüller, 1975
    (b) banaticola E.I. Nyárády & Zahn
 14 flagellare Willd.
    (a) flagellare                36      Europe Gustaffson, 1933
                                  45      Po     Skalińska, 1967
    (b) bicapitatum (P.D. Sell & C. West)
        P.D. Sell                 54      Br     Finch, ined.
 15 x flagellariforme G. Schneider
 16 lactucella Wallr.             18      Po     Skalińska, 1967
                                  27      Co     Contandriopoulos,
                                                   1962
                                  36
    (a) nanum (Scheele) P.D. Sell
    (b) lactucella
    (c) magnauricula (Naegeli & Peter) P.D. Sell
```

283

17 x auriculiforme Fries	18		
	36		
18 x schultesii F.W. Schultz			
19 x paragogum Naegeli & Peter			
20 x sulphureum Doll			
21 koernickeanum (Naegeli & Peter) Zahn			
22 vahlii Froelich	18	Hs	Merxmüller, 1975
23 glaciale Reyn.	18	Ga	Favarger, 1969
24 x niphostribes Peter			
25 sphaerocephalum Froelich			
26 breviscapum DC.			
27 alpicola Schleicher ex Gaudin	36	He	Favarger, 1959
28 piloselloides Vill.			
(a) piloselloides	36	Europe	Finch, ined.
(b) megalomastix (Naegeli & Peter) P.D. Sell			
pavichii	18		
29 praealtum Vill. ex Gochnat			
(a) praealtum	36		
	45		
(b) anadenium (Naegeli & Peter) P.D. Sell			
(c) bauhinii (Besser) Petunnikov	45	Cz	Uhríková, 1970
(d) thaumasium (Peter) P.D. Sell			
30 cymosum L.	36		
	54 63	Ca	Gadella & Kliphuis, 1970
(a) sabinum (Sebastiani & Mauri) Naegeli & Peter			
(b) cymosum			
(c) cymigerum (Reichenb.) Peter	36	Su	G. & B. Turesson, 1963
31 x fallax Willd.			
32 x sciadophorum Naegeli & Peter			
33 x anchusoides (Arvet-Touvet) Arvet-Touvet			
34 x spurium Chaix ex Froelich			
35 x fallacinum F.W. Schultz			
36 x zizianum Tausch			
37 x densiflorum Tausch			
38 caespitosum Dumort.	18	Po	Skalińska & Kubien, 1972
	27	Po	Skalińska, 1967
	36	Po	Skalińska & Kubien, 1972
	45	Po	Skalińska, 1967
(a) caespitosum	45	Su	G. & B. Turesson, 1963
(b) colliniforme (Peter) P.D. Sell			
(c) brevipilum (Naegeli & Peter) P.D. Sell	18	Po	Skalińska et al., 1972
39 x ambiguum Ehrh.	36		
	45		
40 x dubium L.	36	Su	G. & B. Turesson,
	45		1963
41 x macranthelum Naegeli & Peter	45	Su	G. & B. Turesson, 1963
42 x floribundum Wimmer & Grab.	27	Po	Skalińska, 1967
islandicum (Lange) Dahlst.	27	Is	Finch, ined.

43 x piloselliflorum Naegeli & Peter
44 x duplex Peter
45 x melinomelas Peter
46 x arvicola Naegeli & Peter
47 x polymastix Peter

48 aurantiacum L.	18	Po	Skalińska, 1970
	27	Po	Skalińska, 1967
	36	Po	Skalińska, 1967
	45	Po	Skalińska, 1967
	54	Po	Skalińska et al., 1968
	63	Po	Skalińska, 1967
	72		
(a) aurantiacum	36	Br	Finch, ined.

(b) carpathicola Naegeli & Peter
49 x fuscatrum Naegeli & Peter
50 x guthnickianum Hegetschw.
51 x plaicense Woloszczak
52 x biflorum Arvet-Touvet

| 53 x fuscum Vill. | 54 | Po | Skalińska, 1967 |

54 x peteranum Kaeser

| 55 x stoloniflorum Waldst. & Kit. | 45 46 | Europe | Finch, ined. |

56 x calomastix Peter

| 57 echioides Lumn. | 36 | Bot. Gard. | Gentscheff, 1937 |
| (a) echioides | 36 | Cz | Májovský et al., 1970 |

(b) procerum (Fries) P.D. Sell
58 x macrotrichum Boiss.
59 x bifurcum Bieb.
60 x heterodoxum (Tausch) Naegeli & Peter
61 x euchaetium Naegeli & Peter
62 x auriculoides A.F. Lang
63 x echiogenes (Naegeli & Peter) Juxip
64 verruculatum Link

65 murorum L.	27 36	Europe	Finch, ined.
grandidens Dahlst.	27	Europe	Finch, ined.
66 glaucinum Jordan	27		
	36		
gougetianum Gren. & Godron	27	Hb	Mills & Stace, 1974
scotostictum Hyl.	27	Europe	Finch, ined.
67 bifidum Kit.	27	Cz	Májovský et al., 1970
	36		

68 subcaesiiforme (Zahn) Zahn
69 fuscocinereum Norrlin
70 incisum Hoppe

| 71 vulgatum Fries | 27 | Br | Morton, 1974 |

72 benzianum J. Murr & Zahn

73 maculatum Sm.	27	Europe	Finch, ined.
74 caesium (Fries) Fries	27 36	Europe	Finch, ined.
caesiomurorum Lindeb.	36	Europe	Finch, ined.

75 hypastrum Zahn

76 ramosum Waldst. & Kit. ex Willd.	27	It	Battaglia, 1947
77 argillaceum Jordan	27	Europe	Finch, ined.
acuminatum Jordan	27	Europe	Finch, ined.

78 rotundatum Kit. ex Schultes

| 79 diaphanum Fries | 27 | Br | Morton, 1974 |

smolandicum group	27		Gustaffson, 1933
80 schmidtii Tausch	27 36	Europe	Finch, ined.
schmidtii Tausch	27	Europe	Finch, ined.
81 hypochoeroides Gibson	27		
	36		
82 sommerfeltii Lindb.			
83 aymericianum Arvet-Touvet			
84 bourgaei Boiss.			
85 stelligerum Froelich			
86 onosmoides Fries	27	Br	Morton, 1974
subrude (Arvet-Touvet) Arvet-Touvet	27	Europe	Finch, ined.
87 saxifragum Fries			
88 scoticum F.J. Hanb.			
89 caledonicum F.J. Hanb.	27	Br	Mills & Stace, 1974
90 laniferum Cav.	18	Hs	Merxmüller, 1975
91 elisaeanum Arvet-Touvet ex Willk.	18		Merxmüller, ined.
92 candidum Scheele			
93 phlomoides Froelich			
94 rupicaprinum Arvet-Touvet & Gaut.			
95 eriopogon Arvet-Touvet & Gaut.			
96 lawsonii Vill.			
97 briziflorum Arvet-Touvet			
98 subsericeum Arvet-Touvet			
99 cordifolium Lapeyr.			
100 sonchoides Arvet-Touvet			
101 purpurascens Scheele ex Willk.			
102 guadarramense Arvet-Touvet			
103 aragonense Scheele			
104 loscosianum Scheele			
105 cerinthoides L.			
106 alatum Lapeyr.	27	Br	Morton, 1974
	36		
anglicum Fries	36	Br	Mills & Stace, 1974
vogesiacum (Kirschléger) Fries	36	Bot. Gard.	Gentscheff, 1937
107 lamprophyllum Scheele			
108 longifolium Schleicher ex Froelich			
109 villosum Jacq.	27	Au	Polatschek, 1966
	36	He	Larsen, 1954
110 pilosum Schleicher ex Froelich	27		
	36		
111 scorzonerifolium Vill.	36		Christoff & Popoff, 1933
112 leucophaeum Gren. & Godron			
113 ctenodon Naegeli & Peter			
114 dentatum Hoppe			
115 chondrillifolium Fries			
116 cryptadenum Arvet-Touvet			
117 valdepilosum Vill.			
118 wilczekianum Arvet-Touvet			
119 chlorifolium Arvet-Touvet			
120 piliferum Hoppe			
121 dasytrichum Arvet-Touvet			
122 aphyllum Naegeli & Peter			
123 cochlearioides Zahn			
124 armerioides Arvet-Touvet			
125 mixtum Froelich			
126 mixtiforme Arvet-Touvet			
127 pictum Schleicher ex Pers.			

128 farinulentiforme Zahn			
129 pulchellum Gren.			
130 caesioides Arvet-Touvet			
131 pseudoprasinops Zahn			
132 cephalotes Arvet-Touvet			
133 leiopogon Gren. ex Verlot			
134 rupestre All.			
135 lanatum Vill.			
136 erioleucum Zahn			
137 jordanii Arvet-Touvet			
138 pellitum Fries			
139 lansicum Arvet-Touvet			
140 verbascifolium Vill.			
141 chaboissaei Arvet-Touvet			
142 pannosum Boiss.	36		Christoff & Popoff, 1933
143 gymnocephalum Griseb. ex Pant.			
144 pichleri A. Kerner			
145 gaudryi Boiss.			
146 waldsteinii Tausch	27 36		
147 dolopicum Freyn & Sint.			
148 guentheri-beckii Zahn			
149 scheppigianum Freyn			
150 mirificissimum Rohlena & Zahn			
151 lazistanum Arvet-Touvet			
152 calophyllum Uechtr.			
153 pilosissimum Friv.			
154 heldreichii Boiss.			
155 sericophyllum Nejc. & Zahn			
156 jankae Uechtr.			
157 sartorianum Boiss. & Heldr.			
marmoreum Pancic & Vis.	27	Bot. Gard.	Gentscheff, 1937
158 alpinum L.	27	Po	Skalińska et al., 1959
159 nigrescens Willd.	36	Br	Mills & Stace, 1974
calenduliflorum Backh.	36	Br	Morton, 1974
hanburyi Pugsley	36	Br	Mills & Stace, 1974
160 pietroszense Degen & Zahn			
161 fritzei F.W. Schultz			
162 arolae J. Murr			
163 senescens Backh.			
164 atratum Fries	27 36		
165 liptoviense Borbas			
166 krasanii Woloszczak			
167 rohacsense Kit. ex Kanitz			
168 bocconei Griseb.			
169 vollmannii Zahn			
lividorubens (Almq.) Elfstr.	27	Da	Jörgensen et al., 1958
paltinae Jav. & Zahn	36		Christoff & Popoff, 1933
170 sudeticum Sternb.			
171 nigritum Uechtr.	27	Da	Rosenberg, 1926
172 chlorocephalum Wimmer			
173 gombense Lagger & Christener			
174 humile Jacq.	27	Da	Rosenberg, 1926
175 cottetii Godet ex Gremli			

```
176 kerneri Ausserdorfer ex Zahn
177 valoddae (Zahn) Zahn
178 amplexicaule L.                    27      It      Gadella & Kliphuis,
                                                        1970
                                        36      Ho      Gadella & Kliphuis,
                                                        1968
    pulmonarioides Vill.               36      Br      Mills & Stace, 1974
    speluncarum Arvet-Touvet           36      Lu      Fernandes & Queirós,
                                                        1971
179 chamaepicris Arvet-Touvet
180 pseudocerinthe (Gaudin) Koch
181 rupicola Jordan
182 cordatum Scheele ex Costa
    hispanicum Arvet-Touvet            18      Hs      Merxmüller, 1975
183 glaucophyllum Scheele
184 pardoanum Arvet-Touvet
185 pedemontanum Burnat & Gremli
186 scapigerum Boiss., Orph. & Heldr.
187 urticaceum Arvet-Touvet
188 viscosum Arvet-Touvet
189 ramosissimum Schleicher ex Hegetschw.
190 arpadianum Zahn
191 picroides Vill.
192 neopicris Arvet-Touvet
193 intybaceum All.                    27      Da      Rosenberg, 1926
194 pallidiflorum Jordan ex Ascherson
195 khekianum Zahn
196 porrifolium L.                     18      Ju      Favarger, 1965
197 bupleuroides C.C. Gmelin           27      Ga      Májovsky et al., 1970
198 glaucum All.                       27      Au      Polatschek, 1966
                                        36      Au      Tischler, 1950
199 sparsiramum Naegeli & Peter
200 falcatum Arvet-Touvet
201 glabratum Hoppe ex Willd.
202 franconicum Griseb.
203 oxyodon Fries
204 fulcratum Arvet-Touvet
205 neyraeanum Arvet-Touvet
206 austriacum Brittinger
207 dollineri Schultz Bip. ex F.W. Schultz
208 calcareum Bernh. ex Hornem.
209 saxatile Jacq.
210 naegelianum Pančić
211 silesiacum Krause
212 sparsum Friv.                      18              Christoff, 1942
213 heterogynum (Froelich) Guterm.
214 macrodon Naegeli & Peter
215 macrodontoides (Zahn) Zahn
216 tommasinii Reichenb. fil.
217 olympicum Boiss.
218 leiocephalum Bartl. ex Griseb.
219 virgicaule Naegeli & Peter
220 pseudobupleuroides Naegeli & Peter
221 umbrosum Jordan
222 viride Arvet-Touvet
223 pinicola Arvet-Touvet
224 rapunculoides Arvet-Touvet
225 pedatifolium Omang
226 prenanthoides Vill.                27      Cz      Májovský et al.,
                                        36              1974

227 juranum Fries
```

228 juraniforme Zahn			
229 pocuticum Woloszczak			
230 cydonifolium Vill.			
231 doronicifolium Arvet-Touvet			
232 cantalicum Arvet-Touvet			
233 turritifolium Arvet-Touvet			
234 sergureum Arvet-Touvet			
235 epimedium Fries	27	Au	Polatschek, 1966
zetlandicum Beeby	27	Br	Mills & Stace, 1974
236 carpathicum Besser	36	Br	Mills & Stace, 1974
237 dovrense Fries	36	No	Finch, ined.
238 plicatum Lindeb.			
239 truncatum Lindeb.	27 36		
vinicaule P.D. Sell & C. West	36	Br	Mills & Stace, 1974
240 latifolium Froelich ex Link			
241 virosum Pallas	36	Rs( )	Christoff & Popoff, 1933
242 robustum Fries			
243 inuloides Tausch	27	Br	Morton, 1974
latobrigorum (Zahn) Roffey	27	Br	Finch, ined.
244 crocatum Fries			
245 lucidum Guss.	18	Si	Merxmüller, 1975
246 racemosum Waldst. & Kit. ex Willd.	27	It	Battaglia, 1947
crinitum Sibth. & Sm.	36		Merxmüller, ined.
247 pseuderiopus Zahn			
248 compositum Lapeyr.			
249 nobile Gren. & Godron			
250 rectum Griseb.			
251 symphytaceum Arvet-Touvet			
252 insuetum Jordan ex Boreau			
253 bracteolatum Sibth. & Sm.			
254 sabaudum L.	18 27	Lu	Queirós, 1973
sabaudum L.	27	Br	Morton, 1974
vagum Jordan	18	Br	Morton, 1974
255 flagelliferum Ravaud			
256 lycopsifolium Froelich			
257 unbellatum L.	18	Cz	Májovský et al., 1970
	27	Br	Morton, 1974
ssp. bichlorophyllum	18	Br	Morton, 1974
258 laevigatum Willd.	27	He	Gadella & Kliphuis, 1966
tridentatum Fries	27	Rs( )	Christoff & Popoff, 1933
259 eriophorum St.Amans	18	Ga	Merxmüller, 1975
260 prostratum DC.			

HELOBIAE

CLXX ALISMATACEAE

1 Sagittaria L.			
1 sagittifiolia L.	22	Po	Skalińska et al., 1966

2 natans Pallas	22	Fe	Vaarama, 1941
3 latifolia Willd.			
4 rigida Pursh			

2 Baldellia Parl.
1 ranunculoides (L.) Parl.	16	Su	Löve & Löve, 1944
	30		
2 alpestris (Cosson) Vasc.			

3 Luronium Rafin.
1 natans (L.) Rafin.	42	Su	Björkquist, 1961a

4 Alisma L.
1 plantago-aquatica L.	14	Ho	Gadella & Kliphuis, 1963
2 lanceolatum With.	26	Su	Löve & Löve, 1944
	27	Su	Pogan, 1968
	28	Da	Hagerup, 1944
3 gramineum Lej.	14	Cz	Májovský et al., 1976
4 wahlenbergii (Holmberg) Juz.	14	Fe	Björkquist, 1961b

5 Caldesia Parl.
1 parnassifolia (L.) Parl.

6 Damasonium Miller
1 alisma Miller	28	Ga	Schotsman, 1970

CLXXI BUTOMACEAE

1 Butomus L.
1 umbellatus L.	26	Po	Skalińska et al., 1961
	39	Ho	Gadella & Kliphuis, 1963

CLXXII HYDROCHARITACEAE

1 Hydrocharis L.
1 morsus-ranae L.	28	Ho	Gadella & Kliphuis, 1973

2 Stratiotes L.
1 aloides L.	24	?Ge	Schürhoff, 1926

3 Ottelia Pers.
1 alismoides (L.) Pers.	22	non-Europe	Rao, 1951

4 Egeria Planchon
1 densa Planchon

5 Elodea Michx
1 canadensis Michx	24	?Br	Heppell, 1945
	48	Po	Skalińska & Pogan, 1973
2 ernstiae St. John			
3 nuttallii (Planchon) St. John			

6 Hydrilla L.C.M. Richard
  1 verticillata (L. fil.) Royle   16   non-Europe   Larsen, 1963

7 Lagarosiphon Harvey
  1 major (Ridley) Moss

8 Blyxa Noronha ex Thouars
  1 japonica (Miq.) Maxim.

9 Vallisneria L.
  1 spiralis L.   20   ?Br cult.   Jørgensen, 1927

10 Halophila Thouars
   1 stipulacea (Forskål) Ascherson

CLXXIII SCHEUCHZERIACEAE

1 Scheuchzeria L.
  1 palustris L.   22   Br   Manton, 1949

CLXXIV APONOGETONACEAE

1 Aponogeton L. fil.
  1 distachyos L. fil.

CLXXV JUNCAGINACEAE

1 Triglochin L.
  1 maritima L.   24   Ga   Lévêque & Gorenflot,
                                                          1969

                                         30   Po   Skalińska et al.,
                                                          1961
                                         36   Ga   Labadie, 1976
                                         48   Po   Skalińska et al.,
                                                          1961

  2 palustris L.   24   Ga   Lévêque & Gorenflot,
                                                          1969

  3 striata Ruiz & Pavón
  4 bulbosa L.
    (a) barrelieri (Loisel.) Rouy   30   Lu   Castro & Fontes,
                                                          1946
                                         32   Bl   Dahlgren et al.,
                                                          1971
    (b) laxiflora (Guss.) Rouy   18   Lu   Gardé & Malheiros
                                                          Gardé, 1953

CLXXVI LILAEACEAE

1 Lilaea Humb. & Bonpl.
  1 scilloides (Poiret) Hauman

CLXXVII POTAMOGETONACEAE

1 Potamogeton L.

1 natans L.	52	Su	Palmgren, 1939
2 polygonifolius Pourret	26	Su	Palmgren, 1939
3 coloratus Hornem.	26	Da	Palmgren, 1939
4 nodosus Poiret			
5 lucens L.	52	Su	Palmgren, 1939
6 gramineus L.	52	Su	Palmgren, 1939
7 alpinus Balbis	26	Hs	Löve & Kjellqvist, 1973
	52	Is	Löve & Löve, 1956
8 praelongus Wulfen	52	Su	Palmgren, 1939
9 ¬erfoliatus L.	52	Su	Palmgren, 1939
10 epihydrus Rafin.			
11 friesii Rupr.	26	Su	Palmgren, 1939
12 rutilus Wolfg.			
13 pusillus L.	26	Su	Palmgren, 1939
14 obtusifolius Mert. & Koch	26	Su	Palmgren, 1939
15 berchtoldii Fieber	26	Su	Löve & Löve, 1956
16 trichoides Cham. & Schlecht.	26	Su	Palmgren, 1939
17 compressus L.	26	Su	Palmgren, 1939
18 acutifolius Link	26	Su	Palmgren, 1942
19 crispus L.	52	Su	Palmgren, 1939
20 filiformis Pers.	78	Is	Löve & Löve, 1956
21 vaginatus Turcz.	c.88	Su	Palmgren, 1939
22 pectinatus L.	78	Su	Palmgren, 1942

2 Groenlandia Gay			
1 densa (L.) Fourr.	30	?Su	Palmgren, 1939

CLXXVIII RUPPIACEAE

1 Ruppia L.			
1 maritima L.	20	Ge	Löve & Löve, 1961b
2 cirrhosa (Petagna) Grande	40	Ge	Löve & Löve, 1961b

CLXXIX POSIDONIACEAE

1 Posidonia C. Konig
   1 oceanica (L.) Delile

CLXXX ZOSTERACEAE

1 Zostera L.			
1 marina L.	12	Po	Skalińska et al., 1974
2 angustifolia (Hornem.) Reichenb.	12	Ge	Wulff, 1937a
3 noltii Hornem.	12	Ge	Wulff, 1937a

CLXXXI ZANNICHELLIACEAE

1 Althenia Petit
   1 filiformis Petit

2 Zannichellia L.			
1 palustris L.	12	Bu	Kozuharov & Kuzmanov, 1964

292

| | 28 | Ge | Scheerer, 1940 |
| | 24 32 34 36 | Ge | Reese, 1963 |

3 Cymodocea C. Konig
  1 nodosa (Ucria) Ascherson

CLXXXII NAJADACEAE

1 Najas L.
  1 marina L.   12   Scand.   Löve & Löve, 1948
  2 flexilis (Willd.) Rostk. & W.L.E.
    Schmidt   24   Br Su   Löve & Löve, 1958
  3 minor All.   24   Ga   Schotsman, 1970
  4 tenuissima (A. Braun) Magnus
  5 gracillima (A. Braun ex Engelm.)
    Magnus
  6 graminea Delile

LILIIFLORAE

CLXXIII LILIACEAE

1 Tofieldia Hudson
  1 pusilla (Michx) Pers.
    (a) pusilla   30   Br   Elkington, 1974
    (b) austriaca H. Kunz
  2 calyculata (L.) Wahlenb.   30   Ju   Lovka et al.,
                                                       1971

2 Narthecium Hudson
  1 ossifragum (L.) Hudson   26   Lu   Fernandes, 1950
  2 reverchonii Celak.   26   Co   Contandriopoulos,
                                                       1962

  3 scardicum Kosanin

3 Zigadenus L.C.M. Richard
  1 sibiricus (L.) A. Gray

4 Veratrum L.
  1 nigrum L.   16   Ju   Lovka et al.,
                                                       1972

  2 album L.   16   error
    lobelianum   32   Ju   Lovka et al.,
                                                       1971

5 Asphodelus L.
  1 fistulosus L.   28   Hs   Rejón, 1976
                                       56   Hs   Rejón, 1976

  2 albus Miller
    (a) albus   28   Gr   Alden, 1976

                                       56   Hs   Rejón, 1976
    (b) villarsii (Verlot ex Billot)
      I.B.K. Richardson & Smythies   78   Co   Lambert, 1969
  3 ramosus L.   28   Hs   Löve & Kjellqvist,
                                                       1973
                                       56   Lu   Barros Neves, 1973

293

	84	Lu	Barros Neves, 1973
4 aestivus Brot.	28 56 84	Lu	Barros Neves, 1973
5 bento-rainhae P. Silva			

6 Asphodeline Reichenb.
1 lutea (L.) Reichenb.	28	Bu	Borhidi, 1968
2 liburnica (Scop.) Reichenb.	14	Bu	Borhidi, 1968
	28	Bu	Popova, 1972
3 taurica (Pallas ex Bieb.) Kunth	28	Bu	Popova, 1972

7 Anthericum L.
1 liliago L.	32	Lu	Fernandes, 1950
	48		
	64	Ge	Tischler, 1934
2 baeticum (Boiss.) Boiss.	30	Hs	Küpfer, 1968
3 ramosum L.	30	Su	Strandhede, 1963
	32	Ga	Parreaux, 1971

8 Paradisea Mazzuc.
1 liliastrum (L.) Bertol.	16	Ga	Küpfer, 1969
2 lusitanica (Coutinho) Samp.	32	Lu	Fernandes, 1950
	64	Lu	Fernandes, 1950

9 Eremurus Bieb.
1 caucasicus Steven
2 tauricus Steven

10 Simethis Kunth
1 planifolia (L.) Gren.	48	Lu	Barros Neves, 1973

11 Aphyllanthes L.
1 monspeliensis L.	32	Hs	Löve & Kjellqvist, 1973

12 Hemerocallis L.
1 lilioasphodelus L.
2 fulva (L.) L.

13 Phormium J.R. & G. Forster
1 tenax J.R. & G. Forster

14 Aloe L.
1 aristata Haw.
2 humilis (L.) Miller
3 brevifolia Miller
4 vera (L.) Burm. fil.
5 ciliaris Haw.
6 perfoliata L.
7 succotrina All.
8 arborescens Miller
9 ferox Miller

15 Androcymbium Willd.
1 europaeum (Lange) K. Richter	18	Hs	Carpio, 1954

2 rechingeri W. Greuter

16 Colchicum L.
1 triphyllum G. Kunze	20 21	It	D'Amato, 1956
diampolis Delip. & Česchm ex Kuzmanov & Kozuharov	18	Bu	Popova & Česchmedjiev, 1966

```
  2 burttii Meikle
  3 hungaricum Janka                            54    Ju        Milan, 1975
  4 cupanii Guss.                               54    It        D'Amato, 1957a
  5 pusillum Sieber
  6 psaridis Heldr. ex Halacsy
  7 boissieri Orph.
  8 parlatoris Orph.
  9 arenarium Waldst. & Kit.                    36    Ju        Milan, 1975
                                                54    Ju        Milan, 1975
 10 alpinum DC.                                 54    It        D'Amato, 1957a
                                                56    Ga        Perrenoud &
    10 x 14                                  46 47    Ga        Favarger, 1971
 11 corsicum Baker
 12 micranthum Boiss.
    macedonicum Kosanin                         54    Ju        Milan, 1975
 13 umbrosum Steven
 14 autumnale L.                                38    Ga He It  D'Amato, 1957b
 15 neapolitanum (Ten.) Ten.                   140    It        D'Amato, 1957b
 16 lusitanum Brot.                            106    It        D'Amato, 1957b
 17 lingulatum Boiss. & Spruner
 18 parnassicum Sart.
 19 turcicum Janka
 20 callicymbium Stearn & Stefanov
 21 macrophyllum B.L. Burtt
 22 variegatum L.
 23 bivonae Guss.                            36 54    Bot.Gard. Levan, 1940
                                                90    Ju        Milan, 1975


17 Bulbocodium L.
   1 vernum L.
   2 versicolor (Ker-Gawler) Sprengel

18 Merendera Ramond
   1 attica (Spruner ex Tommasini) Boiss.
     & Spruner
   2 androcymbioides Valdes
   3 sobolifera C.A. Meyer                      54    Ju        Lovka et al.,
                                                                1971
   4 pyrenaica (Pourret) P. Fourn.              60    Lu        Fernandes, 1950
                                        54 + 0 - 6B   Hs        Löve &
                                                                Kjellqvist, 1973
   5 filifolia Camb.                            54    Bl        Dahlgren et al.,
                                                                1971

19 Lloydia Salisb. ex Reichenb.
   1 serotina (L.) Reichenb.                    24    He        Larsen, 1954

20 Gagea Salisb.
   1 pratensis (Pers.) Dumort.                  36    Cz        Májovský et al.,
                                                                1976
                                             48 60    Cz        Mešiček & Hrouda,
                                                                1974
     transversalis                             48    Cz        Mešiček & Hrouda,
                                                                1974
   2 pusilla (F.W. Schmidt) Schultes &
     Schultes fil.                             24     Cz        Mešiček & Hrouda,
                                                                1974
   3 lutea (L.) Ker-Gawler                      72    Cz        Mešiček & Hrouda,
                                                                1974
   4 minima (L.) Ker-Gawler                     24    Cz        Mešiček & Hrouda,
                                                                1974
   5 reticulata (Pallas) Schultes & Schultes
     fil.
```

6 fibrosa (Desf.) Schultes & Schultes
   fil.
7 taurica Steven
8 bulbifera (Pallas) Schultes &
   Schultes fil.
9 graeca (L.) A. Terracc.
10 trinervia (Viv.) W. Greuter

11 spathacea (Hayne) Salisb.	106	Cz	Mesiček & Hrouda, 1974

12 fistulosa (Ram. ex DC.) Ker-Gawler

13 arvensis (Pers.) Dumort.	48	Cz	Mesiček & Hrouda, 1974

14 dubia A. Terracc.
15 granatellii (Parl.) Parl.
16 peduncularis (J. & C. Presl) Pascher

17 nevadensis Boiss.	36	Sa	Martinoli, 1950

18 amblyopetala Boiss. & Heldr.
19 foliosa (J. & C. Presl) Schultes &
   Schultes fil.
20 polymorpha Boiss.

21 bohemica (Zauschner) Schultes &	60	Cz	Mesiček & Hrouda,
Schultes fil.			1974

   (a) bohemica
   (b) nebrodensis (Tod. ex Guss.) I.B.K.
      Richardson
   (c) gallica (Rouy) I.B.K. Richardson
22 saxatilis (Mert. & Koch) Schultes &
   Schultes fil.
23 szovitzii (A.F. Lang) Besser

?1 Erythronium L.			
1 dens-canis L.	24	Cz	Hruby, 1934

22 Tulipa L.				
1 sylvestris L.				
(a) sylvestris		36	?S. Europe	Hall, 1937
(b) australis (Link) Pamp.		24	Lu	Barros Neves, 1973
2 orphanidea Boiss. ex Heldr.	24 36 48		Bu	Popova, 1972
3 goulimyi Sealy & Turrill				
4 saxatilis Sieber ex Sprengel		24		
		36	Cr	Hall, 1937
5 cretica Boiss. & Heldr.		24	Cr	Hall, 1937
6 biflora Pallas		24	Ju	Milan, 1975
7 clusiana DC.		48		
		56		
8 gesnerana L.				
hungarica Borbas		24	Rm	Tarnavschi & Lungeanu, 1970
urumoffii Hayek		24	Bu	Popova, 1972
9 boeotica Boiss. & Heldr.		24	Ju	Lovka et al., 1971
10 praecox Ten.		36	It	Hall, 1937
11 agenensis DC.		24	?Ga	Hall, 1937
		36	Bot. Gard.	Woods & Bamford, 1937

23 Fritillaria L.

1 meleagris L.	24	Ho	Gadella & Kliphuis, 1973
(a) meleagris			
(b) burnatii (Planchon) Rix			
2 tubiformis Gren. & Godron			
(a) tubiformis			
(b) moggridgei Boiss. & Reuter ex Rix			
3 macedonica Bornm.			
4 pontica Wahlenb.			
5 involucrata All.			
6 messanensis Rafin.			
(a) messanensis	24	Si	Gori, 1958
(b) gracilis (Ebel) Rix	24	Ju	Milan, 1975
7 epirotica Turrill ex Rix			
8 gussichiae (Degen & Dörfler) Rix			
9 graeca Boiss. & Spruner	24	Gr	Phitos, ined.
(a) graeca			
(b) thessala (Boiss.) Rix			
10 davisii Turrill			
11 rhodocanakis Orph. ex Baker	24	Gr	Phitos, ined.
12 pyrenaica L.			
13 lusitanica Wikström	24	Lu	Fernandes, 1950
14 ruthenica Wikström			
15 orientalis Adams	18 + 0 - 9B	Ju	Milan, 1975
16 meleagroides Patrin ex Schultes fil.			
17 obliqua Ker-Gawler	24	Gr	Phitos, ined.
18 tuntasia Heldr. ex Halácsy			
19 drenovskii Degen & Stoj.			
20 conica Boiss.			
21 euboeica Rix			
22 stribrnyi Velen.	24	Bu	Popova, 1972
23 ehrhartii Boiss. & Orph.			
24 Lilium L.			
1 martagon L.	24 + 0 - 2B	Rs(W)	Van Loon & Oudemans, 1976
2 candidum L.	24	Sa	Arrigoni & Mori, 1976
3 bulbiferum L.	24	Po	Skalińska et al., 1964
4 pyrenaicum Gouan			
5 pomponium L.			
6 carniolicum Bernh. ex Koch	24	It	Zucconi, 1956
7 chalcedonicum L.			
8 rhodopaeum Delip.			
9 monadelphum Bieb.			
10 lancifolium Thunb.			
25 Ornithogalum L.			
1 pyrenaicum L.	16	It	Chiappini & Scrugli, 1972
	18	Rm	Lungeanu, 1971
2 sphaerocarpum A. Kerner	16	Au	Leute, 1974
3 visianicum Tommasini			
4 creticum Zahar.			
5 narbonense L.	54	Hs	Rejón, 1976
6 ponticum Zahar.	16	Rs(K)	Agapova, 1976
7 pyramidale L.	24	Ge	Lungeanu, 1971
8 prasinantherum Zahar.			

9 reverchonii Lange			
10 fischeranum Krasch.	22	Rs(K)	Agapova, 1976
	24	Rs(W)	Lungeanu, 1971
11 oligophyllum E.D. Clarke	24	Bu	Markova et al., 1972
12 montanum Cyr.	12	Bu	Lungeanu, 1971
	14	Bu	Markova et al., 1972
	18	Gr Si	Ratter, 1967
	20 + 3B	Si	Garbari & Tornadore, 1972
13 atticum Boiss. & Heldr.			
14 oreoides Zahar.	18	Rm	Lungeanu, 1971
15 fimbriatum Willd.	12 + 0 - 3B	Bu	Markova et al., 1972
16 collinum Guss.			
17 costatum Zahar.			
18 woronowii Krasch.	16	Rs(K)	Agapova, 1976
19 amphibolum Zahar.	18	Rm	Lungeanu, 1971
20 wiedemannii Boiss.			
21 sibthorpii W. Greuter	14	Bu	Markova et al., 1972
	18 24	Rm	Lungeanu, 1971
	28	Bu	Markova et al., 1972
22 exscapum Ten.	18	It	Garbari & Tornadore, 1972
23 comosum L.	14	Bu	Markova et al., 1972
	18	It	Garbari & Tornadore, 1972
24 umbellatum L.	18-20	Lu	Mesquita, 1964
	27 28	Br	Czapik, 1968
	35 36	Gr	Ratter, 1967
	42	Cz	Májovský et al., 1976
	44	Bu	Markova et al., 1972
	45	It	Garbari & Tornadore, 1972
	54	Br	Czapik, 1968
	72 90 108	It	Garbari & Tornadore, 1971
25 orthophyllum Ten.	14 + 0 - 3B	It	Garbari & Tornadore, 1972
	18-20 27-29	Po	Czapik, 1965
	36	Bu	Markova et al., 1972
(a) psammophilum (Zahar.) Zahar.			
(b) orbelicum (Velen.) Zahar.			
(c) baeticum (Boiss.) Zahar.	18	Hs	Löve & Kjellqvist, 1971
	27		
(d) orthophyllum			
(e) acuminatum (Schur) Zahar.			
(f) kochii (Parl.) Zahar.			
26 armeniacum Baker			
27 refractum Kit. ex Schlecht.	54	Bu	Markova et al., 1972
	56	Rm	Lungeanu, 1971

	72	Bu	Markova et al., 1972
28 divergens Boreau	18	It	Chiappini & Serugli, 1972
	42	Rm	Lungeanu, 1971
	54	Ju	Sušnik & Lovka, 1973
29 exaratum Zahar.			
30 nutans L.	45	It	Garbari & Tornadore, 1972
31 boucheanum Ascherson	28	Rm	Lungeanu, 1971
32 arabicum L.			
33 unifolium (L.) Ker-Gawler	34	Lu	Barros Neves, 1973
34 concinnum (Salisb.) Coutinho	36	Lu	Barros Neves, 1973

26 Urginea Steinh.

1 undulata (Desf.) Steinh.	20	Hs	Sañudo & Rejón, 1975
	60		
2 fugax (Moris) Steinh.	20 + 1B 4B	Sa	Battaglia, 1957b
3 maritima (L.) Baker	20	Bl It Si	Battaglia, 1957
	30	It	Battaglia, 1957
	40	Bl It Sa	Battaglia, 1957
	60	Hs	Rejón. 1974
	64	Lu	Barros Neves, 1973

27 Scilla L.

1 bifolia L.	18	Ju	Sušnik & Lovka, 1973
	36	Ju	Sušnik & Lovka, 1973
	54	Gr	Speta, 1976
2 messeniaca Boiss.			
3 siberica Haw.			
4 amoena L.			
5 bithynica Boiss.			
6 verna Hudson	20	Lu	Barros Neves, 1973
7 ramburei Boiss.	20	Lu	Barros Neves, 1973
8 odorata Link			
9 monophyllos Link	20 40	Lu	Barros Neves, 1973
10 litardierei Breistr.	28	Ju	Lovka et al., 1971
11 hyacinthoides L.	20	Lu	Barros Neves, 1973
12 lilio-hyacinthus L.			
13 peruviana L.	16	Lu	Barros Neves, 1973
cupanii Guss.	14	Si	Maugini, 1952
hughii Tineo ex Guss.	16	Si	Maugini, 1952
14 beirana Samp.			
15 autumnalis L.	14	Hs It Si	Battaglia, 1957
	28	Bl Ga It Lu	Battaglia, 1957
	42	Hu	Baksay, 1956

28 Hyacinthoides Medicus

1 non-scripta (L.) Chouard ex Rothm.	16 24	Br	Wilson, 1956
2 hispanica (Miller) Rothm.	16 24	Br Lu	Wilson, 1959
3 italica (L.) Rothm.	16	Lu	Barros Neves, 1973
reverchonii Degen & Hervier	16	Hs	Löve & Kjellqvist, 1973

29 Chionodoxa Boiss.
  1 nana (Schultes & Schultes fil.)   18   Cr   Speta, 1976
     Boiss. & Heldr.
  2 cretica Boiss. & Heldr.

30 Hyacinthus L.
  1 orientalis L.

31 Brimeura Salisb.
  1 amethystina (L.) Chouard
  2 fastigiata (Viv.) Chouard   28   Sa   Chiarugi, 1950

32 Strangweia Bertol.
  1 spicata (Sibth. & Sm.) Boiss.   8   Gr   Phitos, ined.

33 Bellevalia Lapeyr.
  1 trifoliata (Ten.) Kunth
  2 dubia (Guss.) Reichenb.   8   It   Chiarugi, 1949
    boissieri   8   Gr   Bothmer & Bentzer,
                                       1973
  3 webbiana Parl.   16   It   Chiarugi, 1949
  4 hackelii Freyn
  5 lipskyi (Miscz.) Wulf
  6 ciliata (Cyr.) Nees
  7 sarmatica (Pallas ex Georgi) Woronow
  8 romana (L.) Reichenb.   8   Ju   Lovka et al.,
                                       1971
  9 brevipedicellata Turrill

34 Hyacinthella Schur
  1 leucophaea (C. Koch) Schur
  2 dalmatica (Baker) Chouard
  3 pallasiana (Steven) Losinsk

35 Dipcadi Medicus
  1 serotinum (L.) Medicus   8 + 0 - 6B Lu   Fernandes &
                                       Queirós, 1971
                     16 32   Hs   Gadella et al.,
                                       1966

36 Muscari Miller
  1 macrocarpum Sweet   18   Gr   Stuart, 1970
  2 moschatum Willd.
  3 cycladicum P.H. Davis & Stuart 36   Gr   Bentzer, 1973
                            54   Gr   Stuart, 1970
  4 gussonei (Parl.) Tod.   18   Si   Garbari & Martino,
                                       1972
  5 spreitzenhoferi (Heldr.) Vierh. 18   Cr   Stuart, 1970
                            36   Cr   Bentzer, 1973
  6 weissii Freyn   18 36   Gr   Stuart, 1970
                            54   Gr   Bentzer, 1973
    dionysicum Rech. fil.   18 36   Cr   Bentzer, 1973
  7 comosum (L.) Miller   18   Gr It   Stuart, 1970

                          27   Gr   Bentzer, 1972
  8 tenuiflorum Tausch   18   It   Garbari, 1973
  9 botryoides (L.) Miller   18   Hu   Pólya, 1950
                          36   It   Garbari, 1969
  10 armeniacum Leichtlin ex Baker 18   Gr   Stuart, 1970

11 neglectum Guss. ex Ten.	18	Bu	Kozuharov & Kuzmanov, 1962/3
	36	Lu	Barros Neves, 1973
	45	Ju	Sušnik & Lovka, 1973
	54	Ga	Guinochet & Logeois, 1962
	72	Rs(K)	Stuart, 1970
12 commutatum Guss.	18	Gr	Stuart, 1970
13 parviflorum Desf.	18	It	Garbari, 1969

37 Agapanthus L'Hér.
  1 praecox Willd.

38 Allium L.

1 angulosum L.	16	Cz	Májovský et al., 1970
2 rubens Schrader ex Willd.			
3 senescens L.	16	It	Renzoni & Garbari, 1971
	24 32	Hu	Baksay, 1956
	32 + 0 - 4B	Cz	Holub et al., 1970
4 saxatile Bieb.	16	?Bot. Gard.	Levan, 1929
5 horvatii Lovrić	16		
6 albidum Fischer ex Bieb.	16	Ju	Dietrich, 1967b
(a) albidum	16	Bot. Gard.	Hasitschka-Jenscke, 1958
(b) caucasicum (Regel) Stearn			
7 stelleranum Willd.			
8 ericetorum Thore	16 + 0 - 1B	Cz	Holub et al., 1970
	32	Ju	Lovka et al., 1972
9 kermesinum Reichenb.	16	Ju	Lovka et al., 1971
10 suaveolens Jacq.	16	Gr	Alden, 1976
11 lineare L.	16	Bot. Gard.	Dietrich, 1970
	48	Cz	Holub et al., 1970
12 palentinum Losa & Montserrat			
13 hymenorhizum Ledeb.			
14 obliquum L.	16	Bot. Gard.	Dietrich, 1967b
15 narcissiflorum Vill.	14	Ga	Delay, 1968
16 insubricum Boiss. & Reuter	14	It	Favarger, 1959
17 inderiense Fischer ex Bunge			
18 schoenoprasum L.	16	Cz	Májovský et al., 1976
19 schmitzii Coutinho	16	Lu	Barros Neves, 1973
20 cepa L.	16	It	Battaglia, 1957
21 fistulosum L.	16	Ga	Battaglia, 1957
22 victorialis L.	16	Po	Mizianty & Frey, 1973
23 roseum L.	16	Ga	Delay, 1970
	32	Ga	Delay, 1970
24 circinnatum Sieber			
25 massaessylum Batt. & Trabut	14	Lu	Barros Neves, 1973
26 phthioticum Boiss. & Heldr. ex Boiss.			
27 breviradium (Halácsy) Stearn			

28 neapolitanum Cyr.	14 21 28	W.Africa	Kollmann, 1973
	31-36	Lu	Barros Neves, 1973
29 longanum Pamp.			
30 subhirsutum L.	14	It Si	Renzoni & Garbari, 1971
	28	Hs	Rejón, 1974
31 trifoliatum Cyr.			
32 subvillosum Salzm. ex Schultes & Schultes fil.	28	Bl	Dahlgren,et al., 1971
33 moly L.	14	Hs	Löve & Kjellqvist, 1973
34 scorzonerifolium Desf. ex DC.	14	Lu	Barros Neves, 1973
35 triquetrum L.	18	Lu	Barros Neves, 1973
36 pendulinum Ten.	14	It	Renzoni & Garbari, 1970
37 paradoxum (Bieb.) G. Don	16	Br	Barling, 1958
38 chamaemoly L.	22	It	Marchi et al., 1974
39 ursinum L.	14	Cz	Májovský et al., 1976
(a) ursinum			
(b) ucrainicum Kleopow & Oxner			
40 moschatum L.	16	It	Billeri, 1954
41 inaequale Janka			
42 bornmuelleri Hayek			
43 rubellum Bieb.			
44 meteoricum Heldr. & Hausskn. ex Halácsy	16	Gr	Stearn, 1978
45 delicatulum Sievers ex Schultes & Schultes fil.			
46 frigidum Boiss. & Heldr.			
47 chrysonemum Stearn	32	Hs	Stearn, 1978
48 grosii Font Quer			
49 cupani Rafin.			
(a) cupani	32	Si	Garbari et al., 1979
(b) hirtovaginatum (Kunth) Stearn	16	Gr	Garbari et al., 1979
50 callimischon Link			
(a) callimischon	16	Gr	Stearn, 1978
(b) haemostictum Stearn	16	Cr	Miceli & Garbari, 1979
51 obtusiflorum DC			
52 parciflorum Viv.	16	Sa	Martinoli, 1955
53 rouyi Gaut.			
54 caeruleum Pallas			
55 sabulosum Steven ex Bunge	16	Asia	Vakhtina, 1964
56 paniculatum L.	16	Hs	Rejón, 1976
	40	It	Marchi et al., 1974
(a) paniculatum	24	Bu	Češchmedjiev, 1970
	32	Co	Contandriopoulos, 1962
(b) fuscum (Waldst. & Kit.) Arcangeli	16	Gr	Diannelidis, 1951
	32	Gr	Johnson, 1982
(c) euboicum (Rech. fil.) Stearn			
(d) villosulum (Halácsy) Stearn	16	Bu	Češchmedjiev, 1970
57 pallens L.	16	Co	Contandriopoulos, 1957
(a) pallens			
(b) tenuiflorum (Ten.) Stearn			
(c) siciliense Stearn			
58 sipyleum Boiss.			

59 rupestre Steven
60 favosum Zahar.
61 macedonicum Zahar.
62 podolicum Blocki ex Racib. & Szafer

63 oleraceum L.	32	Fe	V. Sorsa, 1962
	40	It	Gadella & Kliphuis, 1970

64 parnassicum (Boiss.) Halácsy
65 staticiforme Sibth. & Sm.
66 pilosum Sibth. & Sm.
67 luteolum Halácsy
68 flavum L.

(a) flavum	16 32	Gr	Alden, 1976
(b) tauricum (Besser ex Reichenb.) Stearn	32	Bu	Ceschmedjiev, 1970
69 melanantherum Pančić	16	Bu	Ceschmedjiev, 1976
	24 + 1B	Bu	Ceschmedjiev, 1976
	32	Bu	Kozuharov & Kuzmanov, 1965a

70 tardans W. Greuter & Zahar.
71 carinatum L.

(a) carinatum	16	Ga	Parreaux, 1971
	24	Ju	Nilsson & Lassen, 1971
	26	Au	Tschermak-Woess, 1947
(b) pulchellum Bonnier & Layens	16	It	Marchi et al., 1974

72 hirtovaginum P. Candargy
73 hymettium Boiss. & Heldr.
74 stamineum Boiss.
75 sativum L.

76 ampeloprasum L.	16	It	Carpineri, 1971
	32 40 48	Gr	Bothmer, 1970
	56	Cr	Bothmer, 1975
	80	Lu	Barros Neves, 1973
var. holmense Ascherson & Graebner	48	Br	Stearn, 1978
porrum L.	32	Cz	Murin, 1964
77 polyanthum Schultes & Schultes fil.	20	Hs	Gadella et al., 1966
	32	Hs	Löve & Kjellqvist, 1973

78 scaberrimum Serres

79 atroviolaceum Boiss.	16	Tu	Johnson, unpub.
	24		
	32	Bu	Ceschmedjiev, 1970
	40 48		

80 bourgeaui Rech. fil.			
(a) bourgeaui	16	Gr	Bothmer, 1970
(b) cycladicum Bothmer	16 24 32	Gr	Bothmer, 1970
(c) creticum Bothmer	32	Cr	Bothmer, 1975
81 commutatum Guss.	16 24 32	Gr	Bothmer, 1970

82 pardoi Loscos
83 talijevii Klokov

84 baeticum Boiss.	32	Lu	Barros Neves, 1973

85 acutiflorum Loisel.
86 pyrenaicum Costa & Vayr.
87 scorodoprasum L.

(a) rotundum (L.) Stearn	16 + 0 - 2B	Bu	Ceschmedjiev, 1976
	32	Hs	Rejón, 1976

	38 + 1B	Bu	Češchmedjiev, 1976
	40 + 1 - 4B	Bu	Češchmedjiev, 1970
	48 + 1B		
(b) waldsteinii (G. Don) Stearn	16		
	32		
	40		
	48		
(c) jajlae (Vved.) Stearn			
(d) scorodoprasum	16	Cz	Májovský et al.,
			1970
	24	Au	Tschermak-Woess,
88 albiflorum Omelczuk			1947
89 pervestitum Klokov			
90 sphaerocephalon L.	16	Lu	Barros Neves, 1973
	16 + 2B	Hs	Rejón, 1976
(a) sphaerocephalon			
(b) arvense (Guss.) Arcangeli			
(c) trachypus (Boiss. & Spruner) Stearn			
91 proponticum Stearn & N. Ozhatay	16	Tu	Stearn & Ozhatay,
92 melananthum Coincy			1977
93 pruinatum Link ex Sprengel	16	Lu	Barros Neves, 1973
94 regelianum A. Becker			
95 vineale L.	16		
	32 40	Lu	Barros Neves, 1973
96 amethystinum Tausch	16 24 32	Gr	Bothmer, 1970
97 guttatum Steven			
(a) guttatum	16	Rm	Buttler, 1969
(b) sardoum (Moris) Stearn	16	Bu	Češchmedjiev, 1970
	32 48	Lu	Barros Neves, 1977
(c) dalmaticum (A. Kerner ex Janchen	24	Tu	C.Inih, 1967
98 dilatatum Zahar.			
99 gomphrenoides Boiss. & Heldr.			
100 jubatum Macbride			
101 rubrovittatum Boiss. & Heldr.			
102 integerrimum Zahar.			
103 chamaespathum Boiss.	16		
104 heldreichii Boiss.	16	Gr	Alden, 1976
105 atropurpureum Waldst. & Kit.	16	Gr	Dietrich, 1969
106 nigrum L.	16	Bu	Češchmedjiev, 1976
107 cyrilli Ten.			
108 decipiens Fischer ex Schultes & Schultes fil.	16	Bu	Češchmedjiev, 1976
109 orientale Boiss.			
110 caspium (Pallas) Bieb.			
39 Nectaroscordum Lindley			
1 siculum (Ucria) Lindley	18	Si	Garbari & Tornadore, 1970
(a) siculum			
(b) bulgaricum (Janka) Stearn			
40 Nothoscordum Kunth			
1 inodorum (Aiton) Nicholson	18	Ga	Van Loon et al., 1971
41 Ipheion Rafin.			
1 uniflorum (R.C. Graham) Rafin.			
42 Convallaria L.			
1 majalis L.	38	Hu	Pólya, 1949

43 Maianthemum Weber

1 bifolium (L.) F.W. Schmidt	36	Cz	Májovský et al., 1970

44 Smilacina Desf.
1 stellata (L.) Desf.

45 Streptopus Michx
1 amplexifolius (L.) DC.

46 Polygonatum Miller

1 verticillatum (L.) All.	28	Cz	Májovský et al., 1970
2 latifolium (Jacq.) Desf.	20	Hu	Pólya, 1949
3 orientale Desf.			
4 multiflorum (L.) All.	18	Cz	Májovský et al., 1970
	32	Fe	Sorsa, 1963
5 odoratum (Miller) Druce	20	Fe Su	Suomalainen, 1947

47 Paris L.

1 quadrifolia L.	20	Fe	V. Sorsa, 1962

48 Asparagus L.

1 asparagoides (L.) Druce			
2 albus L.	20	Lu	Barros Neves, 1973
3 acutifolius L.	40	It	Bozzini, 1959
4 aphyllus L.	40	It	Bozzini, 1959
5 stipularis Forskål	20	It	Bozzini, 1959
6 verticillatus L.			
7 bresleranus Schultes			
8 persicus Baker			
9 brachyphyllus Turcz.	40	?Rs	Zakharyeva & Makushenko, 1969
10 kasakstanicus Iljin			
11 maritimus (L.) Miller	40	It	Bozzini, 1959
12 litoralis Steven			
13 officinalis L.			
(a) officinalis	20	Ho It	Bozzini, 1959
(b) prostratus (Dumort.) Corb.	40	Br	Davies & Willis, 1959
14 pseudoscaber Grec.			
15 tenuifolius Lam.	20	It	Bozzini, 1959

49 Ruscus L.

1 aculeatus L.	40	Lu	Barros Neves, 1973
2 hypophyllum L.			
3 hypoglossum L.	40	Sa	Martinoli, 1951

50 Smilax L.

1 aspera L.	32	It	Marchi, 1971
2 excelsa L.			
3 canariensis Brouss. ex Willd.			

CLXXXIV AGAVACEAE

1 Agave L.
1 americana L.
2 atrovirens Karwinski ex Salm-Dyck

2 Furcraea Vent.
1 foetida (L.) Haw.

3 Yucca L.
  1 filamentosa L.
  2 gloriosa L.

CLXXXV AMARYLLIDACEAE

1 Amaryllis L.
  1 bella-donna L.

2 Sternbergia Waldst. & Kit.
  1 colchiciflora Waldst. & Kit.   20   Ju   Bedalov & Sušnik,
                                                         1970
                                          26   Rm   Tarnavschi &
                                                         Lungeanu, 1970
  2 lutea (L.) Ker-Gawler ex Sprengel
    (a) lutea   22   Ju   Bedalov & Sušnik,
                                                         1970
    (b) sicula (Tineo ex Guss.) D.A. Webb 18   Gr   Kuzmanov et al.,
                                                         1968

3 Leucojum L.
  1 autumnale L.   14   Lu   Barros Neves, 1939
  2 roseum F. Martin   16   Co Sa Stern, 1956
  3 longifolium (Gay ex M.J. Roemer) Gren. 16   Co   Contandriopoulos,
                                                         1962
  4 trichophyllum Schousboe   14   Lu   Barros Neves, 1939
  5 nicaeense Ardoino   18   Ga   Stern, 1956
  6 valentinum Pau   16   Gr   Damboldt & Phitos,
                                                         1975
  7 vernum L.   20   Po   Skalińska et al.,
                                                         1968
                                          22   Ge   Stern, 1956
  8 aestivum L.
    (a) aestivum   22   Ju   Sušnik & Lovka,
                                                         1973
                                          24   Hu   Pólya, 1949
    (b) pulchellum (Salisb.) Briq.   22   Sa   Garbari &
                                                         Tornadore, 1970

4 Galanthus L.
  1 nivalis L.
    (a) nivalis   24   Ge   Stern, 1956
    (b) reginae-olgae (Orph.) Gottl.-Tann.
  2 plicatus Bieb.   24   Rs(K) Van Loon &
                                                         Oudemans, 1976
  3 elwesii Hooker fil.   24   Rm   Tarnavschi &
                                                         Lungeanu, 1970
                                          48   Gr   Stern, 1956
    (a) elwesii
    (b) minor D.A. Webb
  4 ikariae Baker

5 Lapiedra Lag.
  1 martinezii Lag.   22   Hs   Gadella & Kliphuis,
                                                         1966

6 Narcissus L.
  1 humilis (Cav.) Traub   28   Hs   A. & R. Fernandes, 1945
  2 serotinus L.   10   Hs   Fernandes, 1968
                                          20   It   Garbari et al.,
                                                         1973

	30	It	Scrugli, 1974

3 tazetta L.
  (a) tazetta

(b) italicus (Ker-Gawler) Baker	22	It	Maugini, 1952b
(c) aureus (Loisel.) Baker	22	It	Maugini, 1953

4 papyraceus Ker-Gawler
  (a) papyraceus
  (b) polyanthus (Loisel.) Ascherson &
      Graebner

(c) panizzianus (Parl.) Arcangeli	22	Lu	Fernandes, 1937a
5 dubius Gouan	50	Hs	Fernandes, 1937b

6 elegans (Haw.) Spach
7 poeticus L.

(a) poeticus	14	It	Garbari & Tornadore, 1970
(b) radiiflorus (Salisb.) Baker	14	Ju	Sušnik & Lovka, 1973
8 jonquilla L.	14	Lu	Fernandes, 1966
9 requienii M.J. Roemer	14	Hs	Fernandes, 1939
10 willkommii (Samp.) A. Fernandes	14	Lu	Fernandes, 1966
11 gaditanus Boiss. & Reuter	14	Lu	Fernandes, 1939
	21	Lu	Fernandes & Queirós, 1971
12 rupicola Dufour	14	Lu	Fernandes, 1939
13 cuatrecasasii Casas, Laínz & Ruiz Rejón	14	Hs	Casas et al., 1973
14 calcicola Mendonca	14	Lu	Fernandes, 1939
15 scaberulus Henriq.	14	Lu	Fernandes, 1939

16 viridiflorus Schousboe
17 triandrus L.

(a) pallidulus (Graells) D.A. Webb	14	Lu Hs	Fernandes, 1950b
(b) triandrus	14	Lu	Fernandes, 1950b
(c) capax (Salisb.) D.A. Webb	14	Ga	Fernandes, 1950b

18 bulbocodium L.

(a) bulbocodium	14 21 28 35 42 49 56	Lu	Fernandes, 1967
(b) obesus (Salisb.) Maire	26	Lu	Fernandes, 1967
19 hedraeanthus (Webb & Heldr.) Colmeiro	14	Hs	Rejón, 1974
20 cantabricus DC.	14	Hs	Fernandes, 1959
21 longispathus Pugsley	14	Hs	Heywood, 1955

22 pseudonarcissus L.
  (a) pseudonarcissus
  (b) pallidiflorus (Pugsley) A.
      Fernandes
  (c) moschatus (L.) Baker

(d) nobilis (Haw.) A. Fernandes	28	Lu	Fernandes, 1950

  (e) major (Curtis) Baker
  (f) portensis (Pugsley) A. Fernandes
  (g) nevadensis (Pugsley) A. Fernandes
23 bicolor L.

24 minor L.	14	Lu	Fernandes, 1931
25 asturiensis (Jordan) Pugsley	14 15	Lu	Fernandes & Fernandes, 1946
26 cyclamineus DC.	14	Lu	Fernandes & Fernandes, 1946

27 x incomparabilis Miller
28 x odorus L.

29 x medioluteus Miller	24	It	Garbari et al., 1973

30 x intermedius Loisel.

7 Pancratium L.
  1 maritimum L.   22   Lu   Rodrigues, 1953
  2 illyricum L.   22   Sa   Martinoli, 1949

CLXXXVI DIOSCOREACEAE

1 Dioscorea L.
  1 balcanica Košanin

2 Borderea Miégeville
  1 pyrenaica Miégeville   24   Ga   Heslot, 1953
  2 chouardii (Gaussen) Heslot   24   Hs   Heslot, 1953

3 Tamus L.
  1 communis L.   48   Bl   Dahlgren et al.,
                                                   1971

CLXXXVII PONTEDERIACEAE

1 Pontederia L.
  1 cordata L.

2 Heteranthera Ruiz & Pavón
  1 reniformis Ruiz & Pavón

3 Eichornia Kunth
  1 crassipes (C.F.P. Mart.) Solms-Laub.

4 Monochoria C. Presl
  1 korsakowii Regel & Maack

CLXXXVIII IRIDACEAE

1 Sisyrinchium L.
  1 bermudiana L.   64   Hb   Löve & Löve, 1958b
                                88   Hb   Ingram, 1967
  2 montanum E.L. Greene   96   Cz   Murín & Májovský,
                                                    1976

  3 californicum (Ker-Gawler) Aiton fil.

2 Hermodactylus Miller
  1 tuberosus (L.) Miller

3 Iris L.
  1 sibirica L.   28   Po   Skalińska et al.,
                                                     1961

  2 tenuifolia Pallas
  3 unguicularis Poiret
  4 ruthenica Ker-Gawler
  5 foetidissima L.   40   It   Mori, 1957
  6 pseudacorus L.   24   Cz   Májovský et al.,
                                                   1970
                                30   Hu   Pólya, 1949
                                32   Ju   Lovka et al.,
                                                    1971
                                34   Hu It   Félfoldy, 1947
                                40   It   Mori, 1957

7 versicolor L.
8 spuria L.
  (a) spuria   22   Da   Westergaard, 1938
  (b) maritima P. Fourn.
  (c) halophila (Pallas) D.A. Webb &
      Chater   20   Rm   Lungeanu, 1973
  (d) ochroleuca (L.) Dykes
9 graminea L.   34   Ju   Sušnik et al.,
                                                1972

10 sintenisii Janka
  (a) sintenisii   16   Bu   Popova &
                          18   Česchmedjiev, 1975
  (b) brandzae (Prodan) D.A. Webb &
      Chater   (not 10) 20   Rm   Tarnavschi, 1938
11 pontica Zapal.
12 humilis Georgi   22   Au   Mitra, 1956
13 pumila L.
  (a) pumila   32   Au   Mitra, 1956
  (b) attica (Boiss. & Heldr.) Hayek 14   Gr   Simonet, 1955
                        16   Ju   Lovka et al.,
                                              1971

14 lutescens Lam.
  (a) lutescens   40   It   Mitra, 1956
  (b) subbiflora (Brot.) D.A. Webb &
      Chater
15 pseudopumila Tineo   16   Si   Simonet, 1955
16 reichenbachiana Heuffel   24   Ju   Lovka et al.,
                                              1971
                      48   Bu   Popova &
                                         Česchmedjiev, 1975
17 suaveolens Boiss. & Reuter 24   Bu   Popova &
                                         Česchmedjiev, 1975
18 aphylla L.   24   Ga   Simonet, 1934
                      40   It   Mitra, 1956
                      48   Hu   Pólya, 1950
19 variegata L.   24   Au   Mitra, 1956
  reginae I. & M. Horvat   24   Ju   Mitra, 1956
20 germanica L.   36   Hs   Löve & Kjellqvist,
                                                1973
                    44   Ju   Lovka et al.,
                                            1971
                    48   Gr   Alden, 1976
21 marsica I. Ricci & Colasante 40   It   Ricci & Colasante,
                                              1975
22 albicans Lange
23 pallida Lam.   24   It   Mitra, 1956
  (a) pallida
  (b) cengialti (Ambrosi) Foster
24 planifolia (Miller) Fiori & Paol. 24   Si   Garbari &
                                              Tornadore, 1972b
25 latifolia (Miller) Voss
26 xiphium L.   34   Lu   Fernandes &
                                              Queirós, 1971
27 filifolia Boiss.
28 juncea Poiret
29 boissieri Henriq.   36   Lu   Fernandes, 1950
30 serotina Willk.

4 Gynandriris Parl.
  1 sisyrinchium (L.) Parl.   24   Lu   Fernandes &
                                              Queirós, 1971

5 Ixia L.
  1 maculata L.
  2 paniculata Delaroche

6 Freesia Ecklon ex Klatt
  1 refracta (Jacq.) Ecklon ex Klatt

7 Crocus L.

1 malyi Vis.	30	Ju	Brighton,et al., 1973
2 imperati Ten.			
(a) imperati	26	It	Brighton et al., 1973
(b) suaveolens (Bertol.) Mathew	26	It	Brighton et al., 1973
3 versicolor Ker-Gawler	26	Ga	Brighton et al., 1973
4 cambessedesii Gay	16	Bl	Brighton et al., 1973
5 minimus DC.	24	Co Sa	Brighton et al., 1973
6 corsicus Vanucci ex Maw	18	Co	Brighton et al., 1973
	22	Co	Contandriopoulos, 1962
7 vernus (L.) Hill			
(a) vernus	8	Au It Ju Si	Brighton, 1976
	10	Rm	Brighton, 1976
	12	Ju	Brighton, 1976
	16	Cz It	Brighton, 1976
	18 19 20 22 23	Ju	Brighton, 1976
(b) albiflorus (Kit.) Ascherson & Graebner	8	Au It Ju	Brighton et al., 1973
x fritschii Derganc	13	Al Ju	Brighton et al., 1973
8 tommasinianus Herbert	16	Ju	Sušnik & Lovka, 1973
9 etruscus Parl.	8	It	Brighton et al., 1973
10 kosaninii Pulević			
11 nevadensis Amo	26	Hs	Rejón, 1974
	28 30	Hs	Brighton et al., 1973
12 carpetanus Boiss. & Reuter	64	Hs Lu	Brighton et al., 1973
13 sieberi Gay	22	Cr Gr	Brighton et al., 1973
14 veluchensis Herbert	26	Bu	Brighton et al., 1973
15 dalmaticus Vis.	24	Ju	Sušnik & Lovka, 1973
16 angustifolius Weston	12	Rs(K)	Brighton et al., 1973
17 reticulatus Steven ex Adams	12	Ju	Brighton et al., 1973
18 cvijicii Košanin	18	Gr	Brighton et al., 1973
19 scardicus Košanin	32	Ju	Sopova, 1972
	34	Ju	Brighton et al., 1973
	35 36	Ju	Sopova, 1972

310

20 pelistericus Pulević
21 flavus Weston   8 Bu Gr Ju Rm   Brighton et al.,
                                                                 1973
22 olivieri Gay   6   Bu Gr   Brighton et al.,
                                                                 1973
23 chrysanthus (Herbert) Herbert   8   Bu Gr Ju   Brighton, ined.
                                             12   Bu Gr   Brighton, ined.
                                             20   Gr Ju   Brighton, ined.
24 biflorus Miller   8   It Ju   Brighton et al.,
                                                                 1973
                                             12   Rs(K)   Brighton et al.,
                                                                 1973
                                             16   Bu   Brighton, ined.
25 crewei Hooker fil.   12   Gr   Brighton et al.,
                                                                 1973
26 cartwrightianus Herbert   16   Cr Gr   Brighton et al.,
                                                                 1973
    sativus L.   24   Bl   Brighton, ined.
27 pallasii Goldb.   14   Bu Rs(K)   Brighton et al.,
                                                                 1973
28 thomasii Ten.   16   It Ju   Brighton et al.,
                                                                 1973
29 hadriaticus Herbert   16   Gr   Brighton et al.,
                                                                 1973
30 niveus Bowles   28   Gr   Brighton et al.,
                                                                 1973
31 longiflorus Rafin.   28   Si   Brighton et al.,
                                                                 1973
32 goulimyi Turrill   12   Gr   Brighton et al.,
                                                                 1973
33 serotinus Salisb.
    (a) serotinus   22 23   Lu   Brighton et al.,
                                                                 1973
    (b) clusii (Gay) Mathew   22   Hs Lu   Brighton et al.,
                                                                 1973
                                             24   Hs Lu   Brighton et al.,
                                                                 1973
    (c) salzmannii (Gay) Mathew   22 24   Hs   Brighton et al.,
                                                                 1973
34 nudiflorus Sm.   48   Ga   Brighton et al.,
                                                                 1973
35 medius Balbis   24 + 0 - 5B   It   Brighton, 1976
36 laevigatus Bory & Chaub.   26   Cr Gr   Brighton et al.,
                                                                 1973
37 boryi Gay   30   Gr   Brighton et al.,
                                                                 1973
38 tournefortii Gay   30   Gr   Brighton et al.,
                                                                 1973
39 cancellatus Herbert   16   Ju   Lovka et al.,
                                                                 1971
40 robertianus C.D. Brickell   20   Gr   Brighton et al.,
                                                                 1973
41 speciosus Bieb.   18   Tu   Brighton et al.,
                                                                 1973
42 pulchellus Herbert   12   Gr Ju Tu   Brighton et al.,
                                                                 1973
43 banaticus Gay   26   Rm   Brighton et al.,
                                                                 1973

8 Romulea Maratti
  1 bulbocodium (L.) Sebastiani & Mauri   34   Lu   Fernandes et al.,
                                                                 1948

311

	36	Hs	Löve & Kjellqvist, 1973
	42	Ju	Sušnik & Lovka, 1973

2 ligustica Parl.
3 requienii Parl.　　　　　　　　　34　　　Co　　　Contandriopoulos, 1962
4 revelierei Jordan & Fourr.
5 linaresii Parl.
6(a) linaresii
　(b) graeca Béguinot
6 melitensis Béguinot
7 ramiflora Ten.　　　　　　　　　36　　　It　　　Ricci, 1961
　(a) ramiflora
　(b) gaditana (G. Kunze) Marais
8 columnae Sebastiani & Mauri
　(a) columnae
　(b) rolii (Parl.) Marais
9 rosea (L.) Ecklon

9 Tritonia Ker-Gawler
　1 crocata (L.) Ker-Gawler
　2 x crocosmiflora (Lemoine) Nicholson

10 Gladiolus L.
　1 illyricus Koch　　　　　　　　　60　　　Br　　　Hamilton, ined.
　　　　　　　　　　　　　　　　　　　90　　　Lu　　　Hernandes, 1950
　2 communis L.
　　(a) communis　　　　　　　　　　120　　　Hs　　　Hamilton, ined.
　　(b) byzantinus (Miller) A. P. Hamilton　90,120　Hc　　Hamilton, ined.
　3 palustris Gaudin　　　　　　　　60　　　Po　　　Skalińska et al., 1974
　4 imbricatus L.　　　　　　　　　60　　　Po　　　Skalińska et al., 1964
　5 italicus Miller　　　　　　　　120　　　Ju　　　Sušnik & Lovka, 1973
　　　　　　　　　　　　　　　　171 $\pm$ 2　Hs　　　Hamilton, 1968

JUNCALES

CLXXXIX JUNCACEAE

1 Juncus L.
　1 maritimus Lam.　　　　　　　　48　　　Gr Su　Snogerup, 1963
　2 rigidus Desf.
　3 acutus L.　　　　　　　　　　　48　　　Ga Gr It　Snogerup, 1963
　　(a) acutus
　　(b) leopoldii (Parl.) Snogerup
　4 littoralis C. A. Meyer　　　　　48　　　Europe　Snogerup, ined.
　5 heldreichianus Marsson ex Parl.　48　　Europe　Snogerup, ined.
　6 jacquinii L.　　　　　　　　　c.170　Europe　Snogerup, ined.
　7 filiformis L.　　　　　　　　　84　　　Scand.　Snogerup, 1971
　8 arcticus Willd.　　　　　　　　84　　　Scand.　Snogerup, 1971
　9 balticus Willd.　　　　　　　　?80　　Da　　　Jørgensen et al., 1958
　　　　　　　　　　　　　　　　　84
　10 pyrenaeus Timb.-Lagr. & Jeanb.
　11 inflexus L.　　　　　　　　　　40　　　Cz　　　Májovský et al., 1974

12 effusus L.	40	Cz	Májovský et al., 1976
	42	Su	Snogerup, 1963
13 conglomeratus L.	42	Su	Snogerup, 1963
14 subulatus Forskål	42	Ga	Labadie, 1976
15 trifidus L.	30	Is	Löve & Löve, 1956
(a) trifidus			
(b) monanthos (Jacq.) Ascherson & Graebner			
16 squarrosus L.	42	Da	Snogerup, 1963
17 compressus Jacq.	44	Da Su	Snogerup, 1963
18 gerardi Loisel.	84	Da	Snogerup, 1963
(a) gerardi			
(b) atrofuscus (Rupr.) Printz			
(c) montanus Snogerup			
(d) soranthus (Schrenk) Snogerup			
(e) libanoticus (J. Thieb.) Snogerup			
19 tenuis Willd.	84	Scand.	Snogerup, 1971
20 dudleyi Wieg.	84	non-Europe	
21 imbricatus Laharpe			
22 tenageia Ehrh.			
23 sphaerocarpus Nees			
24 foliosus Desf.	26	Br Hb	Cope, 1976
25 sorrentini Parl.	28	Madeira	Cope, 1976
26 bufonius L.	108 (not 100-110)	Be Ga Hb	Cope, 1976
27 minutulus Albert & Jahandiez	c.70	Europe	Snogerup, ined.
28 ranarius Song. & Perr.	34	Hb	Cope, 1976
29 hybridus Brot.	34	Az Bl Hs Si	Cope, 1976
30 planifolius R. Br.			
31 capitatus Weigel	18	Ga	Snogerup, 1963
32 ensifolius Wikström	40	N. America	Snogerup, 1963
33 canadensis Gay	80	Canada	Snogerup, 1963
34 subnodulosus Schrank	40	Su	Snogerup, 1963
35 pygmaeus L.C.M. Richard	40	Scand.	Snogerup, 1972
36 tingitanus Maire & Weiller			
37 heterophyllus Dufour			
38 emmanuelis A. Fernandes & Garcia			
39 bulbosus L.	40	Po	Skalińska et al., 1971
40 atratus Krocker	40	Cz	Snogerup, 1963
41 thomasii Ten.			
42 acutiflorus Ehrh. ex Hoffm.	40	Scand.	Snogerup, 1972
43 valvatus Link			
44 striatus Schousboe ex E.H.F. Meyer			
45 fontanesii Gay			
(a) fontanesii			
(b) pyramidatus (Laharpe) Snogerup	40	Anatolia	Snogerup, ined.
46 alpinus Vill.			
(a) alpinus	40	Br	Elkington, 1974
(b) nodulosus (Wahlenb.) Lindman	40	Fe	Hämet-Ahti & Virrankoski, 1970
(c) alpestris (Hartman) A. & D. Löve	40	Scand.	Snogerup, 1972
47 anceps Laharpe	40	Ge	Wulff, 1937
48 requienii Parl.			
49 articulatus L.	80	Br	Fearn, 1977
50 biglumis L.	60	No	Knaben & Engelskjon, 1967
51 triglumis L.	50 c.134	Is	Löve & Löve, 1956
52 stygius L.			
53 castaneus Sm.	40	Is	Löve & Löve, 1956

2 Luzula DC.
1 campestris (L.) DC.

1 campestris (L.) DC.	12	Su	Nordenskjold, 1951
2 multiflora (Retz.) Lej.			
(a) multiflora	24	Ga	Kliphuis & Wieffering, 1972
	36	Ga	Lambert & Giesi, 1967
(b) frigida (Buchenau) V. Krecz.	36	No	Knaben & Engelskjon, 1967
(c) congesta (Thuill.) Hyl.	48	Is	Löve & Löve, 1956
3 pallescens Swartz	12	Fe	Hamet-Ahti & Virrankoski, 1970
4 sudetica (Willd.) DC.	36	Ga	Lambert & Giesi, 1967
	48	Fe	Halkka, 1964
5 arcuata Swartz	36	No	Knaben, 1950
	42		
6 confusa Lindeb.	48	Is	Löve & Löve, 1956
7 arctica Blytt	24	No	Knaben, 1950
8 spicata (L.) DC.			
(a) spicata	24	Su	Nordenskjold, 1951
(b) mutabilis Chrtek & Křísa	12 14	Au	Nordenskjold, 1951
9 italica Parl.			
10 pindica (Hausskn.) Chrtek & Křísa			
11 hispanica Chrtek & Křísa			
12 nutans (Vill.) Duval-Jouve	12	Ga	Lambert & Giesi, 1967
13 caespitosa Gay			
14 nodulosa (Bory & Chaub.) E.H.F. Meyer			
15 sylvatica (Hudson) Gaudin			
(a) sylvatica	12	Ga	Lambert & Giesi, 1967
(b) henriquesii (Degen) P. Silva	c.84	Lu	Fernandes, 1950
16 sieberi Tausch			
17 lutea (All.) DC.	12	Hs	Noronha-Wagner, 1949
18 purpureospoendens Seub.			
19 nivea (L.) DC.	12	Ga	Lambert & Giesi, 1967
20 lactea (Link) E.H.F. Meyer	12	Hs	Noronha-Wagner, 1949
21 pedemontana Boiss. & Reuter			
22 luzuloides (Lam.) Dandy & Wilmott	12	Ga	Lambert & Giesi, 1967
(a) luzuloides			
(b) cuprina (Rochel ex Ascherson & Graebner) Chrtek & Křísa			
23 alpinopilosa (Chaix) Breistr.	12	Au	Nordenskjold, 1951
(a) alpinopilosa			
(b) candollei (E.H.F. Meyer) Rothm.			
(c) obscura Fröhner	12	Cz	Májovský et al., 1970
(d) velenovskyi (Kozuharov) Chrtek & Křísa			
24 glabrata (Hoppe) Desv.			
25 desvauxii Kunth	12	Ga	Lambert & Giesi, 1967
26 wahlenbergii Rupr.	24	No	Knaben & Engelskjon, 1967
	36	Su	Löve & Löve, 1944
27 parviflora (Ehrh.) Desv.	24	No	Knaben & Engelskjon, 1967

28 elegans Lowe   6   Lu
29 pilosa (L.) Willd.   66   Fe
                                                  72
30 forsteri (Sm.) DC.   24   Ga
31 luzulina (Vill.) Dalla Torre & Sarnth.   24   Au

BROMELIALES

CXC BROMELIACEAE
1 Fascicularia Mez
  1 pitcairniifolia (Verlot) Mez

COMMELINALES

CXCI COMMELINACEAE

1 Tradescantia L.
  1 virginiana L.
  2 fluminensis Velloso
2 Tripogandra Rafin.
  1 multiflora (Swartz) Rafin.

3 Commelina L.
  1 communis L.

CXCII ERIOCAULACEAE

1 Eriocaulon L.
  1 aquaticum (Hill) Druce   64   Hb
  2 cinereum R. Br.

GRAMINALES

CXCIII GRAMINEAE

Tribe Arundinarieae Steudel

1 Sasa Makino & Shibata
  1 palmata (Burbidge) E.G. Camus

2 Arundinaria Michx
  1 anceps Mitf.
  2 fastuosa (Latour-Marliac ex Mitf.)
    Lehaie
  3 japonica Siebold & Zucc. ex Steudel
  4 simonii (Carriere) A. & C. Riviere
  5 vagans Gamble

3 Phyllostachys Siebold & Zucc.

Tribe Poeae

4 Festuca L.
  1 triflora Desf.
  2 caerulescens Desf.   14

3 paniculata (L.) Schinz & Thell.
  (a) paniculata   14   Ga   Küpfer, 1971
  (b) spadicea (L.) Litard.   42   Ga   Van Loon et al.,
  (c) baetica (Hackel) Markgr.-Dannenb.   1971
4 durandii Clauson   28   Lu   Fernandes & Queirós,
5 spectabilis Jan   42   1969
  (a) spectabilis
  (b) affinis (Boiss. & Heldr. ex Hackel)
    Hackel
6 altissima All.   14   Cz   Váchová & Feráková,
                            1977
                     42   Br   Hubbard, 1954
7 drymeja Mert. & Koch   14   Po   Mizianty & Frey,
                            1973
8 gigantea (L.) Vill.   42   Cz   Májovský et al.,
                            1970
9 pratensis Hudson   14   Cz   Májovský et al.,
                            1974
                     42
  (a) pratensis   14
  (b) apennina (De Not.) Hegi
10 arundinacea Schreber   28   Ge   Tischler, 1934
                     42   Cz   Uhríková &
                            Májovský, 1977
                     70   Ge   Tischler, 1934
  (a) arundinacea   42   Hs   Talavera, 1978
  (b) uechtritziana (Wiesb.) Hegi
  (c) atlantigena (St-Yves) Auquier   56
  (d) fenas (Lag.) Arcangeli   28
  (e) orientalis (Hackel) Tzvelev
11 luxu nust   14
                     28   Au   Gervais, 1965
                     42   Ju   Lovka et al., 1972
12 dimorpha Guss.   28   Ga   Tombal, 1968
13 calabrica Huter
14 pseudeskia Boiss.
15 scariosa (Lag.) Ascherson & Graebner   14
16 carpatica F.G. Dietr.   28   Cz   Májovský et al.,
                            1974
17 pulchella Schrader   14   Au   Gervais, 1965
18 eskia Ramond ex DC.   14   Hs   Küpfer, 1969
19 alpestris Roemer & Schultes
20 elegans Boiss.   28   Lu   Fernandes & Queirós,
                            1969
21 burnatii St-Yves   14   Hs   Küpfer, 1971
22 valida (Uechtr.) Penzes
  (a) valida
  (b) leilaensis Markgr.-Dannenb.
23 gautieri (Hackel) K. Richter   14
                     28
24 flavescens Bellardi   14   It   Gervais, 1965
25 scabriculmis (Hackel) K. Richter
  (a) scabriculmis
  (b) luedii Markgr.-Dannenb.
26 sardoa (Hackel ex Barbey) K. Richter   14   Co   Litardière, 1950
27 quadriflora Honckeny   14
28 versicolor Tausch
  (a) versicolor
  (b) brachystachys (Hackel) Markgr.-   14   Au   Litardière, 1950b
    Dannenb.
  (c) pallidula (Hackel) Markgr.-Dannenb.
  (d) dominii Krajina
29 acuminata Gaudin
30 varia Haenke

31 calva (Hackel) K. Richter
32 xanthina Roemer & Schultes
33 balcanica (Acht.) Markgr.-Dannenb.
   (a) balcanica
   (b) neicevii (Acht.) Markgr.-Dannenb.
   (c) rhodopensis Markgr.-Dannenb.
34 bosniaca Kummer & Sendtner
   (a) bosnaica
   (b) chlorantha (G. Beck) Markgr.-Dannenb.
   (c) pirinensis (Acht.) Markgr.-Dannenb.
35 penzesii (Acht.) Markgr.-Dannenb.
36 pindica (Markgr.-Dannenb.) Markgr.-Dannenb.
37 graeca (Hackel) Markgr.-Dannenb.
   (a) graeca
   (b) pawlowskiana Markgr.-Dannenb.
38 rechingeri Markgr.-Dannenb.
39 cyllenica Boiss. & Heldr.
   (a) cyllenica
   (b) pangaei Markgr.-Dannenb.
   (c) thasia Markgr.-Dannenb.
40 adamovicii (St-Yves) Markgr.-Dannenb.
41 galicicae I. Horvat ex Markgr.-Dannenb.
42 ampla Hackel   28   Lu   Fernandes & Queiros, 1969
             42
             56
43 petraea Guthnick ex Seub.
44 jubata Lowe
45 morisiana Parl.
46 henriquesii Hackel
47 querana Litard.
48 capillifolia Dufour
49 porcii Hackel
50 tatrae (Csakó) Degen   14   Cz   Baksay, 1957b
51 peristerea (Vetter) Markgr.-Dannenb.
52 korabensis (Jav. ex Markgr.-Dannenb.)
53 borderi (Hackel) K. Richter   14   Ga   Litardière, 1950b
54 norica (Hackel) K. Richter
55 amethystina L.
   (a) amethystina   14   ?He   Litardière, 1950b
   (b) kummeri (G. Beck) Markgr.-Dannenb.
   (c) ritschlii (Hackel) Lemke ex Markgr.-Dannenb.
   (d) orientalis Krajina
56 heterophylla Lam.   28   Cz   Májovský et al., 1974
57 plicata Hackel
58 violacea Schleicher ex Gaudin
   (a) violacea   14   Ga   Zickler, 1968
   (b) macrathera (Hackel ex G. Beck) Markgr.-Dannenb.
   (c) handelii Markgr.-Dannenb.
59 nitida Kit.
   (a) nitida
   (b) flaccida (Schur) Markgr.-Dannenb.
60 iberica (Hackel) K. Richter
61 puccinellii Parl.

317

62 clementei Boiss.	14	Hs	Küpfer, 1968
63 picta Kit.	14	Cz	Májovský et al. 1974
64 baffinensis Polunin	28	As	Zhukova, 1965
65 nigrescens Lam.			
(a) nigrescens	28		
	42	Br	Maude, 1940
(b) microphylla (St-Yves) Markgr.-Dannenb.			
66 rubra L.	14	Lu	Rodrigues, 1953
	28 70	Is	Löve & Löve, 1942
	42	Su	Löve & Löve, 1944
	56	Hu	Pólya, 1949
(a) rubra			
(b) litoralis (G.F.W. Meyer) Auquier	42	Be	Auquier & Rammeloo, 1973
(c) arenaria (Osbeck) Syme	56	Da	Kjellqvist, 1964
(d) pruinosa (Hackel) Piper	42		
(e) juncea (Hackel) Soó	42	Be	Auquier & Rammeloo, 1973
(f) thessalica Markgr.-Dannenb.			
(g) asperifolia (St-Yves) Markgr.-Dannenb.			
67 richardsonii Hooker	42	Is	Löve & Löve, 1956
68 pyrenaica Reuter	28		
69 trichophylla (Ducros ex Gaudin) K. Richter	42		
70 pseudotrichophylla Patzke	42		
71 cyrnea (Litard. & St-Yves) Markgr.-Dannenb.			
72 oelandica (Hackel) K. Richter	42		
73 juncifolia St-Amans	56	Be	Auquier & Rammeloo, 1977
74 diffusa Dumort.	42		
	56	Be	Auquier & Rammeloo, 1973
75 rivularis Boiss.	14 42		
76 rothmaleri (Litard.) Markgr.-Dannenb.			
77 nevadensis (Hackel) Markgr.-Dannenb.	70		
78 cretacea T. Popov & Proskorj.			
79 brachyphylla Schultes	28		
	42	Asia	Zhukova, 1965
80 hyperborea Holmen	28	Is	Löve & Löve, 1956
81 pirinica I. Horvat ex Markgr.-Dannenb.			
82 intercedens (Hackel) Ludi ex Becherer			
83 riloensis (Hackel ex Hayek) Markgr.-Dannenb.			
84 halleri All.			
(a) halleri	14	He	Scholte, 1977
(b) scardica (Griseb.) Markgr.-Dannenb.			
85 glacialis (Miégeville ex Hackel) K. Richter	14		
86 rupicaprina (Hackel) A. Kerner	14	Bot. Gard.	Brandberg, 1948
87 bucegiensis Markgr.-Dannenb.			
88 frigida (Hackel) K. Richter			
89 alpina Suter	14	Ga	Parreaux, 1972
90 olympica Vetter			
91 vizzavonae Ronniger			
92 horvatiana Markgr.-Dannenb.			
93 jeanpertii (St-Yves) Markgr.			
(a) jeanpertii			

(b) achaica (Markgr.-Dannenb.) Markgr.-
    Dannenb.
(c) campana (N. Terracc.) Markgr.-
    Dannenb.

94	stenantha (Hackel) K. Richter	14		
95	pseudodura Steudel	28		
96	circummediterranea Patzke	14	Ga	Parreaux, 1972
97	costei (St-Yves) Markgr.-Dannenb.	28	Ga	Parreaux, 1972
98	hervieri Patzke	14	Ga	Litardière, 1950
99	patzkei Markgr.-Dannenb.			
100	hystrix Boiss.			
101	reverchonii Hackel			
102	tenuifolia Sibth.	14 28	Ga	Bidault, 1972
103	ovina L.	14	Is	Löve & Löve, 1956
104	ophioliticola Kerguélen	28	Ga	Kerguélen, 1975
105	niphobia (St-Yves) Kerguélen	28	Ga	Kerguélen, 1976
106	airoides Lam.	14	Cz	Hadde & Haskova, 1956
		28		
		35		
107	armoricana Kerguélen	28	Ga	Kerguélen, 1975
108	eggleri Tracey	28	Au	Tracey, 1977
109	sipylea (Hackel) Markgr.-Dannenb.			
110	lapidosa (Degen) Markgr.-Dannenb.			
111	occitanica (Litard.) Auquier & Kerguélen	14	Ga	Kerguélen, 1975
112	keteropachys (St-Yves) Patzke ex Auquier	28	Be	Auquier & Rammeloo, 1973
113	guestfalica Boenn. ex Reichenb.	28		
114	lemanii Bast.	42	Be	Auquier & Rammeloo, 1973
115	arvernensis Auquier, Kerguélen & Markgr.-Dannenb.	28	Ga	Auquier & Kerguélen, 1977
116	longifolia Thuill.	14	Br	Hubbard, 1954
117	inops De Not.			
118	vaginata Waldst. & Kit. ex Willd.	14	Hu	Baksay, 1956
	(a) vaginata			
	(b) dominii (Krajina) Soó			
119	psammophila (Hackel ex Čelak.) Fritsch			
120	pannonica Wulfen ex Host			
121	beckeri (Hackel) Trautv.			
122	polesica Zapal.	14	Da	Bücher, 1947
123	javorkae Májovský			
124	arenicola (Prodan) Soó			
125	cinerea Vill.	28	Ga	Bidault, 1968
126	pallens Host			
	(a) pallens	14 28	Cz	Májovský et al., 1974
	(b) scabrifolia (Hackel ex Rohlena) Zielonk.			
	(c) treskana Markgr.-Dannenb.			
127	huonii Auquier	42	Ga	Auquier, 1973
128	gracilior (Hackel) Markgr.-Dannenb.	14	Ga	Parreaux, 1972
129	liviensis (Verguin) Markgr.-Dannenb.	14	Ga	Auquier & Kerguélen, 1977
130	ochroleuca Timb.-Lagr.	28		
	(a) ochroleuca			
	(b) gracilior (Verguin) Kerguélen			
	(c) bigorrensis (St-Yves) Kerguélen			
	(d) heteroidea (Verguin) Markgr.-Dannenb.			
131	brigantina (Markgr.-Dannenb.) Markgr.-Dannenb.			

132 valentina (St-Yves) Markgr.-Dannenb.
133 degenii (St-Yves) Markgr.-Dannenb.   28   Ga   Bidault, 1968
134 oviniformis Vetter
135 hirtovaginata (Acht.) Markgr.-Dannenb.
136 ticinensis (Markgr.-Dannenb.
137 grandiaristata Markgr.-Dannenb.
138 thracica (Acht.) Markgr.-Dannenb.
139 hercegovinica Markgr.-Dannenb.
140 robustifolia Markgr.-Dannenb.
141 panciciana (Hackel) K. Richter
142 apuanica Markgr.-Dannenb.
143 koritnicensis Hayek & Vetter
144 pohleana Allexeev
145 macedonica Vetter
146 polita (Halacsy) Tzvelev
147 curvula Gaudin
    (a) curvula   56 + 0 - 1B Ga   Bidault, 1968
    (b) crassifolia (Gaudin) Markgr.-   56 + 1B
        Dannenb.
    (c) cagiriensis (Timb.-Lagr.) Markgr.
        Dannenb.
148 pachyphylla Degen ex E.I. Nyárády
149 igoschiniae Tzvelev
150 vivipara (L.) Sm.   28   Is   Löve & Löve, 1956
                                         21 35 42 49
                                            56
151 glauca Vill.   42
152 vasconensis (Markgr.-Dannenb.)   42   Ga Hs   Kerguélen, 1976
    Auquier & Kerguélen
153 durissima (Hackel) Kerguélen   42 56   Ga   Kerguélen, 1975
154 indigesta Boiss.
    (a) indigesta   42   Hs   Küpfer, 1968
    (b) hackeliana (St-Yves) Markgr.-   42
        Dannenb.
    (c) aragonensis (Willk.) Kerguélen
    (d) alleizettei (Litard.) Kerguélen
    (e) molinieri (Litard.) Kerguélen
    (f) litardierei (St-Yves) Kerguélen
155 pseudovina Hackel ex Wiesb.   14   Hu   Pólya, 1948
                                            28   Hu   Felföldy, 1947
156 illyrica Markgr.-Dannenb.
157 wagneri (Degen, Thaisz & Flatt)
    Krajina   28   Hu   Horánszky et al.,
158 makutrensis Zapal   1972
159 valesiaca Schleicher ex Gaudin   14   Cz   Baksay, 1958
                                            28   Hu   Felföldy, 1947b
160 carnuntina Tracey   42   Au   Tracey, 1977
161 rupicola Heuffel   42   Hu   Pólya, 1949
    (a) rupicola
    (b) saxatilis (Schur) Rauschert
162 taurica (Hackel) A. Kerner ex Trautv.
163 dalmatica (Hackel) K. Richter   28
164 pseudodalmatica Krajina ex Domin   28   Cz   Baksay, 1957
165 callieri (Hackel ex St-Yves) Markgraf
166 wolgensis P. Smirnov
167 stricta Host   42   Hu   Felföldy, 1947
168 trachyphylla (Hackel) Krajina   42   Ga   Bidault, 1972
169 brevipila Tracy   42   Au   Tracey, 1977
170 duvalii (St-Yves) Stohr   28

x Festulolium Ascherson & Graebner
1 loliaceum (Hudson) P. Fourn.   14   Ho   Gadella & Kliphuis,
                                                                    1966
2 braunii (K. Richter) A. Camus   14   Br   Peto, 1933

5 Lolium L.
1 perenne L.   14 + 1 - 3B
2 multiflorum Lam.   14   Lu   Fernandes & Queirós,
                                                                    1969
3 rigidum Gaudin
  (a) rigidum   14   Lu   Fernandes & Queirós,
                                                                    1969
  (b) lepturoides (Boiss.) Sennen &
      Mauricio
4 temulentum L.   14   Bl   Dahlgren et al.,
                                                                    1971
5 remotum Schrank   14 + 1B   Br   Hovin & Hill, 1966

6 Vulpia C.C. Gmelin
1 sicula (C. Presl) Link   14   It Si Cotton & Stace,
                                                                    1976
2 geniculata (L.) Link   14   Lu   Cotton & Stace,
                                                                    1976
3 ligustica (All.) Link   14   Si   Cotton & Stace,
                                                                    1976
4 alopecuros (Schousboe) Dumort.   14   Lu   Fernandes & Queirós,
                                                                    1969
5 fasciculata (Forskål) Samp.   28   Bl   Dahlgren et al.,
                                                                    1971
6 membranacea (L.) Dumort.   14   Ga Hs Cotton & Stace,
                                                                    1976
7 fontquerana Melderis & Stace   14   Hs   Cotton & Stace,
                                                                    1976
8 bromoides (L.) S.F. Gray   14   Br   I. & O. Hedberg, 1961
9 muralis (Kunth) Nees   14   Lu   Fernandes & Queirós,
                                                                    1969
10 myuros (L.) C.C. Gmelin   42 Be Br Co Ga   Cotton & Stace,
                                                      Gr It Ju Si   1976
11 ciliata Dumort.
   (a) ciliata   28 Br Co Cr Ga   Cotton & Stace,
                                                      Gr Hs Ju Lu   1976
                                                      Rm Si
   (b) ambigua (Le Gall) Stace & Auquier   28 Be Br Ga   Cotton & Stace,
                                                                    1976
12 unilateralis (L.) Stace   14   Bu   Kozuharov &
                                                                    Petrova, 1973

7 Ctenopsis De Not.
1 delicatula (Lag.) Paunero   14   Lu   Fernandes & Queirós,
                                                                    1969
2 gypsophila (Hackel) Paunero

8 Wangenheimia Moench
1 lima (L.) Trin.

9 Micropyrum Link
1 tenellum (L.) Link   14   Lu   Fernandes & Queirós,
                                                                    1969
2 patens (Brot.) Rothm. ex Pilger   14   Lu   Fernandes & Queirós,
                                                                    1969

321

10 Castellia Tineo
  1 tuberculosa (Moris) Bor

11 Narduroides Rouy
  1 salzmannii (Boiss.) Rouy   14

12 Desmazeria Dumort.
  1 sicula (Jacq.) Dumort.
  2 marina (L.) Druce   14 28   Lu   Fernandes &
                                             Queirós, 1969
  3 rigida (L.) Tutin   14   Bu   Kozuharov &
                                             Petrova, 1974
    (a) rigida
    (b) hemipoa (Delile ex Sprengel) Stace

13 Cutandia Willk.
  1 maritima (L.) W. Barbey   14   Lu   Fernandes &
                                           Queirós, 1969
  2 stenostachya (Boiss.) Stace
  3 divaricata (Desf.) Ascherson ex W.
    Barbey
  4 memphitica (Sprengel) K. Richter

14 Sphenopus Trin.
  1 divaricatus (Gouan) Reichenb.   12 N. Africa Litardière, 1950b

15 Vulpiella (Trabut) Burollet
  1 tenuis (Tineo) Kerguelen

16 Poa L.
  1 annua L   28   B   Tutin, 1957
  2 informa Kunth   14   Ga Tutin, 1957
  3 supina Schrader   14   He Tutin, 1957
  4 trivialis L.   14   Hs Löve & Kjellqvist,
                                           1973
                   14 + 0 - 5B Su Bosemark, 1957
                      28   Bu Kozuharov &
                                       Kuzmanov, 1970
    (a) trivialis
    (b) sylvicola (Guss.) H. Linb. fil.
  5 feratiana Boiss. & Reuter
  6 alpigena (Fries) Lindman   35   Sb Flovik, 1940
                        43   Sb Flovik, 1940
                      56   Rs(N) Sokolovskaya &
                   70-72 Rs(N) Strelkova, 1960
                    84   Rs(N) Sokolovskaya &
                                     Strelkova, 1941
  7 subcaerulea Sm.   38-117 Is Löve & Löve, 1956
  8 pratensis L.   42   Su Almgaard, 1960
                   50-78 Is Löve & Löve, 1956
                    91   Is Löve & Löve, 1942
                    98   Bu Christov & Nikolov,
                                     1962
  9 angustifolia L.   46-63 Is Löve & Löve, 1956
  10 arctica R. Br.   38   No Nannfeldt, 1940
                    56   Sb Flovik, 1940
                    72   Rs(N) Sokolovskaya &
                                   Strelkova, 1941
                    76   Is Löve & Löve, 1956
  11 granitica Br.-Bl.

(a) granitica	64 67 71 72 94	Po	Skalińska et al., 1957
(b) disparilis (E.I. Nyárády) E.I. Nyárády)			
12 cenisia All.			
(a) sardoa E. Schmid	56	Co	Contandriopoulos, 1962
(b) cenisia	28	He	Favarger, 1959
	49	Ga	Bajon & Greber, 1971
	50-56	Ga Ge	Favarger, 1959
13 longifolia Trin.			
14 sibirica Roshev.			
15 chaixii Vill.	14	Cz	Májovský et al., 1974
16 hybrida Gaudin	14	He	Favarger, 1959
17 remota Forselles	14	Su	Nannfeldt, 1942
18 stiriaca Fritsch & Hayek	14	Cz	Edmondson, ined.
19 flexuosa Sm.	42	Is	Löve & Löve, 1956
20 laxa Haenke	21 28	He	Nygren, 1956
21 pirinica Stoj. & Acht.	14	Bu	Stoeva, 1977
22 minor Gaudin	28	He	Nygren, 1956
23 compressa L.	42	Cz	Májovský et al., 1974
	45 49 56	Is	Löve & Löve, 1948
24 flaccidula Boiss. & Reuter			
25 palustris L.			
26 glauca Vahl	42	Is	Löve & Löve, 1956
	44	No	Knaben & Engelskjon, 1967
	49 56 63	Is	Löve & Löve, 1956
	65	Su	Aakerberg, 1942
	c.70	Sb	Flovik, 1940
27 nemoralis L.	28	Po	Skalińska et al., 1957
	33	Ga	Guinochet, 1943
	42	Is	Löve & Löve, 1956
	56	Br	Hooper, 1961
28 rehmannii (Ascherson & Graebner) Woloszczak	14	Rm	Lungeanu, 1972
29 pannonica A. Kerner			
(a) pannonica			
(b) scabra (Ascherson & Graebner) Soó			
30 versicolor Besser			
31 sterilis Bieb.			
32 stepposa (Krylov) Roshev.			
33 abbreviata R. Br.			
34 trichophylla Heldr. & Sart. ex Boiss.	14	Gr	Edmondson, ined.
35 bulbosa L.	21	Lu	Fernandes & Queirós, 1969
	28	Al	Strid, 1971;
		Co Ga	Litardière, 1949
	39 42	Po	Skalińska et al., 1957
	42	It	Tutin, 1961
36 perconcinna J.R. Edmondson	14	Bu Ga He	Nygren, 1962
37 ligulata Boiss.	14	Hs	Löve & Kjellqvist, 1973
38 timoleontis Heldr. ex Boiss.	14	Ju	Nygren, 1957
39 media Schur	14 + 0 - 2B	Ju	Edmondson, ined.
40 pumila Host	14	Au	Nygren, 1962

41 molinerii Balbis 14 He Nygren, 1955
42 badensis Haenke ex Willd. 14 + 2 - 8B Au Ge Muntzing & Nygren, 1955
18 Cz Holub et al., 1971
21 28 Ga Ge Hu Nygren, 1962
43 alpina L. 22 26 28 Po Skalińska, 1950
33-35
37-40 Su Muntzing, 1966
42 Rs(N) Sokolovskaya & Strelkova, 1960
43-45 Su Muntzing, 1966
33-46 No Knaben & Engelskjon, 1967
56 Bu Stoeva, 1977
44 jubata A. Kerner
45 flabellata (Lam.) Hooker fil.
46 x hartzii Gand.
47 x jemtlandica (Almq.) K. Richter c.36 Br Hedberg, 1958
48 x nobilis Skalińska c.80 Po Skalińska, 1950
49 x taurica H. Pojark.

17 Bellardiochloa Chiov.
   1 violacea (Bellardi) Chiov. 14 He Gervais, 1965

18 Eremopoa Roshev.
   1 altaica (Trin.) Roshev.

19 Catabrosella (Tzvelev) Tzvelev
   1 humilis (Bieb.) Tzvelev

20 Arctophila Rupr. ex N.J. Andersson
   1 fulva (Trin.) N.J. Andersson 42 Rs(N) Sokolovskaya & Strelkova, 1960

21 Puccinellia Parl.
   1 distans (L.) Parl.
     (a) distans 42 Cz Váchová & Feráková, 1978
     (b) limosa (Schur) Jáv. 28 Hu Pólya, 1948
     (c) hauptiana (Trin. ex V. Krecz.) 28 Asia
         W.E. Hughes
                                       42 Asia
     (d) borealis (Holmberg) W.E. Hughes 42 Is Löve & Löve, 1956
   2 tenella (Lange) Holmberg
     svarlbadensis Rønning 42 Sb Rønning, 1962
   3 fasciculata (Torrey) E.P. Bicknell
     (a) fasciculata 28 Br Rutland, 1941
     (b) pungens (Pau) W.E. Hughes
   4 maritima (Hudson) Parl. 56 Ga Delay, 1969
   5 phryganodes (Trin.) Scribner & Merr. 28 Rs(N) Sokolovskaya & Strelkova, 1960
   6 festuciformis (Host) Parl. 35 Ga Labadie, 1976
     (a) festuciformis
     (b) convoluta (Hornem.) W.E. Hughes
     (c) intermedia (Schur) W.E. Hughes
     (d) tenuifolia (Boiss. & Reuter) W.E. Hughes
   7 dolicholepis (V. Krecz.) V. Krecz. ex Pavlov
   8 gigantea (Grossh.) Grossh.

9 poecilantha (C. Koch) Grossh.
10 tenuissima (V. Krecz.) V. Krecz. ex Pavlov
11 angustata (R. Br.) Rand & Redf.   42
12 rupestris (With.) Fernald & Weatherby   42   Br   Rutland, 1941
13 vahliana (Liebm.) Scribner & Merr.

x Pucciphippsia Tzvelev
  1 vacillans (Th. Fries) Tzvelev

22 Phippsia R. Br.
  1 algida (Solander) R. Br.   28   No   Knaben, 1950
  2 concinna (Th. Fries) Lindeb.   28   No   Knaben, 1950

23 Sclerochloa Beauv.
  1 dura (L.) Beauv.   14   Cz   Vachova &
                                                            Ferakova, 1977

24 Dupontia R. Br.
  1 fisheri R. Br.   88   Rs(N)   Sokolovskaya &
                                                            Strelkova, 1960
  2 psilosantha (Rupr.) Griseb.   44   Rs(N)   Sokolovskaya &
                                                            Strelkova, 1960

25 Arctagrostis Griseb.
  1 latifolia (R. Br.) Griseb.   56 62   Rs(N)   Sokolovskaya &
                                                            Strelkova, 1960

26 Dactylis L.
  1 glomerata L.   14   Su   Lundqvist, 1965
                                            28   Is   Löve & Löve, 1956
                                      28 + 0 - 4B   Lu   Fernandes & Queirós,
                                                            1969

27 Beckmannia Host
  1 eruciformis (L.) Host   14   Cz   Májovský et al.,
                                                            1970
    (a) eruciformis
    (b) borealis Tzvelev
  2 syzigachne (Steudel) Fernald

28 Cynosurus L.
  1 cristatus L.   14   Lu   Fernandes & Queirós,
                                                            1969
  2 echinatus L.   14   Bu   Kozuharov &
                                                            Kuzmanov, 1970
  3 elegans Desf.   14   Lu   Fernandes & Queirós,
                                                            1969
  4 callitrichus W. Barbey

29 Lamarckia Moench
  1 aurea (L.) Moench   14   Lu   Fernandes & Queirós,
                                                            1969

30 Catabrosa Beauv.
  1 aquatica (L.) Beauv.   20   Is   Löve & Löve, 1956

31 Cinna L.
  1 latifolia (Trev.) Griseb.

32 Apera Adanson
  1 spica-venti (L.) Beauv.   14   Ho   Gadella & Kliphuis,
                                                            1968

```
    (a) spica-venti
    (b) maritima (Klokov) Tzvelev
  2 interrupta (L.) Beauv.                    14      Br      Maude, 1940
  3 intermedia Hackel


33 Psilurus Trin.
  1 incurvus (Gouan) Schinz & Thell.         28      Lu      Fernandes & Queirós,
                                                             1969

34 Mibora Adanson
  1 minima (L.) Desv.

35 Briza L.
  1 media L.                                  14      Lu      Fernandes & Queirós,
                                                             1969
                                             28      Bu      Kozuharov et al.,
    (a) media                                                1974
    (b) elatior (Sibth. & Sm.) Rohlena
  2 maxima L.                                 14      Lu      Fernandes & Queirós,
                                                             1969
  3 minor L.                                  10      Lu      Fernandes & Queirós,
                                                             1969
  4 humilis Bieb.                             14      Bu      Kozuharov et al.,
                                                             1974

  Tribe Seslerieae Koch

36 Sesleria Scop.
  1 ovata (Hoppe) A. Kerner
  2 leucocephala DC.
  3 ophrosa...phala Mdl                       14      It      Favarger & Humb,
                                                             1964
  4 tenuifolia Schrader
    (a) tenuifolia                            28      Al      Ujhelyi, 1959
                                             42
    (b) kalnikensis (Jav.) Deyl              56      Ju      Ujhelyi, 1960b
  5 comosa Velen.                             28      Bu      Ujhelyi, 1959
  6 tenerrima (Fritsch) Hayek                 28      Gr      Ujhelyi, 1960b
  7 sadlerana Janka
    (a) sadlerana                             56      Hu      Ujhelyi, 1959
    (b) tatrae (Degen) Deyl                   56      Po      Rychlewski, 1955
  8 heuflerana Schur
    (a) heuflerana                            28      Cz      Murín &
                                                             Májovský, 1976
    (b) hungarica (Ujhelyi) Deyl             56      Hu      Ujhelyi, 1959
  9 bielzii Schur                             56      Rm      Ujhelyi, 1959
 10 coerulans Friv.                           56      Bu      Ujhelyi, 1959
 11 korabensis (Kummerle & Jav.) Deyl
 12 vaginalis Boiss. & Orph.
 13 doerfleri Hayek
 14 wettsteinii Dörfler & Hayek              56      Al      Ujhelyi, 1959
 15 robusta Schott, Nyman & Kotschy
    (a) robusta                               56      Al      Ujhelyi, 1959
    (b) skanderbeggii (Ujhelyi) Dehl
 16 nitida Ten.                               28      Bot. Gard. Ujhelyi, 1960b
 17 alba Sibth. & Sm.
 18 latifolia (Adamovic) Degen               28      Ju      Ujhelyi, 1959
 19 autumnalis (Scop.) F.W. Schultz          28      Ju      Ujhelyi, 1959
 20 anatolica Deyl
 21 argentea (Savi) Savi                      28      Hs      Löve & Kjellqvist,
                                                             1973
 22 albicans Kit. ex Schultes                28      Cz      Murín & Májovský,
                                                             1976
```

```
    (a) albicans
    (b) angustifolia (Hackel & G. Beck) Deyl
23 caerulea (L.) Ard.                    28      Ga      Parreaux, 1971
24 taygetea Hayek                        28      Gr      Ujhelyi, 1960
    serbica                              28      Ju      Ujhelyi, 1959
25 rigida Heuffel ex Reichenb.
    (a) rigida
    (b) achtarovii (Deyl) Deyl
26 insularis Sommier                     28      Co      Litardière, 1949
    (a) insularis                        28      Co      Litardière, 1949
    (b) italica (Pamp.) Deyl
    (c) sillingeri (Deyl) Deyl

37 Oreochloa Link
    1 disticha (Wulfen) Link             14      Au      Reese, 1953
    2 blanka Deyl                        14
    3 seslerioides (All.) K. Richter     14      Ga      Gervais, 1965
    4 confusa (Coincy) Rouy

38 Ammochloa Boiss.
    1 palaestina Boiss.

39 Echinaria Desf.
    1 capitata (L.) Desf.                18      Bu      Kozuharov &
                                                         Petrova, 1974


Tribe Meliceae Reichenb.

40 Melica L.
    1 nutans L.                          18      Fe      V. Sorsa, 1962
    2 picta C. Koch                      18              Doulat, 1943
    3 uniflora Retz.                     18      Ho      Gadella & Kliphuis,
                                                         1963
    4 rectiflora Boiss. & Heldr.
    5 minuta L.                          18      Ga      Delay, 1969
                                         36      Lu      Fernandes & Queirós,
                                                         1969
    6 altissima L.                       18              Doulat, 1943
    7 ciliata L.
    (a) ciliata                          18      Ga      Delay, 1969
    (b) taurica (C. Koch) Tzvelev        18
    (c) monticola (Prokudin) Tzvelev
    (d) magnolii (Gren. & Godron)
    8 transsilvanica Schur               18      Po      Frey, 1971
    (a) transsilvanica
    (b) klokovii Tzvelev
    9 cupanii Guss.
   10 bauhinii All.

41 Schizachne Hackel
    1 callosa (Turcz. ex Griseb.) Ohwi

Tribe Glycerieae Endl.

42 Glyceria R. Br.
    1 triflora (Korsh.) Komarov
    2 arundinacea Kunth
    3 maxima (Hartman) Holmberg          60      Ho      Gadella & Kliphuis,
                                                         1963
    4 lithuanica (Gorski) Gorski
    5 striata (Lam.) A.S. Hitchc.
```

```
  6 nemoralis (Uechtr.) Uechtr. &        20      Cz      Holub et al.,
    Koernicke                                            1970
  7 declinata Breb.                      20      Lu      Fernandes & Queirós,
                                                         1969

  8 fluitans (L.) R. Br.                 40      Hu      Pólya, 1950
  9 spicata (Biv.) Guss.
 10 plicata (Fries) Fries                40      Hs      Löve & Kjellqvist,
                                                         1973

 11 x pedicellata Townsend

43 Pleuropogon R. Br.
  1 sabinei R. Br.


Tribe Bromeae Dumort.

44 Bromus L.
  1 diandrus Roth                        56      Hs      Löve & Kjellqvist,
                                                         1973

  2 rigidus Roth                         42      Ga      Cugnac & Simonet, 1941b
  3 sterilis L.                          14      Cz      Májovský et al.,
                                                         1974
  4 tectorum L.                          14      Hs      Löve & Kjellqvist,
                                                         1973

  5 madritensis L.                       28      Bu      Kozuharov et al.,
                                                         1974

                                        42
  6 fasciculatus C. Presl                14
  7 rubens L.                            28      Hs      Björkqvist et al.,
                                                         1969

  8 inermis Leysser                      56      Su      Löve & Löve, 1944
  9 sibiricus Drobov
 10 ramosus Hudson                       14      Br      Hubbard, 1954
                                         28      Hu      Pólya, 1950
                                         42      Cz      Májovský et al.,
                                                         1976
 11 benekenii (Lange) Trimen             28      Cz      Májovský et al.,
                                                         1976
 12 erectus Hudson                       42      Hu      Baksay, 1961
                                         56      Hu      Pólya, 1949
    (a) erectus
    (b) condensatus (Hackel) Ascherson &
        Graebner
    (c) stenophyllus (Link) Ascherson &
        Graebner
    (d) transsilvanicus (Steudel) Ascherson
        & Graebner
 13 pannonicus Kummer & Sendtner
    (a) pannonicus
    (b) monocladus (Domin) P.M. Sm.       28      Cz      Májovský et al.,
                                                         1976

 14 moellendorffianus (Ascherson & Graebner)
    Hayek
 15 pindicus Hausskn.
 16 tomentellus Boiss.
 17 moesiacus Velen.
 18 riparius Rehmann                      70
 19 cappadocicus Boiss. & Balansa         42
    (a) cappadocicus
    (b) lacmonicus (Hausskn.) P.M. Sm.
 20 arvensis L.                           14      Bu      Kozuharov et al.,
                                                         1974
```

21 pseudosecalinus P.M. Sm.	14	Br	Smith, 1967	
22 secalinus L.	28	Su	Löve & Löve, 1944	
23 bromoideus (Lej.) Crépin	28	Ga	Cugnac & Simonet, 1941a	
24 grossus Desf. ex DC.	28	Be	Tournay, 1968	
25 commutatus Schrader	28	Bu	Kozuharov et al., 1974	

   (a) commutatus
   (b) neglectus (Parl.) P.M. Sm.

26 racemosus L.	28	Is	Löve & Löve, 1956	
27 hordeaceus L.	28	Hs	Löve & Kjellqvist, 1973	

   (a) hordeaceus
   (b) molliformis (Lloyd) Maire & Weiller
   (c) ferronii (Mabille) P.M. Sm.
   (d) thominii (Hard.) Maire & Weiller

28 interruptus (Hackel) Druce	28	Br	Hubbard, 1954	
29 lepidus Holmberg	28	Br	Hubbard, 1954	
30 scoparius L.	14	Bu	Kozuharov et al., 1974	
31 alopecuros Poiret	14			

   (a) alopecuros
   (b) caroli-henrici (W. Greuter) P.M. Sm.

32 lanceolatus Roth	28	Lu	Gardé & Malheiros-Gardé, 1954	
33 intermedius Guss.	14			
34 japonicus Thunb.	14	Bu	Kozuharov et al., 1974	

   (a) japonicus
   (b) anatolicus (Boiss. & Heldr.) Penzes

35 squarrosus L.	14	Bu	Jasiewicz & Mizianty, 1975	
36 willdenowii Kunth	42	Lu	Fernandes & Queirós, 1969	
37 carinatus Hooker & Arnott	56	Br	Hubbard, 1954	

Tribe Brachypodieae C.O. Harz

45 Brachypodium Beauv.

1 sylvaticum (Hudson) Beauv.	18	Br	Hedberg & Hedberg, 1961	
	28 42 + 2B 56	Bu	Kozuharov et al., 1974	

   (a) sylvaticum

(b) glaucovirens Murb.	16	Bu	Kozuharov et al., 1974	
2 pinnatum (L.) Beauv.	14 16 28	Bu	Kozuharov et al., 1974	
(a) pinnatum	28	Bu	Kozuharov et al., 1974	
(b) rupestre (Host) Schübler & Martens	14	Bu	Kozuharov & Nicolova, 1975	
3 retusum (Pers.) Beauv.	18			
4 phoenicoides (L.) Roemer & Schultes	28	Lu	Fernandes & Queirós, 1969	
5 distachyon (L.) Beauv.	10	Bu	Kozuharov et al., 1974	
	28	Ga	Kliphuis & Wieffering, 1972	
	30	Lu	Fernandes & Queiros, 1969	

Tribe Triticeae Dumort.

46 Festucopsis (C.E. Hubbard) Melderis
  1 serpentini (C.E. Hubbard) Melderis   14   Al   Jones, 1955
  2 sancta (Janka) Melderis

47 Leymus Hochst.
  1 arenarius (L.) Hochst.   56   Is   Löve & Löve, 1956
    mollis   28   Is   Löve & Löve, 1956
  2 racemosus (Lam.) Tzvelev
    (a) racemosus   28   ?Europe Bowden, 1957
    (b) sabulosus (Bieb.) Tzvelev 28   Asia   Bowden, 1957
    (c) klokovii Tzvelev
  3 karelinii (Turcz.) Tzvelev   56
  4 akmolinensis (Drobov) Tzvelev 28   Rs(E)
  5 paboanus (Claus) Pilger
  6 multicaulis (Kar. & Kir.) Tzvelev
  7 ramosus (Trin.) Tzvelev

48 Elymus L.
  1 sibiricus L.   28
  2 panormitanus (Parl.) Tzvelev 28   It   Schulz-Schaeffer
                               & Jurasits, 1962
  3 trachycaulus (Link) Gould ex Shinners 28
  4 mutabilis (Drobov) Tzvelev   28
  5 caninus (L.) L.   28   Lu   Fernandes & Queirós,
                                1969
  6 uralensis (Nevski) Tzvelev
    (a) uralensis
    (b) viridiglumis (Nevski) Tzvelev
  7 fibrosus (Schrenk) Tzvelev   ?8
  8 macrourus (Turcz.) Tzvelev
    (a) macrourus   28
    (b) turuchanensis (Reverdatto) Tzvelev 28   Rs(N) Sokolovskaya, 1970
  9 alaskanus (Scribner & Merrill) A. Löve
    (a) borealis (Turcz.) Melderis   28
    (b) scandicus (Nevski) Melderis 28   Su   Runemark & Heneen,
                                1968
    (c) subalpinus (Neuman) Melderis
  10 reflexiaristatus (Nevski) Melderis
    (a) reflexiaristatus
    (b) strigosus (Bieb.) Melderis 14 27 Rs(K) Petrova, 1968
  11 stipifolius (Czern. ex Nevski)
    Melderis   28   Rs(K) Petrova, 1968
  12 bungeanus (Trin.) Melderis
    (a) scythicus (Nevski) Melderis
    (b) pruiniferus (Nevski) Melderis
  13 nodosus (Nevski) Melderis
    (a) nodosus   28
    (b) corsicus (Hackel) Melderis 28   Co   Contandriopoulos,
                                1962
  14 flaccidifolius (Boiss. & Heldr.) Melderis
  15 curvifolius (Lange) Melderis
  16 elongatus (Host) Runemark
    (a) elongatus   14   Bu   Kozuharov &
                                Kuzmanov, 1970
    (b) ponticus (Podp.) Melderis 70
                         56   Rm   Tarnavschi, 1948
  17 pycnanthus (Godron) Melderis 42   Br   Löve & Löve, 1961b
  18 repens (L.) Gould
    (a) repens   42   Hs   Löve & Kjellqvist,
                                1973

	28		
	56		
(b) pseudocaesius (Pacz.) Melderis	42		
(c) arenosus (Petif) Melderis			
(d) elongatiformis (Drobov) Melderis	42		
(e) calcareus (Černjavski) Melderis			
repens x truncatus	42	Rs(K)	Petrova, 1968
repens x Hordeum secalinum	49		
19 pungens (Pers.) Melderis			
(a) pungens			
(b) campestris (Godron & Gren.) Melderis	56		
(c) fontqueri Melderis			
20 hispidus (Opiz) Melderis			
(a) hispidus	42	Cz	Májovský et al., 1974
	28	Co	Litardière, 1947
	43		
(b) barbulatus (Schur) Melderis	42		
	28		
(c) graecus Melderis			
(d) varnensis (Velen.) Melderis			
(e) pouzolzii (Godron) Melderis			
21 lolioides (Kar. & Kir.) Melderis	58	Iran	
22 farctus (Viv.) Runemark ex Melderis			
(a) farctus	42	Ga	Labadie, 1976
	56		
(b) boreali-atlanticus (Simonet & Guinochet) Melderis	28	Ho Su	Heneen, 1962
(c) bessarabicus (Savul. & Rayss) Melderis	14		
(d) rechingeri (Runemark) Melderis	28	Gr	Heneen & Runemark, 1962
22b x 17	35		
22b x 18	35		
	49		
49 Agropyron Gaertner			
1 cimmericum Nevski			
2 dasyanthum Ledeb.	28	Rs(W)	Petrova, 1968
3 tanaiticum Nevski			
4 fragile (Roth) P. Candargy	28	Rs(E)	Knowles, 1955
	14		
5 desertorum (Fischer ex Link) Schultes	28		Knowles, 1955
	29		Knowles, 1955
	32		Knowles, 1955
6 cristatum (L.) Gaertner			
(a) pectinatum (Bieb.) Tzvelev	14	Au	Titz, 1966
	28	Rm	Lungeanu, 1972
	42		
(b) sabulosum Lavrenko			
(c) brandzae (Panţu & Solac.) Melderis			
(d) ponticum (Nevski) Tzvelev			
(e) sclerophyllum Novopokr. ex Tzvelev			
50 Eremopyrum (Ledeb.) Jaub. & Spach			
1 triticum (Gaertner) Nevski	14		
2 orientale (L.) Jaub. & Spach	28		
3 distans (C. Koch) Nevski	14		
	28		

51 Crithopsis Jaub. & Spach
1 delileana (Schultes) Roshev.

52 Aegilops L.

1 speltoides Tausch	14	cult. Br	Percival, 1926
2 ventricosa Tausch	28	Bl	Dahlgren et al., 1971
3 cylindrica Host	28	Rm	Priadcencu et al., 1967
4 dichasians (Zhuk.) Humphries	14		
5 comosa Sibth. & Sm.	14		
(a) comosa			
(b) heldreichii (Boiss.) Eig			
6 uniaristata Vis.	28	Asia	
7 triuncialis L.	28	Lu	Fernandes & Queirós, 1969
8 lorentii Hochst.	28	Asia	
9 geniculata Roth	28	Bl	Dahlgren et al., 1971
10 neglecta Req. ex Bertol.	28	Al	Strid, 1971
	42		

53 Triticum L.

1 baeoticum Boiss.			
2 monococcum L.	14	Lu	Câmara &
3 dicoccon Schrank	28	It Lu	Coutinho, 1939
4 durum Desf.	28	It	Giorgi & Bozzini, 1969
5 turgidum L.	28	It	Giorgi & Bozzini, 1969
6 polonicum L	28	It	Câmara & Coutinho, 1939
7 spelta L.	42		Upadhya & Swaminathan, 1963
8 aestivum L.	42	Lu	Câmara & Coutinho, 1939
9 compactum Host			

54 Dasypyrum (Cosson & Durieu) T. Durand

1 villosum (L.) P. Candargy	14	Al	Strid, 1971
2 hordeaceum (Cosson & Durieu) P. Candargy			

55 Secale L.
1 sylvestre Host
2 montanum Guss.
3 cereale L.

56 Psathyrostachys Nevski

1 juncea (Fischer) Nevski	14	Rs(N)	Bowden, 1964

57 Hordeum L.

1 spontaneum C. Koch			
2 distichon L.	14	Br	Gale & Rees, 1970
3 vulgare L.	14	Br	Gale & Rees, 1970
4 murinum L.			
(a) murinum L.	28	Al	Strid, 1971
(b) glaucum (Steudel) Tzvelev	14		
(c) leporinum (Link) Arcangeli	28	Bl	Dahlgren et al., 1971
5 marinum Hudson	14	Hu	Kliphuis, 1977
	28	Lu	Castro & Fontes, 1946

6 hystrix Roth	14	Hu	Pólya, 1948
	28		
7 bulbosum L.	14	It	Morrison, 1959
	28	Ju	Katznelson &
8 bogdanii Wilensky			Zohary, 1967
9 brevisubulatum (Trin.) Link			
10 secalinum Schreber	14	Ga	Labadie, 1976
	28	Lu	Fernandes & Queiros, 1969
11 jubatum L.	28	Br Ge Lu Po	Morrison, 1959

58 Hordelymus (Jessen) C.O. Harz

1 europaeus (L.) C.O. Harz	28	Cz	Májovský et al., 1970

59 Taeniatherum Nevski

1 caput-medusae (L.) Nevski	14	Lu	Fernandes & Queirós, 1969

Tribe Aveneae

60 Avena L.
1 clauda Durieu
2 eriantha Durieu

3 longiglumis Durieu	14	Hs It	Ladizinsky &
4 barbata Pott ex Link			Zohary, 1971
(a) barbata	14 28	Lu	Fernandes & Queiros, 1969
(b) atherantha (C. Presl) Rocha Afonso	14		
5 saxatilis (Lojac.) Rocha Afonso	14		
6 prostrata Ladizinsky	14		
7 strigosa Schreber			
(a) strigosa	14	Hs It	Ladizinsky & Zohary,
(b) agraria (Brot.) Tab. Mor.			1971
8 brevis Roth	14		
9 nuda L.	14		

10 fatua L.
11 sativa L.
(a) sativa
(b) praegravis (Krause) Tab. Mor.
(c) macrantha (Hackel) Rocha Afonso
12 byzantina C. Koch

13 sterilis L.			
(a) sterilis	42	Bl	Dahlgren et al., 1971
(b) ludoviciana (Durieu) Nyman	42	Lu	Fernandes & Queirós, 1969
14 murphyi Ladizinsky	28	Hs	Ladizinsky, 1971

61 Helictotrichon Besser

1 decorum (Janka) Henrard	14	Rm	Litardière, 1950
2 sedenense (DC.) J. Holub	14 + 0 - 2B	Ga	Gervais, 1966
	28 + 0 - 2B	Ga	Gervais, 1966
3 planifolium (Willk.) J. Holub	28 + 0 - 2B		
4 parlatorei (J. Woods) Pilger	14	Ga It	Gervais, 1966
5 setaceum (Vill.) Hendrard	14	Ga	Gervais, 1965
6 petzense Melzer	14	Alps	Sauer, 1971
7 sempervirens (Vill.) Pilger	42 + 0 - 5B	Ga Gr It	Gervais, 1966
8 convolutum (C. Presl) Henrard			
(a) convolutum	14	Gr	Gervais, 1968
(b) heldreichii (Parl.) Gervais			

9 sarracenorum (Gand.) J. Holub	14 28 70	Hs	Gervais, 1972
10 filifolium (Lag.) Henrard	70	Hs	Löve & Kjellqvist, 1973
	84 + 0 - 1B	Hs	Gervais, 1968, 1972
11 murcicum J. Holub			
12 cantabricum (Lag.) Gervais	84 98 98	Ga	Gervais, 1973
13 desertorum (Less.) Nevski			
(a) desertorum			
(b) basalticum (Podp.) J. Holub	14	Cz	Holub et al., 1971

62 Avenula (Dumort.) Dumort.

1 pubescens (Hudson) Dumort.	14	Br No Su	Hedberg & Hedberg, 1961
	14 + 2B	Cz	Gervais, 1966
(a) pubescens			
(b) laevigata (Schur) J. Holub			
2 versicolor (Vill.) Lainz			
(a) versicolor	14 + 0 - 3B	Au Ga	Gervais, 1966
(b) pretutiana (Parl. ex Arcangeli) J. Holub	14		
3 schelliana (Hackel) W. Sauer & Chmelitschek	14	?Rs	Sauer & Chmelitschek, 1976
4 marginata (Lowe) J. Holub	14	Ga	Gervais, 1966
(a) marginata			
(b) sulcata (Gay ex Delastre) Franco			
(c) myrenaica J. Holub			
5 albinervis (Boiss.) Lainz	28	Lu	Fernandes & Queirós, 1969
6 occidentalis (Gervais) J. Holub			
(a) occidentalis	42		
(b) stenophylla Franco			
7 delicatula Franco			
8 levis (Hackel) J. Holub	14	Hs	
9 compressa (Heuffel) W. Sauer & Chmelitschek	14	Hu	Gervais, 1966
10 aetolica (Rech. fil.) J. Holub			
11 blavii (Ascherson & Janka) W. Sauer & Chmelitschek			
12 mirandana (Sennen) J. Holub	98		
13 gonzaloi (Sennen) J. Holub			
14 requienii (Mutel) J. Holub			
15 pungens (Sennen ex St-Yves) J. Holub			
16 pratensis (L.) Dumort.	126	Br Su	Hedberg & Hedberg, 1961
(a) pratensis			
(b) hirtifolia (Podp.) J. Holub			
17 praeusta (Reichenb.) J. Holub	126	It	Sauer & Chmelitschek, 1976
18 planiculmis (Schrader) W. Sauer & Chmelitschek	126	Po	Gervais, 1966
(a) planiculmis			
(b) angustior J. Holub			
19 bromoides (Gouan) H. Scholz	14 + 0 - 2B	Ga Hs	Gervais, 1968
20 murcica J. Holub			
21 cincinnata (Ten.) J. Holub	14	Bu	Kozuharov & Petrova, 1975
22 peloponnesiaca J. Holub			
23 cycladum (Rech. fil. & Scheffer) W. Greuter			
24 gervaisii J. Holub			

25 pruinosa (Batt. & Trabut) J. Holub
26 crassifolia (Font Quer) J. Holub
27 hackelii (Henriq.) J. Holub

63 Danthoniastrum (J. Holub) J. Holub
  1 compactum (Boiss. & Heldr.) J.
    Holub

64 Arrhenatherum Beauv.

1 elatius (L.) Beauv. ex J. & C. Presl			
(a) elatius	28	Is	Löve & Löve, 1956
(b) bulbosum (Willd.) Schübler & Martens	28	Lu	Fernandes & Queirós, 1969
sardoum (Schmid) Gamisans	14	Co	Gamisans, 1974
2 palaestinum Boiss.			
3 album (Vahl) W.D. Clayton	28	Lu	Fernandes & Queirós, 1971

65 Pseudarrhenatherum Rouy

1 longifolium (Thore) Rouy	14		
2 pallens (Link) J. Holub			

66 Gaudinia Beauv.

1 fragilis (L.) Beauv.	14 + 0 - 2B	Lu	Fernandes & Queirós, 1969
2 hispanica Stace & Tutin	14		
3 coarctata (Link) T. Durand & Schinz			

67 Ventenata Koeler

1 dubia (Leers) Cosson	14	Ju	Larsen, 1960
2 macra (Bieb.) Boiss.			

68 Koeleria Pers.

1 vallesiana (Honckeny) Gaudin	14 + 0 - 2B	Ga	Küpfer & Favarger, 1967
	28 42	Ga Hs	Küpfer, 1971
	43		
	112 126		
2 cenisia Reuter ex Reverchon	14 + 0 - 2B	Ga	Favarger, 1969
3 hirsuta Gaudin	14 + 0 - 5B	He	Favarger, 1969
4 asiatica Domin	28	Rs(C)	Zhukova & Tikhonova, 1971
5 macrantha (Ledeb.) Schultes	14	Hu	Ujhelyi, 1961
	28 + 0 - 6B	Au	Larsen, ined.
	42 70	Hu	Ujhelyi, 1962
6 glauca (Schrader) DC.	14	Ga	Delay, 1969
	28	Da	Böcher, 1943
	42 70	Ge Hu	Ujhelyi, 1960b
7 lobata (Bieb.) Roemer & Schultes	28	Da	Larsen, ined.
8 splendens C. Presl	28	Al Hu	Ujhelyi, 1962
9 caudata (Link) Steudel			
10 crassipes Lange	14	Lu	Fernandes & Queirós, 1969
11 nitidula Velen.			
12 eriostachya Pančić			
13 pyramidata (Lam.) Beauv.	42	Europe	Ujhelyi, 1970
	56	Au Ju	Larsen, ined.
	70	Europe	Ujhelyi, 1972
	84	Da	Böcher, 1943

69 Lophochloa Reichenb.
   1 cristata (L.) Hyl.   26   Lu   Fernandes & Queirós,
   2 hispida (Savi) Jonsell   1969
   3 pumila (Desf.) Bor
   4 pubescens (Lam.) H. Scholz
   5 salzmannii (Boiss. & Reuter) H. Scholz

70 Trisetum Pers.
   1 velutinum Boiss.
   2 macrotrichum Hackel
   3 laconicum Boiss. & Orph.
   4 distichophyllum (Vill.) Beauv.   28   Ga   Favarger, 1959
                                                 56   He   Favarger, 1959
   5 argenteum (Willd.) Roemer & Schultes   28   It   Favarger, 1959
   6 glaciale (Bory) Boiss.   14   Hs   Küpfer, 1968
   7 gracile (Moris) Boiss.   14   Co   Gamisans, 1973
   8 bertolonii Jonsell
   9 hispidum Lange
  10 spicatum (L.) K. Richter
      (a) spicatum   28   Is No Su Jonsell et al.,
                                                                     1975
      (b) ovatipaniculatum Hultén ex Jonsell   28   Au   Jonsell et al.,
                                                                     1975
      (c) pilosiglume (Fernald) Hultén   42   Is   Jonsell et al.,
                                                                     1975
  11 subalpestre (Hartman) Neuman   28   No   Knaben & Engelskjon,
                                                                     1967
  12 baregense Laffitte & Miégeville   14
  13 flavescens (L.) Beauv.
      (a) flavescens   28   Gr   Májovský et al.,
      (b) splendens (C. Presl) Arcangeli   1974
      (c) purpurascens (DC.) Arcangeli   12
  14 alpestre (Host) Beauv.   14   Po   Skalińska et al.,
                                                                     1971
  15 tenuiforme Jonsell
  16 fuscum (Kit. ex Schultes) Schultes
  17 sibiricum Rupr.   14   Rs(N)   Sokolovskaya &
                                                                     Strelkova, 1960
      (a) sibiricum
      (b) litorale (Rupr.) Roshev.
  18 paniceum (Lam.) Pers.   14   Lu   Fernandes & Queirós,
                                                                     1969
  19 aureum Ten.
  20 parviflorum (Desf.) Pers.
  21 scabriusculum (Lag.) Cosson ex Willk.
  22 ovatum (Cav.) Pers.   14   Lu   Fernandes & Queirós,
                                                                     1969
  23 loeflingianum (L.) C. Presl
  24 macrochaetum Boiss.

71 Trisetaria Forskål
   1 dufourei (Boiss.) Paunero

72 Parvotrisetum Chrtek
   1 myrianthum (Bertol.) Chrtek

73 Lagurus L.
   1 ovatus L.   14   Co Gr   Gadella & Kliphuis,
                                                                     1968

74 Deschampsia Beauv.

1 cespitosa (L.) Beauv.
(a) cespitosa	24 26	Bu	Kovuharov & Kuzmanov, 1962
	26 + 0 - 7B	Ge	Albers, 1972
(b) alpina (L.) Tzvelev	52	Is	Löve & Löve, 1956
	56	Br	Tutin, 1961
(c) obensis			
(d) bottnica (Wahlenb.) G.C.S. Clarke	26 26 + 2B	Ge	Albers, 1972
(e) borealis (Trautv.) Tzvelev	28	Rs(N)	Sokolovskaya & Strelkova, 1960
(f) orientalis Hultén			
(g) paludosa (Schübler & Martens) G.C.S. Clarke	26 + 0 - 2B	Ge	Albers, 1972
littoralis (Gaudin) Reuter	56		
2 media (Gouan) Roemer & Schultes	26		
3 setacea (Hudson) Hackel	14	Da	Hagerup, 1939
4 foliosa Hackel			
5 flexuosa (L.) Trin.	26	Ga	Delay, 1970
	28	Fe Ge No	Albers, 1972
	56	Lu	Albers, 1972

75 Vahlodea Fries
1 atropurpurea (Wahlenb.) Fries ex Hartman	14	No	Knaben & Engelskjon, 1967

76 Aira L.
1 praecox L.	14	Br	Hedberg & Hedberg, 1961
2 caryophyllea L.			
(a) caryophyllea	28	Ga	Albers, 1973
(b) multiculmis (Dumort.) Bonnier & Layens	28	Da Ga Lu	Bücher & Larsen, 1958
3 cupaniana Guss.	14	Co	Albers, 1973
4 uniaristata Lag. & Rodr.	14		
5 scoparia Adamović			
6 elegantissima Schur	14	Bl	Dahlgren et al., 1971
7 tenorii Guss.	14	Ga	Albers & Albers, 1973
8 provincialis Jordan	28	Ga	Albers, 1973

77 Molineriella Rouy
1 minuta (L.) Rouy			
2 laevis (Brot.) Rouy	8	Lu	Fernandes & Queirós, 1969

78 Airopsis Desv.
1 tenella (Cav.) Ascherson & Graebner	8	Lu	Fernandes & Queirós, 1969

79 Periballia Trin.
1 involucrata (Cav.) Janka	14	Lu	Fernandes & Queirós, 1969

80 Antinoria Parl.
1 agrostidea (DC.) Parl.
2 insularis Parl.

81 Hierochloë R. Br.

```
  1 australis (Schrader) Roemer & Schultes 14      Au Fe Ge   Weimarck, 1971
                                                      It Rm
  2 odorata (L.) Beauv.
    (a) odorata                              28         Po      Skalińska et al.,
                                                                  1974
                                             42         Fe      Sorsa, 1963
    (b) baltica G. Weim.                     42         Fe Su   Weimarck, 1971
  3 hirta (Schrank) Borbás                   56       Fe No Su  Weimarck, 1971
    (a) hirta
    (b) arctica (C. Presl) G. Weim.
  4 repens (Host) Simonkai                   28         Cz Rm   Weimarck, 1971
                                         28 + 2B        Cz      Weimarck, 1971
  5 stepporum P. Smirnov
  6 pauciflora R.Br.
  7 alpina (Willd.) Roemer & Schultes        56        Rs(N)    Sokolovskaya &
                                                                  Strelkova, 1960

 82 Anthoxanthum L.
  1 odoratum L.                              10     Au He No Su I. Hedberg, 1970
                                             20         Cz      Májovský et al.,
                                                                  1970
     alpinum                                 10         Is      Löve & Löve, 1956
  2 pauciflorum Adamović
  3 amarum Brot.                             90         Lu      Fernandes & Queirós,
                                                                  1969
  4 aristatum Boiss.
    (a) aristatum                     10 + 0 - 2B      Hs      Valdés, 1973
    (b) macranthum Valdés             10 + 0 - 1B      Hs      Valdés, 1973
  5 ovatum Lag.                              10         Hs      Valdés, 1973
  6 gracile Biv.

 83 Holcus L.
  1 lanatus L.                        14 + 0 - 2B      Po      Skalińska et al.,
                                                                  1968
  2 setiglumis Boiss. & Reuter              14         Lu      Queirós, 1973
    (a) setiglumis
    (b) duriensis P. Silva
  3 notarisii Nyman
  4 mollis L.                             14 28        Lu      Fernandes & Queirós,
                                                                  1969
                                             35         Po      Walter, 1973
                                          42 49        Br      Jones, 1958
    1 x 4                                   21         Br      Jones, 1958
  5 rigidus Hochst.
  6 gayanus Boiss.                            8         Lu      Fernandes, 1950
  7 grandiflorus Boiss. & Reuter
  8 caespitosus Boiss.                       14         Hs      Küpfer, 1969

 84 Corynephorus Beauv.
  1 canescens (L.) Beauv.                    14         Lu      Fernandes & Queirós,
                                                                  1969
  2 divaricatus (Pourret) Breistr.          14         Co      Litardière, 1949
  3 fasciculatus Boiss. & Reuter            14         Lu      Fernandes & Queirós,
                                                                  1969
  4 macrantherus Boiss. & Reuter

 85 Avellinia Parl.
  1 michelii (Savi) Parl.

 86 Triplachne Link
  1 nitens (Guss.) Link
```

87 Agrostis L.

1 canina L.	14	Br	Hubbard, 1954
	16	Ge	Heitz, 1967
	28	Bu	Kozuharov & Nicolova, 1975
2 vinealis Schreber	28	No	Knaben & Engelskjon, 1967
3 curtisii Kerguelen	14	Br Ga Lu	Björkman, 1960
4 mertensii Trin.	56	Fe No Su	Björkman, 1960
5 alpina Scop.	14	Ga He	Björkman, 1960
6 schleicheri Jordan & Verlot	42	Ga He	Björkman, 1960
7 rupestris All.	14 + 0 - 3B	Ga Po	Björkman, 1960
	21	Ga	Björkman, 1960
	28	Au Ga He	Björkman, 1960
8 nevadensis Boiss.	42 + 0 - 10B	Hs	Björkman, 1960
9 agrostiflora (G. Beck) Rauschert	28	He	Scholte, 1977
10 clavata Trin.	42	Su	Björkman, 1960
11 scabra Willd.			
12 pourretii Willd.	14 + 0 - 2B	Lu	Björkman, 1960
	16 18 19 21	Lu	Fernandes & Queirós, 1969
13 tenerrima Trin.	14	Lu	Björkman, 1960
14 juressi Link	14	Lu	Björkman, 1960
15 nebulosa Boiss. & Reuter			
16 delicatula Pourret ex Lapeyr.	14	Lu	Björkman, 1960
17 durieui Boiss. & Reuter ex Willk.			
18 reuteri Boiss.	14	Lu	Björkman, 1960
19 capillaris L.	28	Su	Hedberg, 1964
20 gigantea Roth	42 + 0 - 2B	Su	Björkman, 1954
(a) gigantea			
(b) maeotica (Klokov) Tzvelev			
21 salsa Korsh.			
22 stolonifera L.	28	Hu	Pólya, 1948
	30 32 35		
	42	Lu	Fernandes & Queirós, 1969
	44 46	Ge	Juhl, 1952
23 congestiflora Tutin & E.F. Warburg			
(a) congestiflora			
(b) oreophila Franco			
24 gracililaxa Franco			
25 castellana Boiss. & Reuter	28 + 0 - 4B	Lu	Björkman, 1954
	42 + 0 - 3B	Lu	Fernandes & Queirós, 1969

88 Gastridium Beauv.

1 ventricosum (Gouan) Schinz & Thell.	14	Lu	Fernandes & Queirós, 1969

89 Polypogon Desf.

1 monspeliensis (L.) Desf.	28	Lu	Fernandes & Queirós, 1969
2 maritimus Willd.	14	Lu	Fernandes & Queirós, 1969
	28	Lu	Rodriguez, 1953
(a) maritimus			
(b) subspathaceus (Req.) Bonnier & Layens	14	Bl	Dahlgren et al., 1969
3 viridis (Gouan) Breistr.	28	Lu	Fernandes & Queirós, 1969

x Agropogon P. Fourn.
  1 littoralis (Sm.) C.E. Hubbard   28   Br   Rutland, 1941

90 Ammophila Host
  1 arenaria (L.) Link   14   Da   Westergaard, 1941
                                           28   Ga   Wulff, 1937a
                                           56   Po   Skalińska et al.,
  (a) arenaria   1957
  (b) arundinacea H. Lindb. fil.

x Ammocalamagrostis P. Fourn.
  1 baltica (Flugge ex Schrader) P. Fourn. 28 42   Po   Skalińska et al.,
                                                           1964

91 Calamagrostis Adanson
  1 epigejos (L.) Roth   28   Ho   Gadella & Kliphuis,
                                                           1967
                                           42   Su   Nygren, 1946
                                           56   Hu   Pólya, 1949
  2 pseudophragmites (Haller fil.) Koeler 28   Po   Kubién, 1968
  3 villosa (Chaix) J.F. Gmelin   56   Cz   Murín & Májovský,
                                                           1976
  4 canescens (Weber) Roth   28   Su   Nygren, 1946
  (a) canescens
  (b) vilnensis (Besser) H. Scholz
  5 purpurea (Trin.) Trin.   28   Fe   V. Sorsa, 1962
                                           56-91   Su   Nygren, 1946
  (a) purpurea
  (b) phragmitoides (Hartman) Tzvelev
  (c) langsdorfii (Link) Tzvelev
  (d) pseudopurpurea (Gewitlaani ox R.
      Heine) G.C.S. Clarke
  6 stricta (Timm) Koeler   28   Su   Nygren, 1946
  7 scotica (Druce) Druce
  8 holmii Lange
  9 deschampsioides Trin.
 10 lapponica (Wahlenb.) Hartman   28   Rs(N)   Sokolovskaya &
                                                           Strelkova, 1960
                                           42-112   Nygren, 1946
 11 arundinacea (L.) Roth   28 + 0 - 2B Cz   Holub et al.,
                                                           1972
 12 varia (Schrader) Host
  (a) varia   28   Hu   Baksay, 1956
  (b) corsica (Hackel) Rouy   28   Co   Litardière, 1949
 13 chalybaea (Laest.) Fries   42   Su   Nygren, 1946
 14 obtusata Trin.

92 Chaetopogon Janchen
  1 fasciculatus (Link) Hayek   14   Lu   Fernandes & Queirós,
                                                           1969

93 Phleum L.
  1 pratense L.
  (a) pratense   42   Hu   Pólya, 1950
                              21 35 36 49 63 Su   Muntzing, 1935, 1937
                                    56 70 84 Su   Muntzing & Prakken,
                                                           1940
  (b) bertolonii (DC.) Bornm.   14   Su   Muntzing, 1935
  2 alpinum L.
  (a) alpinum   28   Su   Muntzing, 1935
  (b) rhaeticum Humphries   14   He   Muntzing, 1935

340

3 echinatum Host	10	It	Ellerstrom & Tjio, 1950
4 phleoides (L.) Karsten	14	Su	Muntzing, 1935
	28 + 1 - 5B	Su	Bosemark, 1956
5 montanum C. Koch			
6 hirsutum Honckeny	14	It	Nordenskjold, 1937
7 paniculatum Hudson	28	Rs(K)	Avdulov, 1928
8 arenarium L.	14	Br Da Hb Ho Ga	Bücher & Larsen, 1958

9 graecum Boiss. & Heldr.
  (a) graecum
  (b) aegaeum (Vierh.) W. Greuter
10 crypsoides (D'Urv.) Hackel
  (a) crypsoides
  (b) sardoum (Hackel) Horn af Rantzien
11 subulatum (Savi) Ascherson & Graebner
  (a) subulatum
  (b) ciliatum (Boiss.) Humphries

94 Alopecurus L.
  1 pratensis L.

(a) pratensis	28	Fe	V. Sorsa, 1962
(b) laguriformis (Schur) Tzvelev			
2 arundinaceus Poiret	28	Su	Johnsson, 1941
3 geniculatus L.	28	Su	Johnsson, 1941
4 aequalis Sobol.	14	Cz	Májovský et al., 1970
5 bulbosus Gouan	14	Br	Maude, 1940
6 himalaicus Hooker fil.			
7 alpinus Sm.			
(a) alpinus	100 117	Br	Fearn, 1977
	c.120	Sb	Flovik, 1938
(b) glaucus (Less.) Hultén			
8 utriculatus Solander			
9 setarioides Gren.			
10 myosuroides Hudson	14	Su	Johnsson, 1941
11 creticus Trin.	14		
12 rendlei Eig	14	Ju	Gadella & Kliphuis, 1972
13 gerardii Vill.	14	Ga	Favarger & Huynh, 1964
14 vaginatus (Willd.) Boiss.			

95 Cornucopiae L.
  1 cucullatum L.

Tribe Hainardieae W. Greuter

96 Pholiurus Trin.

1 pannonicus (Host) Trin.	14	Hu	Pólya, 1948

97 Parapholis C.E. Hubbard

1 incurva (L.) C.E. Hubbard	38	Lu	Fernandes & Queirós, 1969
2 marginata Runemark	14	Gr	Runemark, 1962
3 filiformis (Roth) C.E. Hubbard	14	Lu	Rodriguez, 1953
4 strigosa (Dumort.) C.E. Hubbard	28	Lu	Fernandes & Queirós, 1969
5 pycnantha (Hackel) C.E. Hubbard			

98 Hainardia W. Greuter

1 cylindrica (Willd.) W. Greuter	26	Lu	Fernandes & Queirós, 1969

Tribe Phalarideae Kunth

99 Phalaris L.
1 arundinacea L.

(a) arundinacea	28	Br	Hedberg & Hedberg, 1961
	42 + 6B	Lu	Fernandes & Queirós, 1969
(b) rotgesii (Fouc. & Mandon ex Husnot) Kerguelen	14	Co	Litardière, 1949
2 truncata Guss. ex Bertol.			
3 aquatica L.	28	Lu	Fernandes & Queirós, 1969
4 canariensis L.	12	Lu	Fernandes & Queirós, 1969
5 minor Retz.	28	Lu	Fernandes & Queirós, 1969
6 brachystachys Link	12	Lu	Fernandes & Queirós, 1969
7 paradoxa L.	14	Lu	Fernandes & Queirós, 1969
8 coerulescens Desf.	14	Lu	Fernandes & Queirós, 1969
	42	Hs	Löve & Kjellqvist, 1973

Tribe Coleantheae Ascherson & Graebner

100 Coleanthus Seidl
1 subtilis (Tratt.) Seidl

Tribe Scolochloeae Tzvelev

101 Scolochloa Link			
1 festucacea (Willd.) Link	28	Su	Löve & Löve, 1944

Tribe Milieae Endl.

102 Milium L.			
1 effusum L.	14	Rs(N)	Sokolovskaya & Strelkova, 1960
	28	Fe	V. Sorsa, 1962
2 vernale Bieb.	8	Ga	Tutin, ined.

103 Zingeria P. Smirnov			
1 biebersteiniana (Claus) P. Smirnov	4	Rs(E)	Tzvelov & Zhukova, 1974
2 pisidica (Boiss.) Tutin			

Tribe Stipeae Dumort.

104 Piptatherum Beauv.			
1 miliaceum (L.) Cosson			
2 virescens (Trin.) Boiss.			
3 paradoxum (L.) Beauv.	24	Hs	Löve & Kjellqvist, 1973
4 coerulescens (Desf.) Beauv.			
5 holciforme (Bieb.) Roemer & Schultes			

105 Stipa L.
 1 pennata L.
　(a) austriaca (G. Beck) Martinovsky &
　　　Skalicky
　(b) lithophila (P. Smirnov) Martinovsky
　(c) eriocaulis (Borbás) Martinovsky　　　44　　　Rm　　　Tarnavschi &
　　　　　　　　　　　　　　　　　　　　　　　　　　　　　　Lungeanu,　1970
　(d) pennata
　(e) kiemii Martinovsky
 2 pulcherrima C. Koch　　　　　　　　　44
 3 mayeri Martinovsky
 4 dasyvaginata Martinovsky
 5 crassiculmis P. Smirnov
 6 bavarica Martinovsky & H. Scholz
 7 epilosa Martinovsky
 8 endotricha Martinovsky
 9 cretacea P. Smirnov
10 syreistschikovii P. Smirnov
11 iberica Martinovsky
　(a) iberica
　(b) pauneroana Martinovsky
12 apertifolia Martinovsky
13 rechingeri Martinovsky
14 austroitalica Martinovsky
15 novakii Martinovsky
16 dasyphylla (Lindem.) Trautv.
17 zalesskii Wilensky
18 ucrainica P. Smirnov　　　　　　　　44
19 pontica P. Smirnov
20 joannis Čelak.
　(a) puberula (Podp. & Suza) Martinovsky
　(b) joannis　　　　　　　　　　　　　44　　　Rm　　　Tarnavschi, 1948
　(c) balcanica Martinovsky
21 borysthenica Klokov ex Prokudin
　(a) borysthenica
　(b) germanica (Endtmann) Martinovsky &
　　　Rauschert
22 styriaca Martinovsky
23 anomala P. Smirnov
24 danubialis Dihoru & Roman　　　　　44　　　Rm　　　Tarnavschi &
　　　　　　　　　　　　　　　　　　　　　　　　　　　　　　Lungeanu,　1970
25 tirsa Steven　　　　　　　　　　　　44
　(a) tirsa
　(b) albanica Martinovsky
26 lessingiana Trin. & Rupr.　　　　　44　　　Rm　　　Lungeanu, 1975
　(a) lessingiana
　(b) brauneri Pacz.
27 capensis Thunb.　　　　　　　　　　36　　　Lu　　　Fernandes & Queirós,
　　　　　　　　　　　　　　　　　　　　　　　　　　　　　　1969
28 barbata Desf.
29 arabica Trin. & Rupr.
30 orientalis Trin.
31 capillata L.　　　　　　　　　　　　44　　　Po　　　Rychlewski, 1968
32 sareptana A. Becker　　　　　　　　44　　　Rs(E)　Titova, 1935
　(a) sareptana
　(b) praecapillata (Alechin) Tzvelev
33 fontanesii Parl.
34 lagascae Roemer & Schultes
35 celakovskyi Martinovsky
36 offneri Breistr.　　　　　　　　　　44　　　Bl　　　Nilsson & Lassen,
　　　　　　　　　　　　　　　　　　　　　　　　　　　　　　1971

37 korshinskyi Roshev			
38 parviflora Desf.	28	Hs	Lorenzo-Andreu, 1950
39 tenacissima L.	40	Hs	Fernandes & Queirós, 1969
40 gigantea Link	96	Lu	Fernandes & Queirós, 1969
41 bromoides (L.) Dörfler	28	Hu	Ujhelyi, 1961
42 trichotoma Nees			

106 Achnatherum Beauv.

1 calamagrostis (L.) Beauv.	24	Ga He	Gervais, 1966
2 splendens (Trin.) Nevski			

Tribe Ampelodesmeae (Concert) Tutin

107 Ampelodesmos Link

1 mauritanica (Poiret) T. Durand & Schinz	48	Bl	Nilsson & Lassen, 1971

Tribe Arundineae Dumort.

108 Arundo L.

1 donax L.	110 112	Ga	Gorenflot et al., 1972a
2 plinii Turra	72	Lu	Fernandes & Queirós, 1969

109 Phragmites Adanson

1 australis (Cav.) Trin. ex Steudel	36 44 46 48 49 50 51 52 96	Ge He Rm	Gorenflot et al., 1972b

Tribe Cortaderieae Zotov

110 Cortaderia Stapf
    1 selloana (Schultes & Schultes fil.)
        Ascherson & Graebner

Tribe Danthonieae (G. Beck) C.E. Hubbard

111 Danthonia DC.

1 decumbens (L.) DC.	24	Ge	Schwarz & Bässler, 1964
	36	Cz	Májovský et al., 1974
	124	Br	Maude, 1940
2 alpina Vest	36	Europe	Packer, 1961

112 Schismus Beauv.
    1 barbatus (L.) Thell.
    2 arabicus Nees

Tribe Molinieae Jirasek

113 Molinia Schrank

1 caerulea (L.) Moench			
(a) caerulea	36	Hu	Pólya, 1950
(b) arundinacea (Schrank) H. Paul	?36	Ga	Guinochet & Lemée, 1950
	90	Hu	Ujhelyi, 1961

Tribe Aristideae C.E. Hubbard ex Bor

114 Aristida L.
    1 adscensionis L.

115 Stipagrostis Nees
    1 pennata (Trin.) De Winter
    2 karelinii (Trin. & Rupr.) Tzvelev

Tribe Nardeae Reichenb.

116 Nardus L.
    1 stricta L.   26   Br   Hedberg & Hedberg,
                                                         1961

Tribe Lygeeae Lange

117 Lygeum Loefl. ex L.
    1 spartum L.   40   Hs   Lorenzo-Andreu &
                                                         Garcia, 1951

Tribe Pappophoreae Kunth

118 Enneapogon Desv. ex Beauv.
    1 persicus Boiss.

Tribe Aeluropodeae Nevski ex Bor

119 Aeluropus Trin.
    1 littoralis (Gouan) Parl.   20 Bot. Gard. Litardière, 1950
    2 lagopoides (L.) Trin. ex Thwaites

Tribe Eragrostideae Stapf

120 Cleistogenes Keng
    1 serotina (L.) Keng
    (a) serotina   40   Cz   Májovský et al.,
                                                         1970
    (b) bulgarica (Bornm.) Tutin   40   Bu   Kozuharov & Petrova,
                                                         1974
    2 squarrosa (Trin.) Keng

121 Eragrostis N.M. Wolf
    1 pilosa (L.) Beauv.   40   Lu   Fernandes & Queirós,
                                                         1969
    2 pectinacea (Michx) Nees
    3 papposa (Dufour) Steudel
    4 cilianensis (All.) F.T. Hubbard   20   Lu   Fernandes & Queirós,
                                                         1969
    5 minor Host   40   Cz   Májovský et al.,
                                                         1974
                                         80   Lu   Fernandes & Queirós,
                                                         1969
    6 barrelieri Daveau
    7 aegyptiaca (Willd.) Link
    8 diarrhena (Schultes & Schultes fil.)
       Steudel
    9 collina Trin.

122 Sporobolus R. Br.
    1 pungens (Schreber) Kunth
    2 indicus (L.) R. Br.   36   Lu   Fernandes & Queirós,
                                                         1969

345

123 Crypsis Aiton
   1 alopecuroides (Piller & Mitterp.)
     Schrader
   2 acuminata Trin.

3 aculeata (L.) Aiton	16	Hu	Pólya, 1948
4 schoenoides (L.) Lam.	36	Lu	Fernandes & Queirós, 1969

   5 turkestanica Eig.

124 Eleusine Gaertner
   1 indica (L.) Gaertner
   2 tristachya (Lam.) Lam.

125 Dactyloctenium Willd.
   1 aegyptium (L.) Beauv.

  Tribe Chlorideae J.G. Agardh

126 Cynodon L.C.M. Richard

1 dactylon (L.) Pers.	36	Hu	Pólya, 1948
	40	Lu	Fernandes & Queirós, 1969

  Tribe Spartineae (Prat) C.E. Hubbard

127 Spartina Schreber

1 maritima (Curtis) Fernald	60	Lu	Fernandes & Queirós, 1969
2 x townsendii H. & J. Groves	62	Br	Marchant, 1968
3 anglica C.E. Hubbard	120 122 124	Br	Marchant, 1968
4 alterniflora Loisel.	62	Br	Marchant, 1968
5 versicolor Fabre			
6 densiflora Brongn.			

  Tribe Zoysieae Miq.

128 Tragus Haller
   1 racemosus (L.) All.

  Tribe Oryzeae Dumort.

129 Oryza L.
   1 sativa L.

130 Leersia Swartz

1 oryzoides (L.) Swartz	48	Lu	Fernandes & Queirós, 1969

131 Zizania L.
   1 aquatica L.
   2 latifolia (Griseb.) Stapf.

  Tribe Paniceae R. Br.

132 Panicum L.
   1 miliaceum L.
   2 capillare L.
   3 dichotomiflorum Michx
   4 implicatum Scribner ex Britton & A. Br.

5 repens L.	54	Lu	Fernandes & Queirós, 1969

6 maximum Jacq.

133 Oplismenus Beauv.
1 undulatifolius (Ard.) Roemer & Schultes

134 Echinochloa Beauv.
1 colonum (L.) Link
2 crus-galli (L.) Beauv.   54   Lu   Fernandes & Queirós, 1969

3 oryzoides (Ard.) Fritsch

135 Brachiaria (Trin.) Griseb.
1 eruciformis (Sibth. & Sm.) Griseb.

136 Eriochloa Kunth
1 succincta (Trin.) Kunth

137 Digitaria Haller
1 debilis (Desf.) Willd.   36   Lu   Fernandes & Queirós, 1969

2 sanguinalis (L.) Scop.   36   Lu   Fernandes & Queirós, 1969

3 ciliaris (Retz.) Koeler
4 ischaemum (Schreber) Muhl.

138 Paspalum L.			
1 dilatatum Poiret	50	Lu	Fernandes & Queirós, 1969
2 urvillei Steudel	20	Lu	Fernandes & Queirós, 1969
3 paspalodes (Michx) Scribner	60	Lu	Fernandes & Queirós, 1969
4 vaginatum Swartz	20	Lu	Fernandes & Queirós, 1969

139 Stenotaphrum Trin.
1 secundatum (Walter) O. Kuntze

140 Setaria Beauv.			
1 pumila (Poiret) Schultes	36	Cz	Májovský et al., 1976
	72	Hs	Björkqvist et al., 1969
geniculata (Lam.) Beauv.	72	Lu	Fernandes & Queirós, 1969
2 verticillata (L.) Beauv.	18	Lu	Fernandes & Queirós, 1969
	36	Cz	Májovský et al., 1974
	c.54	Al	Strid, 1971
3 viridis (L.) Beauv.	18		
2 x 3	36	Lu	Fernandes & Queirós, 1969
4 italica (L.) Beauv.			

141 Pennisetum L.C.M. Richard
1 setaceum (Forskål) Chiov.
2 villosum R. Br. ex Fresen.   45   Lu   Fernandes & Queirós, 1969

142 Cenchrus L.

```
      1 ciliaris L.
      2 incertus M.A. Curtis

143 Tricholaena Schrader
      1 teneriffae (L. fil.) Link

    Tribe Andropogoneae Dumort.
```

144 Imperata Cyr.
    1 cylindrica (L.) Raeuschel   50 60   Lu   Fernandes & Queirós, 1969

145 Saccharum L.
    1 spontaneum L.
    2 officinarum L.
    3 ravennae (L.) Murray
    4 strictum (Host) Sprengel

146 Sorghum Moench
    1 halepense (L.) Pers.   40 41 42 43   Lu   Fernandes & Queirós, 1969

    2 sudanense (Piper) Stapf
    3 bicolor (L.) Moench

147 Chrysopogon Trin.
    1 gryllus (L.) Trin.

148 Dichanthium Willemet
    1 insculptum (A. Richard) W.D. Clayton
    2 ischaemum (L.) Roberty

149 Andropogon L.
    1 distachyos L.   40   Ga   Saura, 1943

150 Hyparrhenia N.J. Andersson ex E. Fourn.
    1 hirta (L.) Stapf   30   Ga   Larsen, 1954
                        40   Lu   Fernandes & Queirós, 1969

151 Heteropogon Pers.
    1 contortus (L.) Beauv. ex Roemer &
      Schultes

152 Phacelurus Griseb.
    1 digitatus (Sibth. & Sm.) Griseb. 20   Bu   Kozuharov & Nicolova, 1975

153 Hemarthria R. Br.
    1 altissima (Poiret) Stapf & C.E. Hubbard

154 Coix L.
    1 lacryma-jobi L.

155 Zea L.
    1 mays L.

PRINCIPES

CXCIV PALMAE

    1 Chamaerops L.

1 humilis L.

2 Phoenix L.
  1 dactylifera L.
  2 canariensis hort. ex Chabaud
  3 theophrasti W. Greuter

SPATHIFLORAE

CXCV ARACEAE

1 Acorus L.
  1 calamus L.

	24	?Fe	Vaarama, 1948
	36	Ho	Gadella & Kliphuis, 1973

2 Calla L.
  1 palustris L.

	36	Ge	Wulff, 1939
	72	Su	Löve & Löve, 1944

3 Lysichiton Schott
  1 americanus Hulten & St. John

4 Zantedeschia Sprengel
  1 aethiopica (L.) Sprengel

5 Colocasia Schott
  1 esculenta (L.) Schott

6 Arum L.
  1 italicum Miller

(a) italicum	84	It	Marchi, 1971
(b) neglectum (Townsend) Prime	84	Br	Prime, 1954
(c) albispathum (Steven) Prime			
(d) byzantinum (Blume) Nyman			
2 maculatum L.	56	Br	Prime, 1954
3 orientale Bieb.			
(a) orientale	28	Au Cz Hu Po Rm	Beuret, 1972
(b) lucanum (Cavara & Grande) Prime	28	It	Beuret, 1972
(c) danicum (Prime) Prime	28	Da	Hagerup, 1944
4 elongatum Steven			
5 petteri Schott			
nigrum var. apulum Carano	56	It	Gori, 1957
6 creticum Boiss. & Heldr.	28	Cr	Marchant, 1972
7 pictum L. fil.	28	Co	Contandriopoulos, 1962

7 Biarum Schott
  1 davisii Turrill

2 tenuifolium (L.) Schott	16	It	Caldo, 1971
	20	Si	Monti & Garbari, 1974
arondandum Boiss. & Reut.	22	Hs	Talavera, 1976
galiani Talavera	26	Hs	Talavera, 1976
3 carratracense (Haenseler ex Willk.) Font Quer	98	Hs	Messia & Rejon, 1976
	c.96	Hs	Talavera, 1976
4 dispar (Schott) Talavera	74	Sa	Chiappini & Scrugli, 1972

5 spruneri Boiss.

8 Arisarum Miller
  1 vulgare Targ.-Tozz.
    (a) vulgare   56   Cr   Marchant, 1972
    (b) simorrhinum (Durieu) Maire &
        Weiller
  2 proboscideum (L.) Savi   28   Hs   Marchant, 1972

9 Ambrosina Bassi
  1 bassii L.

10 Dracunculus Miller
  1 vulgaris Schott   32   Bu   Popova &
                                                               Češchmedjiev, 1978
  2 muscivorus (L. fil.) Parl.   56   Sa   Scrugli &
                                                               Bocchieri, 1977

CXCVI LEMNACEAE

1 Wolffia Horkel ex Schleiden
  1 arrhiza (L.) Horkel ex Wimmer   50   Po   Wcislo, 1970

2 Lemna L.
  1 trisulca L.   40 Br Ga Ge Hb Urbanska-
                                                               Worytkiewicz, 1975
                                               44   Br   Blackburn, 1933
  2 gibba L.   40 50 Br Ga Ge Urbanska-
                                                   Hs It Lu   Worytkiewicz, 1975
                                               64   Br   Blackburn, 1933
  3 minor L.   40 42 Br Gr Hb Urbanska-
                                                   Hs It Lu   Worytkiewicz, 1975
                                               50   Cz   Murín & Májovský,
                                                               1978
  4 perpusilla Torrey   40   Ga   Urbanska-
                                                               Worytkiewicz, 1975
  5 valdiviana Philippi   40   Ga   Urbanska-
                                                               Worytkiewicz, 1975

3 Spirodela Schleiden
  1 polyrhiza (L.) Schleiden   40   Po   Wcislo, 1970

PANDANALES

CXCVII SPARGANIACEAE

1 Sparganium L.
  1 erectum L.
    (a) erectum
    (b) microcarpum (Neuman) Domin
    (c) neglectum (Beeby) Schinz & Thell.   30   Da   Larsen, 1965
  2 gramineum Georgi   30   Scand. Lohammon, 1942
  3 emersum Rehmann   30   Cz   Májovský et al.,
                                                               1976
  4 angustifolium Michx   30   Is   Löve & Löve, 1956
  5 glomeratum Laest. ex Beurl.   30   Scand. Lohammon, 1942
  6 minimum Wallr.   30   Is   Löve & Löve, 1956
  7 hyperboreum Laest. ex Beurl.   30   Is   Löve & Löve, 1956

350

CXCVIII TYPHACEAE

1 Typha L.
  1 angustifolia L.   30   Ho   Gadella & Kliphuis, 1963

  2 domingensis (Pers.) Steudel   30   Hs   Löve & Kjellqvist, 1973

  3 minima Funck
  4 laxmannii Lepechin
  5 latifolia L.   30   Da   Larsen, 1954
  6 shuttleworthii Koch & Sonder

CYPERALES

CXCIX CYPERACEAE

1 Scirpus L.
  1 sylvaticus L.   62   Cz   Májovský et al., 1976
                         64   Su   Ehrenberg, 1945
  2 radicans Schkuhr
  3 maritimus L.   80   Lu   Rodriguez, 1953
                 c.104   Cz   Májovský et al., 1976
    (a) maritimus
    (b) affinis (Roth) T. Norlindh
  4 lacustris L.
    (a) lacustris   42   Fe   Sorsa, 1963
    (b) tabernaemontani (C.C. Gmelin) Syme 38   Hu   Pólya, 1949
                         42   Cz   Májovský et al., 1976
                         44   Ge   Wulff, 1938
  5 mucronatus L.
  6 pungens Vahl
  7 litoralis Schrader
  8 triqueter L.
  9 supinus L.
  10 juncoides Roxb.
  11 holoschoenus L.
  12 setaceus L.   28   Ge   Scheerer, 1940
  13 pseudosetaceus Daveau
  14 cernuus Vahl
  15 prolifer Rottb.
  16 fluitans L.   60   Ge   Scheerer, 1940
  17 hudsonianus (Michx) Fernald
  18 cespitosus L.   104   Ge   Scheerer, 1940
    (a) cespitosus
    (b) germanicus (Palla) Broddeson
  19 pumilus Vahl

2 Blysmus Panzer
  1 compressus (L.) Panzer ex Link
  2 rufus (Hudson) Link

3 Eriophorum L.
  1 angustifolium Honckeny   58   Ho   Gadella & Kliphuis, 1966

    triste (Th. Fries) A. Löve & Hadac 60   Is   Löve & Löve, 1950
  2 latifolium Hoppe   54   Ge   Scheerer, 1940
                       72 non-Europe Löve & Löve, 1948

```
    3 gracile Koch ex Roth
    4 vaginatum L.                              58        Fe        Sorsa, 1963
    5 brachyantherum Trautv. & C.A. Meyer
    6 russeolum Fries ex Hartman
    7 scheuchzeri Hoppe


4 Eleocharis R. Br.
    1 quinqueflora (F.X. Hartmann) O.
      Schwarz                                  132        Su        Strandhede &
                                               134        Su          Dahlgren, 1968
    2 parvula (Roemer & Schultes) Link ex
      Bluff, Nees & Schauer                     10        Ge        Wulff, 1937a
    3 acicularis (L.) Roemer & Schultes         20        Is        Löve & Löve, 1956
    4 bonariensis Nees
    5 ovata (Roth) Roemer & Schultes
    6 obtusa (Willd.) Schultes
    7 atropurpurea (Retz.) C. Presl
    8 palustris (L.) Roemer & Schultes
      (a) palustris                             16        Cz        Májovský et al.,
                                                                      1974
      (b) vulgaris Walters                      38        Fe No     Strandhede, 1965
                                               37 39      Fe Su     Strandhede, 1965
                                                40        Po        Pogan, 1972
    9 mitracarpa Steudel
   10 austriaca Hayek                           16        Au Ge No  Strandhede, 1961
   11 mamillata H. Lindb. fil.                  16        Fe No Su  Strandhede, 1965
   12 uniglumis (Link) Schultes                 46        Fe No Su  Strandhede, 1965
                                                40        Br        Davies, 1956a
                                                44        Br        Walters, 1949
                                               54 92      Su        Sorsa Viv., 1963
   13 oxylepis (Holmsh.) B. Feddech.
   14 multicaulis (Sm.) Desv.                   20        Br        Davies, 1956a
   15 carniolica Koch                           20        Hu        Baksay, 1958


5 Fuirena Rottb.
    1 pubescens (Poiret) Kunth


6 Fimbristylis Vahl
    1 bisumbellata (Forskål) Bubani
    2 annua (All.) Roemer & Schultes
    3 squarrosa Vahl
    4 ferruginea (L.) Vahl


7 Cyperus L.
    1 papyrus L.
    2 auricomus Sieber ex Sprengel
    3 longus L.
    4 rotundus L.
    5 esculentus L.
    6 glomeratus L.
    7 glaber L.
    8 eragrostis Lam.                           42        Ga        Kliphuis &
                                                                      Wieffering, 1972
    9 textilis Thunb.
   10 alternifolius L.
   11 fuscus L.                                 72        Su        Löve & Löve, 1944
   12 difformis L.
   13 capitatus Vandelli                        72        Lu        Rodriguez, 1953
   14 michelianus (L.) Link
      (a) michelianus
      (b) pygmaeus (Rottb.) Ascherson & Graebner
```

```
15 cerotinus Rottb.
16 pannonicus Jacq.
17 laevigatus L.
   (a) laevigatus
   (b) distachyos (All.) Maire & Weiller
18 flavidus Retz.
19 polystachyos Rottb.
20 mundtii (Nees) Kunth
21 flavescens L.
22 strigosus L.
23 congestus Vahl
24 hamulosus Bieb.
25 squarrosus L.
26 ovularis (Michx) Torrey
27 brevifolius (Rottb.) Hassk.


8 Cladium Browne
   1 mariscus (L.) Pohl                        36        Europe    Löve, 1954a
                                             c.60        Da        Larsen, 1965


9 Rhynchospora Vahl
   1 alba (L.) Vahl                            26        Ge        Scheerer, 1940
                                               42        Su        Löve & Löve, 1942
   2 fusca (L.) Aiton fil.                     32        Ge        Scheerer, 1940
   3 rugosa (Vahl) S. Gale


10 Schoenus L.
   1 nigricans L.                              44        Ga        Labadie, 1976
                                               54        Lu        Rodriguez, 1953
   2 ferrugineus L.                            76        Scand.    Löve & Löve, 1944


11 Kobresia Willd.
   1 myosuroides (Vill.) Fiori                 60        Is        Löve & Löve, 1956
   2 simpliciuscula (Wahlenb.) Mackenzie       76        No        Knaben & Engelskjon,
                                                                     1967


12 Carex L.
   1 phyllostachys C.A. Meyer
   2 distachya Desf.                           74        Hs        Löve & Kjellqvist,
                                                                     1973
   3 paniculata L.                       60 62 64        Su        Lövqvist, 1963
     (a) paniculata
     (b) lusitanica (Schkuhr) Maire
   4 appropinquata Schumacher                  64        Su        Lövqvist, 1963
   5 diandra Schrank                           60        Is        Löve & Löve, 1956
   6 vulpina L.                                68        Su        Lövqvist, 1963
   7 otrubae Podp.                             58        It        Dietrich, 1972
                                               60        Su        Lövqvist, 1963
   8 vulpinoidea Michx
   9 spicata Hudson                            58        Ge        Dietrich, 1972
                                               60        Ju        Druskevic, ined.
  10 muricata L.                             c.56        Su        Lövqvist, 1963
                                               58        Ge        Dietrich, 1972

     (a) muricata
     (b) lamprocarpa Čelak.
  11 divulsa Stokes                            56        It        Dietrich, 1972
                                               58        Br        Davies, 1956b
     (a) divulsa
     (b) leersii (Kneucker) Walo Koch
  12 arenaria L.                               58        Lu        Rodriguez, 1953
```

353

	64	Ge	Wulff, 1937b
13 ligerica Gay	58	Ge	Dietrich, 1972
14 reichenbachii Bonnet			
15 disticha Hudson	62	Br	Davies, 1956
16 repens Bellardi	70	Au	Dietrich, 1972
17 praecox Schreber	58	Ge	Dietrich, 1972
18 brizoides L.	58	Ge	Dietrich, 1964
19 chordorrhiza L. fil.	60	Su	Lövqvist, 1963
	c.66	No	Knaben & Engelskjon, 1967
20 divisa Hudson	60 62	Hs	Löve & Kjellqvist, 1973
21 stenophylla Wahlenb.	60	It	Dietrich, 1972
22 physodes Bieb.			
23 foetida All.	58	He	Favarger, 1959
24 maritima Gunnerus	60	Is	Löve & Löve, 1956
25 remota L.	62	It	Dietrich, 1972
26 bohemica Schreber	80	Ge	Dietrich, 1972
27 ovalis Good.	64	Hs	Kjellqvist & Löve, 1963
	66	Ge	Dietrich, 1972
	68	Ge	Wulff, 1939
28 macloviana D'Urv.	86	Is	Löve & Löve, 1956
29 echinata Murray	56 58	Su	Lövqvist, 1963
30 dioica L.	52	Su	Lövqvist, 1963
	60		
31 parallela (Laest.) Sommerf.			
(a) parallela	44	Sb	Flovik, 1943
(b) redowskiana (C.A. Meyer) Egorova			
32 davalliana Sm.	18	Ge	Dietrich, 1972
33 elongata L.	56	Ju	Druskevic, ined.
34 lachenalii Schkuhr	58	Su	Davies, 1956
	62	Is	Löve & Löve, 1956
	74		
35 glareosa Wahlenb.	66	Sb	Flovik, 1943
36 ursina Dewey	64	Sb	Flovik, 1943
37 mackenziei V. Krecz.	64	Is	Löve & Löve, 1956
38 heleonastes L. fil.	c.64	Is	Löve & Löve, 1956
39 marina Dewey			
40 curta Good.	54	Su	Lövqvist, 1963
	56	No	Knaben & Engelskjon, 1967
41 lapponica O.F. Lang			
42 brunnescens (Pers.) Poiret	56	Is	Löve & Löve, 1956
43 loliacea L.	54	Su	Lövqvist, 1963
44 tenuiflora Wahlenb.			
45 disperma Dewey			
46 baldensis L.	88	It	Favarger, 1959
	90	It	Dietrich, 1972
47 curvula All.	c.86	Au	Reese, 1953
(a) curvula			
(b) rosae Gilomen			
48 hirta L.	112	Br	Davies, 1956
49 atherodes Sprengel			
50 lasiocarpa Ehrh.	c.56	Ge	Reese, 1952
51 acutiformis Ehrh.	c.38	Ge	Reese, 1952
	78	Ge	Dietrich, 1972
52 melanostachya Bieb. ex Willd.			
53 riparia Curtis	72	Cz	Májovský et al., 1976

54 pseudocyperus L.	66	Su	Heilborn, 1924
55 rostrata Stokes	60	Ju	Druskevič, ined.
	72-74	Su	Lövqvist, 1963
	76	Cz	Murin & Májovský, 1976
	82	Su	Lövqvist, 1963
56 rhynchophysa Fischer, C.A. Meyer & Ave-Lall.			
57 vesicaria L.	70	Ju	Druskevic, ined.
	74	Su	Lövqvist, 1963
	82	Su	Heilborn, 1924
	86	Ge	Dietrich, 1972
	88	Ju	Druskevič, ined.
58 saxatilis L.	80	No	Knaben & Engelskjon, 1967
59 rotundata Wahlenb.			
60 stenolepis Less.			
61 mollissima Christ			
62 pendula Hudson	58	Hs	Kjellqvist & Löve, 1963
	60	Su	Lövqvist, 1963
	62	Ju	Druskevič, ined.
63 microcarpa Bertol. ex Moris	60	Co	Contandriopoulos, 1962
64 sylvatica Hudson			
(a) sylvatica	58	Cz	Májovský et al., 1976
(b) paui (Sennen) A. & O. Bolos			
65 arnellii Christ			
66 debilis Michx			
67 capillaris L.			
(a) capillaris	54	Su	Löve et al., 1957
(b) fuscidula (V. Krecz. ex Egorova) A. & D. Löve			
68 krausei Boeckeler	36	Is	Löve & Löve, 1956
69 ledebourana C.A. Meyer ex Trev.			
70 strigosa Hudson	66	Su	Lövqvist, 1963
71 vulcani Hochst.			
72 flacca Schreber			
(a) flacca	76	Su	Lövqvist, 1963
	90	Hs	Kjellqvist & Löve, 1963
(b) serratula (Biv.) W. Greuter	76	Ju	Druskevič, ined.
	90	Hs	Löve & Kjellqvist, 1973
73 hispida Willd.	38	Hs	Love & Kjellqvist, 1973
	42	Hs	Kjellqvist & Löve, 1963
74 panicea L.	32	Is	Löve & Löve, 1956
75 vaginata Tausch	32	No	Knaben & Engelskjon, 1967
76 livida (Wahlenb.) Willd.	32	Su	Lövqvist, 1963
77 asturica Boiss.	46	Hs	Löve & Kjellqvist, 1973
78 depauperata Curtis ex With.	44	Ge	Dietrich, 1972
79 brevicollis DC.			
80 michelii Host	62	Ju	Druskevič, ined.
	c.70	Cz	Murin & Májovský, 1976

355

81 pilosa Scop.	44	Ge	Dietrich, 1972
82 hordeistichos Vill.			
83 secalina Willd. ex Wahlenb.			
84 laevigata Sm.	72	Ge	Dietrich, 1972
85 camposii Boiss. & Reuter	68	Lu	Dietrich, 1964
86 hochstetterana Gay ex Seub.			
87 binervis Sm.	74	Be	Dietrich, 1972
88 distans L.	72	It	Dietrich, 1972
	74	Hs	Löve & Kjellqvist, 1973
89 cretica Gradstein & Kern			
90 punctata Gaudin	68	Ju	Druskević, ined.
91 diluta Bieb.			
92 extensa Good.	60	Ga	Labadie, 1976
93 mairii Cosson & Germ.	68	Hs	Kjellqvist & Löve, 1963
	70	It	Dietrich, 1972
94 hostiana DC.	56	Ju	Druskević, ined.
95 durieui Steudel	52	Hs	Dietrich, 1972
96 flava L.	30 33	Ga	Bidault, 1974
	60	Br	Davies, 1956b
	64	?Su	Heilborn, 1918
97 jemtlandica (Palmgren) Palmgren			
98 lepidocarpa Tausch	58	Ju	Druskević, ined.
	68	Br	Davies, 1956b
99 nevadensis Boiss. & Reuter	68	Hs	Löve & Kjellqvist, 1963
100 bergrothii Palmgren			
101 demissa Hornem.	70	Br	Davies, 1956
102 serotina Mérat			
(a) serotina	35	Ga	Bidault, 1974
	68	Au	Dietrich, 1972
	70	No	Heilborn, 1924
(b) pulchella (Lonnr.) Van Ooststr.	70	Su	Heilborn, 1928
103 pallescens L.	62	Ge	Dietrich, 1972
	64	Br	Davies, 1956b
	66	Su	Lövqvist, 1963
104 hallerana Asso	50	It	Dietrich, 1964
	52	It	Dietrich, 1972
	54	Hs	Löve & Kjellqvist, 1973
105 rorulenta Porta	48-50	Bl	Dietrich, 1972
106 digitata L.	48 50 52	Br	Davies, 1956b
	54	Rm	Dietrich, 1972
107 ornithopoda Willd.			
(a) ornithopoda	c.46	Su	Heilborn, 1924
	52	No	Knaben & Engelskjon, 1967
	54	Su	Lövqvist, 1963
(b) ornithopodioides (Hausm.) Nyman	54	Au	Dietrich, 1972
	56	Ju	Druskević, ined.
108 pediformis C.A. Meyer			
(a) pediformis			
(b) rhizodes (Blytt) H. Lindb. fil.			
(c) macroura (Meinsh.) Podp.			
109 humilis Leysser	36 38	Ju	Druskević, ined.
110 glacialis Mackenzie	34	No	Knaben & Engelskjon, 1967
111 caryophyllea Latourr.	62 64 66 68	Su	Lövkvist, 1963
112 umbrosa Host			
(a) umbrosa	62	Ge	Dietrich, 1964
	66	Ge	Dietrich, 1972

(b) huetiana (Boiss.) Soó
(c) sabynensis (Less. ex Kunth) Kuk.
113 depressa Link
(a) depressa
(b) transsilvanica (Schur) Egorova
114 oedipostyla Duval-Jouve
115 illegitima Cesati
116 globularis L.

117 tomentosa L.	48	Ge	Dietrich, 1972
118 grioletii Roemer	48	It	Dietrich, 1964
119 ericetorum Pollich	30	Su	Lövqvist, 1963

120 melanocarpa Cham. ex Trautv.

121 montana L.	38	Ge	Dietrich, 1972
122 fritschii Waisb.	30	He	Dietrich, 1972

123 markgrafii Kuk.
124 pilulifera L.

(a) pilulifera	18	Ge	Dietrich, 1972

(b) azorica (Gay) Franco & Rocha Afonso
125 tricolor Velen.
126 amgunensis Friedrich Schmidt Petrop.

127 olbiensis Jordan	46	It	Dietrich, 1972
128 alba Scop.	54	Ge	Dietrich, 1964
129 supina Willd. ex Wahlenb.	38	It	Dietrich, 1964

130 liparocarpos Gaudin

(a) liparocarpos	38	It	Dietrich, 1972

(b) bordzilowskii (V. Krecz.) Egorova

131 sempervirens Vill.	30 31	Ge	Dietrich, 1967
	32	It	Dietrich, 1967
	34	Rm	Dietrich, 1967
	68	Ga	Lazare, 1976
132 firma Host	34	Ge	Dietrich, 1967
	34 68	Ju	Druskevic, ined.
133 kitaibeliana Degen ex Becherer	36	It	Dietrich, 1967
134 macrolepis	36 37	It	Dietrich, 1967
135 mucronata All.	34	Au	Dietrich, 1964
	36	Au Ge It	Dietrich, 1967

136 fuliginosa Schkuhr

(a) fuliginosa	40	Ge Rm	Dietrich, 1967
(b) misandra (R. Br.) Nyman	40	No	Knaben & Engelskjon, 1967

137 ferruginea Scop.

(a) ferruginea	39 40	Au Ge He It	Dietrich, 1967
(b) australpina (Becherer) W. Dietr.	38	It	Dietrich, 1964
	40	It	Dietrich, 1967
(c) caudata (Kuk.) Pereda & Laínz			
(d) macrostachys (Bertol.) Arcangeli	40	It	Dietrich, 1967
(e) tendae W. Dietr.	38	Ga It	Dietrich, 1967
138 brachystachys Schrank	40	Ge It Ju	Dietrich, 1967
139 fimbriata Schkuhr	42	He	Dietrich, 1967
140 atrofusca Schkuhr	38	No	Knaben & Engelskjon, 1967
	40	He	Dietrich, 1967
141 frigida All.	56	Au Ga He	Dietrich, 1967
	58	Ga	Dietrich, 1964
142 limosa L.	56 62	Su	Lövqvist, 1963
	64	Ge	Wulff, 1939
143 rariflora (Wahlenb.) Sm.	c.52	No	Knaben & Engelskjon, 1967
	54	Is	Löve & Löve, 1956
144 magellanica Lam.	58	He	Favarger, 1959

145 laxa Wahlenb.
146 atrata L.

(a) atrata	52 54	Ju	Druskević, ined.
(b) aterrima (Hoppe) Čelak.			
(c) caucasica (Steven) Kuk.			

147 parviflora Host   54   Au   Dietrich, 1972
148 buxbaumii Wahlenb.

(a) buxbaumii	c.74 c.100	Su	Lövqvist, 1963
(b) alpina (Hartman) Liro	106	Is	Löve & Löve, 1956

149 hartmanii Cajander
150 norvegica Retz.   56   Is   Löve & Löve, 1956

   (a) norvegica
   (b) inferalpina (Wahlenb.) Hultén
   (c) pusterana (Kalela) Chater

151 holostoma Drejer
152 stylosa C.A. Meyer

153 bicolor All.	16	He	Reese, 1953
	c.48	Is	Löve & Löve, 1956
	50	He	Davies, 1956
	52	No	Knaben & Engelskjon, 1967
154 rufina Drejer	60	Is	Löve & Löve, 1956
155 paleacea Schreber ex Wahlenb.	72 73	No	Faulkner, 1972
156 lyngbyei Hornem.	72	Is	Löve & Löve, 1956

157 vacillans Drejer

158 recta Boott	73 c.74 75	Br	Faulkner, 1972

159 halophila F. Nyl.
160 salina Wahlenb.

161 subspathacea Wormsk.	42	Is	Löve & Löve, 1956
162 aquatilis Wahlenb.	76	Br Fe No	Faulkner, 1972
	77	Br	Faulkner, 1972
	84	Ge	Dietrich, 1972
163 bigelowii Torrey ex Schweinitz	70	Is	Löve & Löve, 1956
(a) bigelowii	68-71	Br Hb	Faulkner, 1972

   (b) rigida Schultze-Motel
   (c) arctisibirica (Jurtzev) A. & D. Löve
   (d) ensifolia (Turcz. ex Gorodkov)
         J. Holub

164 elata All.	74	Br	Davies, 1956
	75 76	Hb	Faulkner, 1972
	76 77	Br	Faulkner, 1972
	78 80	Su	Lovqvist, 1963

   (a) elata
   (b) omskiana (Meinsh.) Jalas

165 cespitosa L.	78-80	Da	Faulkner, 1972

166 buekii Wimmer

167 nigra (L.) Reichard	82 84	Su	Lövqvist, 1963
	83-85	Br	Faulkner, 1972
vulgaris ssp. juncella (Fries) Nyman	76	Is	Löve & Löve, 1956
	84	No	Knaben & Engelskjon, 1967
168 acuta L.	72 74-76 78	Su	Ehrenberg, 1945
	82-85	Br	Faulkner, 1972

169 trinervis Degl.

170 microglochin Wahlenb.	48	He	Dietrich, 1972
	58	Is	Löve & Löve, 1956
171 pauciflora Lightf.	46 76	Ju	Druskević, ined.
172 rupestris All.	50	He	Favarger, 1959
	52	No	Knaben & Engelskjon, 1967

```
173 obtusata Liljeblad
174 nardina Fries                    70      Is     Löve & Löve, 1956
175 capitata L.                      50      Is     Löve & Löve, 1956
176 scirpoidea Michx.
177 pyrenaica Wahlenb.
178 pulicaris L.                   58 60     Su     Lövqvist, 1963
```

SCITAMINEAE

CC MUSACEAE

```
1 Musa L.
  1 cavandishii Lamb. ex Paxton
```

CCI ZINGIBERACEAE

```
1 Hedychium Koenig
  1 gardneranum Sheppard ex Ker-Gawler
```

CCII CANNACEAE

```
1 Canna L.
  1 indica L.
```

MICROSPERMAE

CCIII ORCHIDACEAE

```
1 Cypripedium L.
  1 calceolus L.                    20      Po     Skalińska et al.,
                                                   1957
  2 macranthos Swartz
  3 guttatum Swartz

2 Epipactis Zinn
  1 palustris (L.) Crantz           40      Su     Löve & Löve, 1944
                             40 44 46 48     Ga     Lévèque &
                                                   Gorenflot, 1969
  2 helleborine (L.) Crantz         36      Ga     Lévèque &
                                                   Gorenflot, 1969
                                    38      Po     Skalińska et al.,
                                                   1961
                                    40      Cz     Uhríková &
                                                   Feráková, 1978
                                    44      Ga     Lévèque &
                                                   Gorenflot, 1969
  3 leptochila (Godfery) Godfery    36      Da     Hagerup, 1947
  4 muelleri Godfery
  5 dunensis (T. & T.A. Stephenson) Godfery
  6 purpurata Sm.                   40      He     Meili-Frei, 1966
  7 phyllanthes G.E. Sm.            36      Da     Young, 1953
  8 atrorubens (Hoffm.) Besser      40      Su     Löve & Löve, 1944
                              40 + 0 - 7B    He     Meili-Frei, 1966
  9 microphylla (Ehrh.) Swartz      40      He     Meili-Frei, 1966

3 Cephalanthera L.C.M. Richard
  1 damasonium (Miller) Druce       32      Da     Hagerup, 1947
  2 longifolia (L.) Fritsch         32      Hs     Löve & Kjellqvist,
                                                   1973
```

```
  3 cucullata Boiss. & Heldr.
  4 epipactoides Fischer & C.A. Meyer
  5 rubra (L.) L.C.M. Richard              48        Au      Titz, 1966

  4 Limodorum Boehmer
  1 abortivum (L.) Swartz                  56        Lu      Coutinho, 1957
    ssp. trabutianum (Batt.) Rouy          60        Lu      Coutinho, 1957

  5 Epipogium R. Br.
  1 aphyllum Swartz

  6 Neottia Ludwig
  1 nidus-avis (L.) L.C.M. Richard         36        He      Meili-Frei, 1966

  7 Listera R. Br.
  1 ovata (L.) R. Br.               34 + 0 - 2B      He      Meili-Frei, 1966
                                         34-38       Ho      Gadella & Kliphuis,
                                                               1963
                                         34 36       Su      Löve & Löve, 1944
                                            38       Da      Löve & Löve, 1944
  2 cordata (L.) R. Br.                     38       Po      Skalińska et al.,
                                                               1961
                                            40       Ho      Gadella & Kliphuis,
                                                               1963

  8 Spiranthes L.C.M. Richard
  1 spiralis (L.) Chevall.                  30       Da      Hagerup, 1944
  2 aestivalis (Poiret) L.C.M. Richard
  3 romanzoffiana Cham.                     60       Hb      Heslop-Harrison,
                                                               1961

  9 Goodyera R. Br.
  1 repens (L.) R. Br.                      30       Su      Löve & Löve, 1942
                                            40       Po      Skalińska et al.,
                                                               1974

 10 Gennaria Parl.
  1 diphylla (Link) Parl.

 11 Herminium Guett.
  1 monorchis (L.) R. Br.                   40       He      Heusser, 1938

 12 Neottianthe Schlechter
  1 cucullata (L.) Schlechter

 13 Platanthera L.C.M. Richard
  1 bifolia (L.) L.C.M. Richard            42        He      Heusser, 1938
  2 chlorantha (Custer) Reichenb.          42        Po      Skalińska et al.,
                                                               1947
  3 micrantha (Hochst.) Schlechter
  4 hyperborea (L.) Lindley                84        Is      Löve & Löve, 1956
  5 obtusata (Pursh) Lindley              126        Su      Löve, 1954a

 14 Chamorchis L.C.M. Richard
  1 alpina (L.) L.C.M. Richard             42        He      Heusser, 1938

 15 Gymnadenia R. Br.
  1 conopsea (L.) R. Br.           40 80 c.97        Au      Groll, 1965
                                   c.117 c.119
  2 odoratissima (L.) L.C.M. Richard       40        Ju      Sušnik & Lovka,
                                                               1973
```

16 Pseudorchis Séguier
  1 albida (L.) Á. & D. Löve
    (a) albida
    (b) straminea (Fernald) Á. & D. Love

(b) straminea (Fernald) Á. & D. Love	40	Rs(N)	Sokolovskaya & Strelkova, 1960
	42	Is	Löve & Löve, 1956

  2 frivaldii (Hampe ex Griseb.) P.F. Hunt

17 Nigritella L.C.M. Richard
  1 nigra (L.) Reichenb. fil.   64   Ju   Lovka et al., 1972

    (a) nigra
    (b) rubra (Wettst.) Beauverd
    (c) corneliana Beauverd

18 Coeloglossum Hartman
  1 viride (L.) Hartman   40   Br   Richards, 1972

19 Dactylorhiza Necker ex Nevski
  1 iberica (Bieb.) Soó

2 sambucina (L.) Soó	40	Ho	Vermeulen, 1947
	42	Fe	Sorsa, 1963

    (a) sambucina

(b) insularis (Sommier) Soó	60	It	Scrugli, 1977

  3 sulphurea (Link) Franco
    (a) sulphurea
    (b) pseudosambucina (Ten.) Franco
    (c) siciliensis (Klinge) Franco

4 incarnata (L.) Soó	40	Ga	Vaucher, 1966

    (a) incarnata
    (b) coccinea (Pugsley) Soó
    (c) pulchella (Druce) Soó

(d) cruenta (O.F. Mueller) P.D. Sell	40	Su	Löve & Löve, 1944

  5 pseudocordigera (Neuman) Soó

6 majalis (Reichenb.) P.F. Hunt & Summerhayes	80	Cz	Májovský et al., 1974

    (a) majalis

(b) occidentalis (Pugsley) P.D. Sell	80	Br	Vermeulen, 1938
(c) alpestris (Pugsley) Senghas			
(d) purpurella (T. & T.A. Stephenson) D. Moresby Moore & Soó	80	Br	Roberts, 1966
(e) praetermissa (Druce) D. Moresby Moore & Soó	80	Br	Roberts, 1966
7 cordigera (Fries) Soó	80	Ho	Vermeulen, 1947

    (a) cordigera
    (b) bosniaca (G. Beck) Soó
    (c) siculorum (Soó) Soó

8 traunsteineri (Sauter) Soó	80	Hs	Löve & Kjellqvist, 1973
	122	Rs(B)	Vermeulen, 1938

    (a) traunsteineri
    (b) curvifolia (Nyl.) Soó
    (c) lapponica (Laest. ex Hartman) Soo

9 russowii (Klinge) J. Holub	120	?Rs(B)	Löve & Löve, 1961a
10 elata (Poiret) Soó	80		Vermeulen, 1947
11 maculata (L.) Soó	40	Su	Löve & Löve, 1944
	40 60 80	Au	Groll, 1965

    (a) maculata

(b) elodes (Griseb.) Soó	40	Br	Maude, 1939
	80	Is	Löve & Löve, 1956

```
   (c) schurii (Klinge) Soó
   (d) islandica (Á. & D. Löve) Soó        80     Is      Löve & Löve, 1956
   (e) transsilvanica (Schur) Soó
12 fuchsii (Druce) Soó
   (a) fuchsii                             40     Ju      Lovka et al., 1971
   (b) psychrophila (Schlechter) J. Holub
   (c) sooana (Borsos) Borsos
13 saccifera (Brongn.) Soó                 40
                                           80     Europe  Vermeulen, 1947


20 Steveniella Schlechter
   1 satyrioides (Steven) Schlechter


21 Comperia C. Koch
   1 comperiana (Steven) Ascherson & Graebner


22 Neotinea Reichenb. fil.
   1 maculata (Desf.) Stearn               40     It      Scrugli et al.,
                                                          1976


23 Traunsteinera Reichenb.
   1 globosa (L.) Reichenb.                42     Po      Skalińska et al.,
                                                          1957


24 Orchis L.
   1 papilionacea L.                       32     Hs      Löve & Kjellqvist,
                                                          1973

   2 boryi Reichenb. fil
   3 morio L.                              36     Hs      Löve & Kjellqvist,
                                                          1973
   (a) morio                               36     It      Scrugli et al.,
                                                          1976
   (b) picta (Loisel.) Arcangeli
   (c) champagneuxii (Barn.) Camus
   4 longicornu Poiret                     36     It      Scrugli et al.,
                                                          1976
   5 coriophora L.                         36     Ju      Sušnik & Lovka,
                                                          1973
   (a) coriophora
   (b) fragrans (Pollini) Sudre            20     Ga      Labadie, 1976
                                           38     It      Scrugli et al.,
                                                          1976
   (c) martrinii (Timb.-Lagr.) Nyman
   6 sancta L.
   7 ustulata L.                           42     Hs      Löve & Kjellqvist,
                                                          1973
   8 tridentata Scop.                      42     It      Scrugli, 1977
   (a) tridentata
   (b) commutata (Tod.) Nyman
   9 lactea Poiret                         42     It      Scrugli, 1977
   10 italica Poiret                       42     It      Del Prete, 1977
   11 simia Lam.                           42     He      Heusser, 1938
   12 militaris L.                         42     Cz      Májovský et al.,
                                                          1976

   13 punctulata Steven ex Lindley
   (a) punctulata
   (b) sepulchralis (Boiss. & Heldr.) Soó
   14 purpurea Hudson                      42     Cz      Májovský et al.,
                                                          1976
   15 saccata Ten.                         36     It      Scrugli et al.,
                                                          1976
```

16 patens Desf.
17 spitzelii Sauter ex Koch

(a) spitzelii	42	Hs	Löve & Kjellqvist, 1973
(b) nitidifolia (Teschner) Soó	40	Cr	Hautzinger, 1976

18 mascula (L.) L. 42 Hs Löve & Kjellqvist, 1973

(a) mascula
(b) wanjkowii (Wulf) Soó
(c) signifera (Vest) Soó   42   Cz   Uhríková & Schwarzova, 1978
(d) olbiensis (Reuter ex Grenier)
    Ascherson & Graebner
(e) hispanica (A. & C. Nieschalk) Soó
19 pallens L.   40   He   Heusser, 1938
20 provincialis Balbis   42   It   Scrugli, 1977
(a) provincialis
(b) pauciflora (Ten.) Camus
21 anatolica Boiss.
22 quadripunctata Cyr. ex Ten.
23 laxiflora Lam.
(a) laxiflora   36   It   Scrugli, 1977
(b) palustris (Jacq.) Bonnier & Layens   42   Cz   Májovský et al., 1976

(c) elegans (Heuffel) Soó

25 Aceras R. Br.
1 anthropophorum (L.) Aiton fil.   42   Ho   Kliphuis, 1963

26 Himantoglossum Koch
1 hircinum (L.) Sprengel   24   He   Heusser, 1915
                           36   Ju   Sušnik & Lovka, 1973

(a) hircinum
(b) calcaratum (G. Beck) Soó
(c) caprinum (Bieb.) Sundermann

27 Barlia Parl.
1 robertiana (Loisel.) W. Greuter   36   Ga   Raynaud, 1971

28 Anacamptis L.C.M. Richard
1 pyramidalis (L.) L.C.M. Richard   36   Ho   Kliphuis, 1963

29 Serapias L.
1 cordigera L.   36   It   Scrugli et al., 1976

2 neglecta De Not.
3 vomeracea (Burm.) Briq.   36   Ju   Sušnik & Lovka, 1973

(a) vomeracea
(b) orientalis W. Greuter
4 lingua L.
5 parviflora Parl.   36   It   Del Prete, 1977

30 Ophrys L.
1 insectifera L.   36   Cz   Májovský et al., 1976

2 speculum Link   36   It   Scrugli, 1977
(a) speculum
(b) lusitanica O. & A. Danesch
3 lutea (Gouan) Cav.

(a) lutea	36	Hs	Löve & Kjellqvist, 1973

(b) murbeckii (Fleischm.) Soó
(c) melena Renz

| 4 fusca Link | c.73 | Bl | Dahlgren et al., 1971 |

(a) durieui (Reichenb. fil.) Soó

| (b) fusca | 72 76 | Bl | Greilhuber & Ehrendorfer, 1975 |
| (c) iricolor (Desf.) O. Schwarz | 36 | It | Scrugli, 1977 |

(d) omegaifera (Fleischm.) E. Nelson
5 pallida Rafin.
6 sphegodes Miller

| (a) sphegodes | 36 | He | Heusser, 1938 |
| (b) litigiosa (Camus) Becherer | 36 | He | Heusser, 1938 |

(c) tommasinii (Vis.) Soó

| (d) atrata (Lindley) E. Mayer | 36 | Bl | Greilhuber & Ehrendorfer, 1975 |

(e) mammosa (Desf.) Soó ex E. Nelson
(f) aesculapii (Renz) Soó
(g) helenae (Renz) Soó & D. Moresby
    Moore
7 spruneri Nyman
(a) spruneri
(b) panormitana (Tod.) Soó
8 ferrum-equinum Desf.
(a) ferrum-equinum
(b) gottfriediana (Renz) E. Nelson

| 9 bertolonii Moretti | 36 | Ju | Sušnik & Lovka, 1973 |
| promontorii | 36-38 | It | Greilhuber & Ehrendorfer, 1975 |

10 lunulata Parl.
11 argolica Fleischm.
12 reinholdii Spruner ex Fleischm.
13 cretica (Vierh.) E. Nelson
14 carmeli Fleischm. & Bornm.
15 scolopax Cav.

| (a) scolopax | 36 | It | Scrugli, 1977 |
| | 38 40 | Lu | Shimoya & Ferlan, 1952 |

(b) oestrifera (Bieb.) Soó

| (c) cornuta (Steven) Camus | 36 | Ju | Sušnik & Lovka, 1973 |

(d) heldreichii (Schlechter) E. Nelson

16 fuciflora (F.W. Schmidt) Moench	36	He	Heusser, 1938
(a) fuciflora	36 37 38	It	Greilhuber & Ehrendorfer, 1975
ssp. sundermannii	36	It	

(b) candica E. Nelson ex Soó
(c) oxyrrhynchos (Tod.) Soó
(d) exaltata (Ten.) E. Nelson

17 arachnitiformis Gren. & Philippe	36 38	It	Scrugli, 1977
18 tenthredinifera Willd.	36	It	Scrugli, 1977
19 apifera Hudson	36	Hs	Löve & Kjellqvist, 1973

(a) apifera
(b) jurana Ruppert

| 20 bombyliflora Link | 36 | It | Scrugli, 1977 |

31 Corallorhiza Chatel.
   1 trifida Chatel.   42   Da   Hagerup, 1941

32 Calypso Salisb.
   1 bulbosa (L.) Oakes

33 Liparis L.C.M. Richard
   1 loeselii (L.) L.C.M. Richard

34 Microstylis (Nutt.) A. Eaton
   1 monophyllos (L.) Lindley

35 Hammarbya O. Kuntze
   1 paludosa (L.) O. Kuntze   28   Ho   Kliphuis, 1963

REFERENCES

Aakerberg, E., 1942. Hereditas, 28: 1-126.

Agapova, N.D., 1976. Taxon, 25: 635-636.

Ahti, T., 1962. Arch. Soc. Zool. Bot. Fenn. 'Vanamo', 16: 22-35.

Albers, A. and Albers, I., 1973. Österr. Bot. Zeits., 122: 293-298.

Albers, F., 1972. Ber. Deutsch. Bot. Gesell. 85 (5-6): 279-285.

Albers, F., 1973. Österr. Bot. Zeits., 121: 251-254.

Alden, B., 1976. Bot. Not., 129: 297-321.

Aleskoovskij, M.K., 1929. Izv. Saratovskogo Gosodarstvenn Instit. Selsk.
    Khozjajstva, 5: 375-382.

Almgaard, G., 1960. Kungl. Lantbrukshögsk. Annal., 26: 77-119.

Ančev, M.E., 1974. In: Ehrendorfer, F. and Ančev, M.E., Bot. Jour. Linn.
    Soc. London, 68: 268-272.

Ančev, M.E., 1975. In: Löve, Á., Taxon, 24: 514.

Ančev, M.E., 1976. Bulg. Acad. Sc., Phytol., 5: 51-56.

Arato, M., 1911. Nuovo Giorn. Bot. Ital., 51: 39-43.

Armenise, V., 1958. Nuovo Giorn. Bot. Ital., 64: 297-318.

Arnoldi, W., 1910. Flora, 100: 121-139.

Arrigoni, P.V. and Mori, B., 1971. Inf. Bot. Ital., 3: 226-233.

Arrigoni, P.V. and Mori, B., 1976. Inf. Bot. Ital., 8: 269-276.

Ashri, A. and Knowles, P.F., 1960. Agron. Jour., 52: 11-17.

Auquier, P., 1973. Candollea, 28: 15-19.

Auquier, P. and Kerguélen, M., 1977. Lejeunia, 89: 1-82.

Auquier, P. and Rammeloo, J., 1973. Bull. Soc. Roy. Bot. Belg., 106: 317-328.

Avdulov, N.P., 1928. Driev. Vsesojuz. Sezda Botanik. Leningrad, 1928: 65-66.

Azevedo Coutinho, L. de and Santos, A.C., 1943. Agron. Lusit., 5: 349-361.

Babcock, E.B., 1947. Univ. Calif. Publs. Bot., 22: 199-1030.

Babcock, E.B., Stebbins, G.L. and Jenkins, J.A., 1937. Cytologia, Fuji
    Jubil. Vol.: 188-210.

Badr, A. and Elkington, T.T., 1977. Pl. Syst. Evol., 128: 23-35.

Baez-Mayer, A.B., 1934. Cavanillesia, 6: 59-103.

Bagnoli, L., Guglielminetti, P. and Loreti, F., 1963. Nuovo Giorn. Bot. Ital., 70: 544-545.

Bajer, A., 1950. Acta Soc. Bot. Polon., 20: 635-646.

Bajon, R. and Greber, F., 1971. Actes Cell. Fl. Veg. Alp. Jur., Paris, 255-270.

Baker, H.G., 1948. New Phytol., 47: 131-145.

Baker, H.G., 1955. In: Darlington, C.D. and Wylie, A.P., Chromosome Atlas of Flowering Plants.

Baker, H.G., 1959. Proc. Bot. Soc. Brit. Is., 3: 288.

Baksay, L., 1951. In: Soó, R. and Javorka, S. A Magyar Növenyvilag Kezikönyve I-II. Budapest.

Baksay, L., 1954. Ann. Hist.-Nat. Mus. Natl. Hung., S.N., 5: 139-148.

Baksay, L., 1955. Ann. Hist.-Nat. Mus. Natl. Hung., S.N., 6: 167-176.

Baksay, L., 1956. Ann. Hist.-Nat. Mus. Natl. Hung., S.N., 7: 321-334.

Baksay, L., 1957b. Ann. Hist.-Nat. Mus. Natl. Hung., S.N., 8: 169-174.

Baksay, L., 1958. Ann. Hist.-Nat. Mus. Natl. Hung., S.N., 50: (N.S.9): 121-125.

Baksay, L., 1961. In: Löve, Á. and Löve, D., Opera Bot., 5.

Ball, P.W., 1961. In: Löve, Á. and Löve, D., Opera Bot., 5.

Ball, P.W. and Tutin, T.G., 1959. Watsonia, 4: 193-205.

Bambacioni-Mezzetti, V., 1936. Anal. Bot., 21: 186-204.

Banach-Pogan, E., 1954. Acta Soc. Bot. Polon., 23: 375-382.

Banach-Pogan, E., 1955. Acta Soc. Bot. Polon., 24: 275-286.

Banach-Pogan, E., 1958. Acta Biol. Cracov., Bot., 1 (2): 91-101.

Barling, D.M., 1958. Proc. Bot. Soc. Brit. Is., 3: 85.

Barros Neves, J., 1939. Bol. Soc. Brot., Sér. 2, 13: 545-572.

Barros Neves, J., 1973. Bol. Soc. Brot., Sér. 2, 47: 157-212.

Battaglia, E., 1947. Atti Soc. Toscana Sci. Nat. Mem., Ser. B, 54: 70-78.

Battaglia, E., 1957. Caryologia, 9: 293-314.

Battaglia, E., 1957b. Caryologia, 10: 1-28.

367

Baumberger, H., 1970. Ber. Schweiz. Bot. Ges., 80: 17-95.

Beatus, R., 1936. Zeitschr. Vererb., 71: 353-381.

Beaudry, J.R., 1970. Natural. Canad., 97: 431-445.

Bedalov, M., 1969. Acta Bot. Croat., 28: 39-41.

Bedalov, M. and Sušnik, F., 1970. Caryologia, 23: 519-524.

Behre, K., 1929. Planta, 7: 208-306.

Bentzer, B., 1972. Bot. Not., 125: 406-418.

Bentzer, B., 1973. Bot. Not., 126: 69-132.

Bernstrüm, P., 1946. Hereditas, 32: 514-520.

Berudye, E.M. and Jandar, E., 1907. New Phytol., 6: 127-134, 167-174.

Beuret, E., 1972. Bull. Soc. Neuchât. Sci. Nat., 95: 35-41.

Bidault, M., 1968. Rév. Cytol. Biol. Veg., 31: 217-356.

Bidault, M., 1972. Ann. Sci. Univ. Besançon, Ser. 3., Bot., 12: 119-123.

Bidault, M., 1974. Compt. Rend. Séances Soc. Biol., 168: 466.

Bijok, K., 1959. Acta Bot. Soc. Polon., 28: 365-371.

Bijok, K   Kronoko, D. and Pawlak, I., 1972. Acta Soc. Bot. Polon., 41:
    433-438.

Billeri, G., 1954. Caryologia, 6: 45-51.

Bir, S.S., 1960. Acta Bot. Neerl., 9: 224-234.

Björkman, S.O., 1954. Hereditas, 40: 254-258.

Björkman, S.O., 1960. Symb. Bot. Upsal., 17: 1-113.

Björkquist, I., 1961a. Bot. Not., 114: 365-367.

Björkquist, I., 1961b. Bot. Not., 114: 281-299.

Björkquist, I., Bothmer, R. von, Nilsson, O. and Nordenstam, B., 1969. Bot.
    Not., 122: 271-283.

Björse, S.A., 1961. In: Löve, Á. and Löve, D., Opera Bot., 5.

Blackburn, K.B., 1923. Nature (Lond.), 112: 687-688.

Blackburn, K.B., 1933. Proc. Univ. Durham Philos. Soc., 9 (2): 84-90.

Blackburn, K.B., 1939. Jour. Bot., 77: 67-71.

Blackburn, K.B., 1950. In: Tischler, G., Chromosomenzahlen der Gefässpflanzen
    Mitteleuropas; Junk. S-Gravenhage.

Blackburn, K.B. and Adams, A.W., 1955. Proc. Bot. Soc. Brit. Is., 1: 380.

Blackburn, K.B. and Heslop-Harrison, J.W., 1921. Ann. Bot., 35: 159-188.

Blackburn, K.B. and Heslop-Harrison, J.W., 1924. Ann. Bot., 38: 361-378.

Blackburn, K.B. and Morton, J.K., 1957. New Phytol., 56: 344-351.

Blaise, S., 1966. Compt. Rend. Acad. Sc. Paris, 262: 103-106.

Blaise, S., 1969. In: Löve, Á., Taxon, 18: 311.

Blaise, S., 1970. In: Löve, Á., Taxon, 19: 102-113.

Böcher, T.W., 1932. Bot. Tidsskr., 42: 183-206.

Böcher, T.W., 1938a. Dansk Bot. Ark., 9 (4): 1-33.

Böcher, T.W., 1938b. Svenska Bot. Tidskr., 32: 346-361.

Böcher, T.W., 1940a. Dansk Vid. Selsk. Biol. Medd., 15 (3): 1-64.

Böcher, T.W., 1947. Bot. Not., 1947: 353-360.

Böcher, T.W., 1954. Bot. Tidsskr., 51: 33-47.

Böcher, T.W. and Larsen, K., 1955. Bot. Tidsskr., 52: 125-132.

Böcher, T.W. and Larsen, K., 1958. Dansk Vid. Selsk. Biol. Medd., 10 (2):
    1-24.

Böcher, T.W. and Larsen, K., 1958b. New Phytol., 57: 311-317.

Böcher, T.W., Larsen, K. and Rahn, K., 1953. Hereditas, 39: 290-303.

Böcher, T.W., Larsen, K. and Rahn, K., 1955. Hereditas, 41: 423-453.

Bocquet, G., Favarger, C. and Zurcher, P.A., 1967. Zeits. Basler Bot. Ges.,
    3 (2): 229-242.

Bonnet, A.L.M., 1958. Nat. Monspel., Ser. Bot., 10: 3-6.

Bonnet, A.L.M., 1961. Nat. Monspel., Ser. Bot., 13: 7-13.

Bonnet, A.L.M., 1966. Nat. Monspel., Ser. Bot., 17: 21-29.

Borhidi, A., 1968. Acta Bot. Acad. Sc. Hung., 14 (3-4): 253-260.

Bosemark, N.O., 1956. Hereditas, 42: 443-467.

Bosemark, N.O., 1957. Hereditas, 43: 236-298.

Bothmer, R. von, 1967. Bot. Not., 120: 202-208.

Bothmer, R. von, 1969. Bot. Not., 122: 38-56.

Bothmer, R. von, 1970. Bot. Not. 123: 52-60.

Bothmer, R. von, 1975. Mitt. Bot. Staatssamm., München, 12: 267-288.

Bothmer, R. von and Bentzer, B., 1973. Ann. Mus. Goulandris, 1: 11-13.

Bowden, W.M., 1957. Can. Jour. Bot., 35: 951-993.

Bowden, W.M., 1964. Can. Jour. Bot., 42: 547-601.

Bozzini, A., 1959. Caryologia, 12: 459-470.

Bradshaw, M.E., 1963. Watsonia, 5: 321-326.

Bradshaw, M.E., 1964. Watsonia, 6: 76-81.

Brandt, J.P., 1952. Bull. Soc. Neuchat. Sci. Nat., 75: 179-188.

Brandt, J.P., 1961. Bull. Soc. Neuchat. Sci. Nat., 84: 35-87.

Bresinskya, A. and Grau, J., 1970. Ber. Bayer. Bot. Ges., 42: 101-108.

Breton-Sintes, S., 1971. These Univ. Clermont-Ferrand, E146: 1-70.

Brett, O.E., 1955. New Phytol., 54: 138-148.

Breviglieri, N. and Battaglia, E., 1954. Caryologia, 6: 271-283.

Brighton, C.A., 1976. Kew Bull., 31: 33-46.

Brighton, C.A., Mathew, B. and Marchant, C.J., 1973. Kew Bull., 28: 451-464.

Brøgger, A., 1960. Blyttia, 10, 00 18.

Brummitt, R.K., 1973. Watsonia, 9: 369-373.

Brunsberg, K., 1965. Bot. Not., 118: 356-381.

Bruun, H.G., 1932. Symb. Bot. Upsal., 1: 1-239.

Burdet, H.M., 1967. Candollea, 22: 107-

Burdet, H.M. and Greuter, W., 1973. In: Greuter, W., Boissiera, 22:
    1-215.

Burdet, H.M. and Miege, J., 1968. Candollea, 23: 109-120.

Buttler, K.P., 1969. Rev. Roum. Biol., Ser. Bot., 14: 275-282.

Caldo, L. del, 1971. Inf. Bot. Ital., 3: 71-73.

Camara, A. and Coutinho, L.A., 1939. Agron. Lusit., 1: 268-314.

Campion-Bourget, F., 1967. Compt. Rend. Acad. Sc. Paris, 264: 2100-2102.

Cardona, M.A., 1973. Acta Phytotax. Barcinon., 14: 1-20.

Carolin, R.C., 1957. New Phytol., 56: 81-97.

Carpineri, R., 1971. Inf. Bot. Ital., 3: 75-79.

Carpio, M.D.A., 1954. Genet. Iber., 6: 101-111.

Carpio, M.D.A., 1957. Genet. Iber., 9: 163-185.

Cartier, D., 1965. Compt. Rend. Acad. Sc. Paris., 261: 4475-4478.

Cartier, D. and Lennoir, A., 1968. Compt. Rend. Acad. Sc. Paris, 266: 119-122.

Casas, J.F., 1973. Cuad. Cienc. Biol. (Granada), 2: 39-41.

Casas, J.F., Lainz, M. and Rejón, M.R., 1973. Cuad. Cienc. Biol. (Granada),
   2: 3-5.

Casper, S.J., 1962. Feddes Rep., 66: 1-148.

Casper, S.J., 1966. Biblioth. Bot., 127/128: 1-20.

Castro, D. de, 1941. Agron. Lusit., 3: 103-113.

Castro, D. de, 1943. Agron. Lusit., 5: 243-249.

Castro, D. de, 1945. Bol. Soc. Brot. Sér. 2, 19: 525-539.

Castro, D. de and Fontes, F.C., 1946. Broteria, 15: 38-46.

Cesca, G., 1961. Caryologia, 14: 79-86.

Cesca, G., 1961. Nuovo Giorn. Bot. Ital., 70: 542-543.

Cesca, G., 1972. Inf. Bot. Ital., 4: 45-66.

Ceschmedjiev, I.V., 1970. Bot. Zhurn., 55: 1100-1110.

Ceschmedjiev, I.V., 1973. Bot. Zhurn., 58: 864-875.

Ceschmedjiev, I.V., 1976. Taxon, 25: 642.

Chiappini, M., 1954. Nuovo Giorn. Bot. Ital., 61: 276-289.

Chiappini, M. and Scrugli, A., 1972. Inf. Bot. Ital., 4: 131-133.

Chiarugi, A., 1928. Nuovo Giorn. Bot. Ital., N.S., 34: 1452-1496.

Chiarugi, A., 1933. Nuovo Giorn. Bot. Ital., N.S., 40: 63-75.

Chiarugi, A., 1935. In: Tischler, G., Tab. Biol., 11: 281-304.

Chiarugi, A., 1949. Caryologia, 1: 362-376.

Chiarugi, A., 1950. Caryologia, 3: 148.

Choudhuri, H.C., 1942. Ann. Bot., N.S., 6: 184-217.

Chouksanova, N.A., Sveshnikova, L.I. and Alexandrova, T.V., 1968.
   Citologija, 10: 381-386.

Christoff, M., 1942. Zeitschr. Indukt. Abst. Vererb., 80: 103-125.

Christoff, M. and Christoff, M.A., 1948. Genetics, 33 (1): 36-42.

371

Christoff, M. and Popoff, A., 1933. Planta, 20: 440-447.

Christov, M. and Nikolov, T., 1962. Izv. Nauch. Inst. Rast (Sofia), 15: 33-41.

Cinčura, F. and Hindakova, M., 1963. Biologia (Bratislava), 18: 184-194.

Clausen, J., 1931. Bot. Tidsskr., 41: 317-335.

Cole, M.J., 1962. Watsonia, 5: 113-116.

Constantinescu, G., Morlova, I., Cosmin, S. and Molea, I., 1964. Res. Roumania Biol. Ser. Bot., 9: 223-233.

Contandriopoulos, J., 1957. Bull. Soc. Bot. Fr., 104: 533-538.

Contandriopoulos, J., 1962. Thèse Fac. Sc. Montpellier, 1-354.

Contandriopoulos, J., 1964a. Rev. Gen. Bot., 71: 361-384.

Contandriopoulos, J., 1964b. Bull. Soc. Bot. France, 111: 222-234.

Contandriopoulos, J., 1966. Bull. Soc. Bot. France, 113: 453-474.

Contandriopoulos, J., 1967. 91e Congres des Soc. Sav. Rennes, 3: 271-280.

Contandriopoulos, J., 1968. Bull. Mus. Hist. Nat. Marseilles, 27: 45 53

Contandriopoulos, J. and Cauwet, A.M., 1968. Nat. Monspel., 19: 29-35.

Contandriopoulos, J. and Martin, D., 1967. Bull. Soc. Bot. Fr., 114 (7-8): 257-275.

Cook, C.D.K., 1962. Watsonia, 5: 123-126.

Cook, C.D.K., 1968. In: Löve, Á., Taxon, 17: 419-422.

Coombe, D.E., 1961. Watsonia, 5: 68-87.

Cope,T.A., 1976. Unpub. Ph.D. thesis, Univ. Leicester.

Cotton, R. and Stace, C.A., 1976. Genetica, 46: 235-255.

Coutinho, L. de A., 1957. Agron. Lusit., 19: 219-231.

Coutinho, L. de A. and Lorenzo-Andreu, A., 1948. An. Aula Dei, 1: 1-32.

Crane, M.B. and Gairdner, A.E.,11923. Jour. Genet., 13: 187-200.

Cugnac, A. de and Simonet, M., 1941a. Compt. Rend. Soc. Biol. (Paris), 135: 728-731.

Cugnac, A. de and Simonet, M., 1941b. Bull. Soc. Bot. Fr., 88: 513-517.

Czapik, R., 1961. Acta Biol. Cracov. Ser. Bot., 4: 97-120.

Czapik, R., 1965. Acta Biol. Cracov., 8: 21-34.

Czapik, R., 1968. Watsonia, 6 (6): 345-349.

Dabrowska, L., 1971. Fragm. Flor. Geobot., 17: 383-386.

Dahlgren, K.V.O., 1952. Hereditas, 38: 314-320.

Dahlgren, R., Karlsson, T. and Lassen, P., 1971. Bot. Not., 124 (2): 249-269.

Daker, M.G., 1965. In: Löve, Å. and Solbrig, O.T., Taxon, 14: 88-89.

D'Amato, F., 1939. Nuovo Giorn. Bot. Ital., N.S., 46: 470-510.

D'Amato, F., 1945. Nuovo Giorn. Bot. Ital., N.S., 52: 86-87.

D'Amato, F., 1946. Nuovo Giorn. Bot. Ital., N.S., 53: 405-437.

D'Amato, F., 1957a. Caryologia, 10: 111-151.

D'Amato, F., 1957b. Caryologia, 9: 315-339.

D'Amato-Avanzi, M.G., 1957. Caryologia 9: 373-375.

Damboldt, J., 1964. Madroño, 17: 266, 268.

Danboldt, J., 1965. Öster. Bot. Zeit., 112: 392-406.

Damboldt, J., 1966. Ber. Deutsch. Bot. Ges., 78 (9): 373-376.

Damboldt, J., 1968. Taxon, 17: 91-104.

Domboldt, J., 1968b. Ber. Deutsch. Bot. Ges., 81: 43-52.

Damboldt, J., 1971. Taxon, 20: 787.

Damboldt, J. and Phitos, D., 1975. Pl. Syst. Evol., 123: 119-131.

Damboldt, J. and Podlech, D., 1963. Ber. Bay. Bot. Ges., 36: 29-32.

Damboldt, J., Graumann, G. and Matthäs, U., 1973. In: Löve, Å., Taxon, 22: 116.

Dangeard, P., 1941. Compt. Rend. Soc. Biol. Paris, 135: 581-583.

Dark, S.O.S., 1932. Ann. Bot., 46: 965-977.

Datta, R.M., 1932. Mem. Mchtr. Lit. Phil. Soc., 76: 85.

Davies, E.W., 1953. Nature (Lond.), 171: 659-660.

Davies, E.W., 1955. Watsonia, 3: 129-138.

Davies, E.W., 1956a. Watsonia, 3: 242-243.

Davies, E.W., 1956b. Hereditas, 42: 349-366.

Davies, E.W. and Willis, A.J., 1959. Proc. Bot. Soc. Brit. Is., 3: 328.

Delay, C., 1947. Rev. Cytol. Cytophys. Végét., 9: 169-223, 10: 103-229.

Delay, J., 1967. Inf. Ann. Caryosyst. Cytogénét., 1: 11-14.

Delay, J., 1968. Inf. Ann. Caryosyst. Cytogénét., 2: 17-22.

Delay, J., 1969. Inf. Ann. Caryosyst. Cytogenet., 3: 6, 7, 13, 17-27.

Delay, J., 1970. Inf. Ann. Caryosyst. Cytogenet., 4: 1-16.

Delay, J., 1971. Inf. Ann. Caryosyst. Cytogenet., 5: 37-40.

Del Prete, C., 1977. Inf. Bot. Ital., 9: 135-140.

Dersch, G., 1964. Ber. Deutsch. Bot. Ges., 76: 351-359.

Diannelidis, T., 1951. Portug. Acta Biol., Ser. A, 3: 151-170.

Dietrich, J., 1967b. Inf. Ann. Caryosyst. Cytogenet., 1: 23-28.

Dietrich, J., 1969. Inf. Ann. Caryosyst. Cytogenet., 3: 33-34.

Dietrich, J., 1970. Inf. Ann. Caryosyst. Cytogenet., 4: 29.

Dietrich, M., 1964. Madrono, 17: 266-267.

Dietrich, M., 1967. Feddes Rep., 75: 1-42.

Dietrich, M., 1972. In: Löve, Á., Taxon, 21: 333-346.

Dittrich, M., 1968. Öesterr. Bot. Zeit., 115: 379-390.

Dolcher, T. and Pignatti, S., 1971. Nuovo Giorn. Bot. Ital., 105: 95-107.

Dominguez, E. and Galiano, E.F., 1974. Lagascalia, 4: 61-84.

Donadille, P., 1967. Compt. Rend. Acad. Sc. Paris, 264: 813-816.

Doulat, E., 1943. Bull. Soc. Sci. Dauphine, 61: 1-232.

Dovaston, H.E., 1955. Notes Roy. Bot. Gard. Edinb., 21: 289-291.

Duckert, M.M. and Favarger, C., 1960. Bull. Soc. Neuchat. Sci. Nat., 83:
    109-119.

Durand, B., 1962. Rev. Cytol. Biol. Veg., 25: 337-341.

Eaton, R.D., 1968. Cytological Studies in the Genus Primula. Ph.D.
    thesis, University of Durham.

Edmonds, J.M., 1977. Bot. Jour. Linn. Soc., London, 75: 141-178.

Edmonds, J.M., Sell, P.D. and Walters, S.M., 1974. Watsonia, 10:
    159-161.

Egolf, D.R., 1962. Journ. Arn. Arbor., 43: 132-172.

Ehrenberg, L., 1945. Bot. Not., 1945: 430-437.

Ehrendorfer, F., 1952. Carnegie Inst. Yrbook, 51: 125-131.

Ehrendorfer, F., 1953. Öesterr. Bot. Zeits., 100: 670-672.

Ehrendorfer, F., 1955. Öesterr. Bot. Zeits., 102 (2-3): 195-234.

Ehrendorfer, F., 1959. Öesterr. Bot. Zeits., 106: 363-368.

Ehrendorfer, F., 1962. Ber. Deutsch. Bot. Ges., 75 (5): 137-152.

Ehrendorfer, F., 1962. Öesterr. Bot. Zeits., 109 (3): 276-343.

Ehrendorfer, F., 1963. In: Turrill, W.B. (Ed.) Vistas in Botany, 4: 158.

Ehrendorfer, F., 1964. Öesterr. Bot. Zeits., 111: 84-142.

Ehrendorfer, F., 1974. In: Ehrendorfer, F. and Krendl, F., Bot. Jour.
    Linn. Soc., London, 68: 268-272.

Ehrendorfer, F., 1975. In: Jordanov, D. et al. (Eds.) Problems of Balkan
    Flora and Vegetation. Sofia, Bulgarian Academy of Science, p. 178-185.

Ehrendorfer, F. and Ančev, M.E., 1975. Pl. Syst. Evol., 124: 1-6.

Eklundh-Enrenberg, C., 1946. Bot. Not.,1946: 529-536.

Elkington, T.T., 1974. Watsonia, 10 (1): 80.

Elkington, T.T. and Woodell, S.R.J., 1963. Jour. Ecol., 51: 769-781.

Ellerström, S. and Tjio, J.H., 1950. Bot. Not., 1950: 463-465.

Ellis, R.P. and Jones, B.M.G., 1970. Watsonia, 8: 45.

Engel, K., 1972. Taxon, 21: 495-500.

Engelskjon, T., 1967. Nytt Mag. Bot., 14: 125-139.

Erben, M., 1978. Mitt. Bot. Staatssamm., München, 14: 361-631.

Erdtman, G. and Nordborg, G., 1961. Bot. Not. 114: 19-21.

Ernet, D., 1972. Taxon, 21: 495-500.

Fabbri, F., 1957. Caryologia, 10: 388-390.

Fabergé, A.C., 1943. Jour. Genet., 45: 139-170.

Fagerlind, F., 1934. Hereditas, 19: 223-232.

Fagioli, A. and Fabbri, F., 1971. Inf. Bot. Ital., 3: 55-62.

Fagerlind, A., 1968. Blyttia, 26: 112-123.

Fahmy, T.Y., 1955. Rec. Trav. Lab. Bot. Geol. Zool. Fac. Sci. Univ.
    Montpellier, 7: 103-114.

Faulkner, J.S., 1972. Bot. Jour. Linn. Soc., London, 65: 271-301.

Faurg, J. and Pietrera, D., 1969. Ann. Fac. Sci. Marseilles, 42: 271-283.

Favarger, C., 1946. Ber. Schweiz. Bot. Ges., 56: 365-467.

Favarger, C., 1949a. Bull. Soc. Neuchât. Sci. Nat., 72: 15-22.

Favarger, C., 1949b. Ber. Schweiz Bot. Ges., 59: 62-86.

Favarger, C., 1952. Ber. Schweiz. Bot. Ges., 62: 244-257.

Favarger, C., 1953. Bull. Soc. Neuchât. Sci. Nat., 76: 133-169.

Favarger, C., 1954. VIII Congr. Internat. Bot. Compt. Rendus Rapp. and
    Comm., 9-10: 51-56.

Favarger, C., 1958. Veröffentl. Geobot. Inst. Rubel in Zurich, 33: 59-80.

Favarger, C., 1959. Bull. Soc. Neuchâtel Sci. Nat., 82: 255-285.

Favarger, C., 1960. Bull. Soc. Bot. France, 107: 94-98.

Favarger, C., 1961. Phyton, 9: 252-256.

Favarger, C., 1962. Bull. Soc. Neuchâtel. Sci. Nat., 85: 53-81.

Favarger, C., 1962b. In: Contandriopoulos, J., Ann. Fac. Sci. Marseilles, 32.
    1 761.

Favarger, C., 1965. Bot. Not., 118: 273-280.

Favarger, C., 1965b. Bull. Soc. Neuchâtel. Sci. Nat., 88: 5-60.

Favarger, C., 1965c. Monde Plantes, 348: 1-4.

Favarger, C., 1969a. Bull. Soc. Neuchâtel. Sci. Nat., 92: 13-30.

Favarger, C., 1969b. Taxon, 18: 434-435.

Favarger, C., 1969c. Acta Bot. Croat., 28: 63-74.

Favarger, C., 1971. Taxon, 20: 351.

Favarger, C., 1975. Anal. Inst. Bot. A.J. Cavanilles, 32: 1209-1243.

Favarger, C., 1975b. In: Löve, Á., Taxon, 24: 144.

Favarger, C. and Contandriopoulos, J., 1961. Ber. Schweiz. Bot. Ges., 71:
    384-408.

Favarger, C. and Huynh, K.L., 1964. Taxon, 13 (6): 201-209.

Favarger, C. and Küpfer, P., 1967. Compt. Rend. Acad. Sci. Paris, 264: 2463.

Favarger, C. and Küpfer, P., 1968. Collect. Bot., 7 (1): 325-357.

Favarger, C. and Küpfer, P., 1970. Bull. Soc. Bot. Suisse, 80: 269-288.

Favarger, C. and Villard, M., 1965. Bull. Soc. Bot. Suisse, 75: 57-79.

Fearn, G.M., 1977. Watsonia, 11: 254.

Felföldy, L.J.M., 1947. Arch. Biol. Hung., 17: 101-103.

Felföldy, L.J.M., 1947b. Rep. Hung. Agric. Exp. Sta., 47-49: 11-16.

Ferakova, V., 1968a. Biologia Bratislava, 23: 317.

Ferakova, V., 1968b. Folia Geobot. Phytotax., Praha, 3: 111-116.

Fernandes, A., 1931. Bol. Soc. Brot., Sér. 2, 7: 3-110.

Fernandes, A., 1937a. Bol. Soc. Brot., Sér. 2, 12: 159-219.

Fernandes, A., 1937b. Bol. Soc. Brot., Sér. 2, 12: 93-116.

Fernandes, A., 1939. Bol. Soc. Brot., Sér. 2, 13: 487-542.

Fernandes, A., 1950. Agron. Lusit., 12: 551-600.

Fernandes, A., 1950b. Rev. Fac. Ci. Univ. Coimbra, 19: 5-39.

Fernandes, A., 1959. Compt. Rend. Acad. Sci., Paris, 248: 3672-3675.

Fernandes, A., 1966. Bol. Soc. Brot., Sér. 2, 40: 207-261.

Fernandes, A., 1967. Portug. Acta Biol., Sér. B, 9: 1-44.

Fernandes, A., 1968. Collect. Bot. (Barcelona), 7: 381-392.

Fernandes, A. and Fernandes, R., 1945. Herbertia, 12: 85-96.

Fernandes, A. and Fernandes, R., 1946. Acta Univ. Conimbrig., 1: 1-33.

Fernandes, A. and Franca, F., 1972. Genet. Iber., 24: 31-36.

Fernandes, A. and Franca, F., 1973. Bol. Soc. Brot., Sér. 2, 46: 465-501.

Fernandes, A., Garcia, J. and Fernandes, R., 1948. Mem. Soc. Brot., 4:
    5-89.

Fernandes, A. and Queirós, M., 1969. Bol. Soc. Brot., Sér. 2, 43: 20-140.

Fernandes, A. and Queirós, M., 1970/1. Mem. Soc. Brot., 21: 343-385.

Fernandes, A. and Queirós, M., 1971. Bol. Soc. Brot., Sér. 2, 45: 5-121.

Fernandes, A. and Santos, M.F., 1971. Bol. Soc. Brot., Sér. 2, 45: 177-226.

Fernandes, R., 1970. Bol. Soc. Brot., Sér. 2, 44: 343-367.

Fernandes, R. and Leitao, M.T., 1969. An. Est. Exp. Aula Dei, 9 (2-4):
    210-222.

Fernandes, R.B., 1969. Bol. Soc. Brot., Ser. 2, 43: 145-158.

Fischer, M., 1967. Öesterr. Bot. Zeits., 144: 189-233.

Fischer, M., 1969. Öesterr. Bot. Zeits., 116: 430-443.

Fischer, M., 1970. Öesterr. Bot. Zeits., 118: 206-215.

Flovik, K., 1938. Hereditas, 24: 265-376.

Flovik, K., 1940. Hereditas, 26: 430-440.

Flovik, K., 1943. Nytt Mag. Naturv., 83: 77-78.

Fothergill, P.G., 1936. Proc. Univ. Durham Phil. Soc. 9: 205-216.

Fothergill, P.G., 1944. New Phytol., 43: 23-35.

Francini, E., 1927. Nuovo Giorn. Bot. Ital., 34 (2): 381-395.

Frey, L., 1969a. Fragm. Flor. Geobot., 15: 179-184.

Frey, L., 1969b. Fragm. Flor. Geobot., 15: 261-267.

Frey, L., 1970. Fragm. Flor. Geobot., 16: 391-394.

Frey, L., 1971. Fragm. Flor. Geobot., 17: 251-256.

Frisendahl, A., 1927. Acta Horti. Getob., 3: 99-142.

Fritsch, R.M., 1973, In· Löve, Á , Taxon, LL. 401.

Fuchs, A., 1938. Öesterr. Bot. Zeits., 87: 1-41.

Furnkranz, D., 1965. Öesterr. Bot. Zeits., 112: 421-423.

Furnkranz, D., 1967. Öesterr. Bot. Zeits., 114: 341-345.

Gadella, T.W.J., 1963. Acta Bot. Neerland., 12: 17-39.

Gadella, T.W.J., 1964. Wentia, 11: 1-104.

Gadella, T.W.J., 1966. Proc. Koni. Nederl. Akad. Wetenschappen., Ser. C,
     69 (4): 502-521.

Gadella, T.W.J., 1972. Bot. Not., 125: 361-369.

Gadella, T.W.J. and Kliphuis, E., 1963. Acta Bot. Neerland., 12: 195-230.

Gadella, T.W.J. and Kliphuis, E., 1966. Proc. Kon. Nederl. Akad. Wetenschappen.,
     Ser. C., 69: 541-556.

Gadella, T.W.J. and Kliphuis, E., 1967. Proc. Kon. Nederl. Akad. Wetenschappen.,
     Ser. C., 70: 7-20.

Gadella, T.W.J. and Kliphuis, E., 1968. Taxon, 17: 199-204.

Gadella, T.W.J. and Kliphuis, E., 1968b. Proc. Kon. Nederl. Akad. Wetenschappen.,
    Ser. C., 71: 168-183.

Gadella, T.W.J. and Kliphuis, E., 1970. Caryologia, 23: 363-379.

Gadella, T.W.J. and Kliphuis, E., 1972. Acta Bot. Croat., 31: 91-103.

Gadella, T.W.J. and Kliphuis, E., 1973. Kon. Nederl. Akad. Wetenschappen.,
    Ser. C., 76: 303-311.

Gadella, T.W.J., Kliphuis, E. and Mennega, E.A., 1966. Acta Bot. Neerl.,
    15 (2): 484-489.

Gagnieu, A., Bouteiller, J. and Linder, R., 1959. Bull. Soc. Bot. Fr.,
    106: 148-154.

Gajewski, W., 1947. Acta Soc. Bot. Polon., 18: 33-44.

Gajewski, W., 1957. Monogr. Bot., 4: 1-416.

Gale, J.S. and Rees, H., 1970. Heredity, 25: 393-410.

Gamisans, J., 1973. Candollea, 28: 39-82.

Gamisans, J., 1974. Candollea, 29: 39-55.

Garbari, F., 1969. Nuovo Giorn. Bot. Ital., 103: 1-9.

Garbari, F., 1969. Nuovo Giorn. Bot. Ital., 103: 613.

Garbari, F., 1973. Webbia, 28: 57-80.

Garbari, F., Greuter, W. and Miceli, P., 1979. Webbia, 34: 459-480.

Garbari, F. and Martino, A. di, 1972. Webbia, 27: 289-297.

Garbari, F. and Tornadore, N., 1970. Inf. Bot. Ital., 2: 79-82.

Garbari, F. and Tornadore, N., 1971. Inf. Bot. Ital., 3: 153-154.

Garbari, F. and Tornadore, N., 1972. Nuovo Giorn. Bot. Ital., 106: 285.

Garbari, F. and Tornadore, N., 1972b. Inf. Bot. Ital., 4: 65.

Garbari, F., Tornadore, N. and Pecori, E., 1973. Inf. Bot. Ital., 5:
    161-169.

Gardé, A., 1952. Genet. Iber., 3: 133-144.

Gardé, A. and Malheiros-Gardé, N., 1949. Agron. Lusit., 11: 91-140.

Gardé, A. and Malheiros-Gardé, N., 1953. Genet. Iber., 5: 115-124.

Gardé, A. and Malheiros-Gardé, N., 1954. Broteria, 23: 5-35.

Gardou, C., 1972. In: Löve, Á., Taxon, 21: 495-500.

Gauger, W., 1937. Planta, 26: 529-531.

Geitler, L., 1929. Züchter, 1: 243-247.

Gentscheff, G., 1937. Planta, 27: 165-195.

Gervais, C., 1965. Bull. Soc. Neuchât. Sci. Nat., 88: 61-64.

Gervais, C., 1966. Bull. Soc. Neuchât. Sci. Nat., 89: 87-100.

Gervais, C., 1968. Bull. Soc. Neuchât. Sci. Nat., 91: 105-117.

Gervais, C., 1972. Bull. Soc. Neuchât. Sci. Nat., 95: 57-61.

Gervais, C., 1973. Denkschr. Schweiz. Naturf. Ges., 88: 1-56.

Giorgi, B. and Bozzini, A., 1969. Caryologia, 22: 249-259.

Gill, J.J.B., 1965. Watsonia, 6: 188-189.

Glasau, F., 1939. Planta, 30: 507-550.

Glendinning, D.R., 1960. Nature (London), 188: 604-605.

Gorenflot, R., 1960. Rév. Cytol. Biol. Vég., 22: 79-108.

Gorenflot, R., Raicu, P. and Cartier, D., 1972. Compt. Rend. Acad. Sci.
    Paris, 274: 391 307

Gorenflot, R., Raicu, P., Cartier, D., et al., 1973b. Compt. Rend. Acad.
    Sci., Paris, 274: 1501-1504.

Gori, C., 1954. Caryologia, 6: 241-246.

Gori, C., 1957. Caryologia, 10: 454-456.

Gori, C., 1958. Caryologia, 11: 28-33.

Grant, W.F., 1959. Canad. Jour. Genet. Cytol., 1: 313-328.

Grau, J., 1964a. Madroño, 17: 266-268.

Grau, J., 1964b. Öesterr. Bot. Zeits., 111: 561-617.

Grau, J., 1967a. Mitt. Bot. Staatssamm., München, 6: 517-530.

Grau, J., 1967b. Öesterr. Bot. Zeits., 114: 66-72.

Grau, J., 1968a. Mitt. Bot. Staatssamm., München, 7: 17-100.

Grau, J., 1968b. Mitt. Bot. Staatssamm., München, 7: 277-294.

Grau, J., 1970. Mitt. Bot. Staatssamm., München, 8: 127-136.

Grau, J., 1971. Mitt. Bot. Staatssamm., München, 9: 177-194.

Grau, J., 1976. Mitt. Bot. Staatssamm., München, 12: 609-654.

Graze, H., 1933. Jahrb. Wiss. Bot., 77: 507-559.

Greilhuber, J. and Ehrendorfer, F., 1975. Pl. Syst. Evol., 124: 125-138.

Griesinger, R., 1937. Ber. Deutsch. Bot. Ges., 55: 556-571.

Griff, V.G., 1965. Bot. Zhur., 50: 1133.

Groll, M., 1965. Österr. Bot. Zeit., 112: 657-700.

Gross, R., 1966. Ber. Schweiz. Bot. Ges., 76: 306-351.

Guinochet, M., 1943. Rév. Cytol. Cytophysiol. Vég., 6: 209-220.

Guinochet, M., 1957. Bull. Soc. Hist. Nat. Afr. Nord., 48: 282-300.

Guinochet, M., 1964. Compt. Rend. Acad. Sci. Paris, 259: 3817-3819.

Guinochet, M., 1967. Compt. Rend. Acad. Sci. Paris, 264: 1623-1625.

Guinochet, M. and Foissac, J., 1962. Rév. Cytol. Biol. Végetal., 25 (3-4):
    373-387.

Guinochet, M. and Lefrane, M., 1972. In: Löve, Á., Taxon, 21: 495-500.

Guinochet, M. and Lemee, G., 1950. Rév. Gén. Bot., 57: 565-593.

Guinochet, M. and Logeois, A., 1962. Rév. Cytol. Biol. Vég., 25: 465-480.

Guittonneau, G., 1964. Bull. Soc. Bot. France, 111: 1-4.

Guittonneau, G., 1965. Bull. Soc. Bot. France, 112: 25.

Guittonneau, G., 1966. Bull. Soc. Bot. France, 111: 3-11.

Guittonneau, G., 1967. Bull. Soc. Bot. France, 114: 32-41.

Gustaffson, Å., 1932. Hereditas, 16: 41-62.

Gustaffson, Å., 1933. Bot. Gaz., 94: 512-533.

Gustaffson, Å., 1935. Hereditas, 20: 1-31.

Gustaffson, Å., 1939. Hereditas, 25: 33-47.

Gustaffson, Å., 1943. Acta Univ. Lund, II, 39 (2): 1-200.

Gustaffson, M. and Snogerup, S., 1972. Bot. Not., 125: 323.

Guşuleac, M. and Tarnavschi, I.T., 1935. Bull. Fac. Stiinţ. Cernauti, 9:
    387-400.

Gutermann, W., 1961. In: Löve, Á. and Löve, D., Opera Bot. 5.

Hadac, E. and Haskova, V., 1956. Biologia (Bratislava), 11: 717-723.

Häfliger, E., 1943. Ber. Schweiz. Bot. Gesell., 53: 317-382.

Hagerup, O., 1928. Dansk Bot. Arkiv., 6: 1-26.

Hagerup, O., 1932. Hereditas, 16: 19-40.

Hagerup, O., 1939. Hereditas, 25: 185-192.

Hagerup, O., 1940. Hereditas, 26: 399-410.

Hagerup, O., 1941a. Planta, 32: 6-14.

Hagerup, O., 1941b. Bot. Tidsskr., 45: 385-395.

Hagerup, O., 1944. Hereditas, 30: 152-160.

Hagerup, O., 1947. Kung. Dansk. Vidensk. Selsk. Biol. Medd., 20 (9): 1-22.

Hakansson, A., 1944. Hereditas, 30: 639-640.

Hakansson, A., 1953. Bot. Not., 1953: 301-307.

Hakansson, A., 1955. Hereditas, 41: 454-482.

Halkka, O., 1964. Hereditas, 52: 81-88.

Hall, A.D., 1937. Bot. Journ. Linn. Soc. London, 50: 181-409.

Halliday, C., 1900. Watsonia, 4: 207-210.

Halliday, G., 1961. In: Löve, A. and Löve, D., Opera Bot., 5.

Hambler, D.J., 1954. Nature (London), 174: 838.

Hambler, D.J., 1955. Proc. Bot. Soc. Brit. Isles, 1: 384-385.

Hambler, D.J., 1958. Bot. Journ. Linn. Soc. London, 55: 772-777.

Hamel, J.L., 1953. Rev. Cytol. et Cytophysiol., 14: 3-4.

Hamel, J.L., 1954. Mem. Soc. Bot. France, 34: 106-121.

Hämet-Ahti, L. and Virrankoski, V., 1970. Ann. Bot. Fenn., 7: 177-181.

Hamilton, A.P., 1968. Gladiolus Annual, 1968: 27-31.

Hanelt, P., 1963. Feddes Rep., 67: 41-180.

Hanelt, P. and Mettin, D., 1966. Kulturpflanze, 14: 137-161.

Harvey, B.L., 1966. Dissert. Abstr., Ser. B., 27: 365.

Hawkes, J.G., 1956. Proc. Linn. Soc., London, 166 (3): 97-144.

Harley, R.M., 1977. In: Harley, R.M. and Brighton, C.A., Bot. Jour. Linn.
    Soc., 74: 71-96.

Haskell, G., 1964. Jour. Roy. Hort. Soc., 89: 504.

Hedberg, O., 1958. Svensk. Bot. Tidskr., 52: 37-46.

Hautzinger, L., 1976. Pl. Syst. Evol., 124: 311-313.

Hedberg, I., 1970. Hereditas, 64: 153-176.

Hedberg, I. and Hedberg, O., 1961. Bot. Not.: 114: 397-99.

Hedberg, I. and Hedberg, O., 1964. Svensk. Bot. Tidskr., 58: 125-128.

Heilborn, O., 1918. Svensk. Bot. Tidskr., 12: 212-220.

Heilborn, O., 1924. Hereditas, 5: 129-216.

Heilborn, O., 1927. Hereditas, 9: 59-68.

Heilborn, O., 1928. Hereditas, 11: 182-192.

Heitz, B., 1967. Inf. Ann. Caryosyst. Cytogén., 1: 22.

Helms, A. and Jorgensen, C.A., 1925. Bot. Tidsskr., 39: 57-134.

Heneen, W.K., 1962. Hereditas, 48: 471-502.

Heneen, W.K. and Runemark, H., 1962. Hereditas, 48: 545-564.

Henin, W., 1960. In: Boulos, L., Bot. Not., 113: 400-420.

Heppell, R., 1945. In: Darlington, C.D. and Janaki-Ammal, E.K., Chromosome
    Atlas of Cultivated Plants, London, Allen and Unwin.

Heslop-Harrison, J., 1961. In: Löve, Á. and Löve, D., Opera Bot. 5.

Heslop-Harrison, Y., 1953. New Phytol., 52: 22-39.

Heslop-Harrison, Y., 1953b. Watsonia, 3: 7-25.

Heslop-Harrison, Y., 1955. Jour. Ecol., 43: 719-734.

Heslot, M.H., 1953. Compt. Rend. Acad. Sci. (Paris), 237: 2477-2479.

Heusser, C., 1915. Beih. Bot. Centralbl., 32: 218-277.

Heusser, C., 1938. Bull. Soc. Bot. Suisse, 48: 562-599.

Heywood, V.H., 1955. In: Turrill, W.B., Bot. Mag., 170: t.246.

Heywood, V.H. and Walker, S., 1961. Nature (London), 189: 604.

Hjelmquist, H., 1959. Bot. Not., 112: 17-64.

Hofelich, A., 1935. Jahrb. Wiss. Bot., 81: 541-572.

Holmen, K., 1961. In: Löve, Á. and Löve, D., Opera Bot. 5.

Holub, J. and Křísa, B., 1971. Folia Geobot. Phytotax. (Praha), 6: 82.

Holub, J., Měsíček, J. and Javůrková, V., 1970. Folia Geobot. Phytotax.,
    Praha, 5: 339-368.

Holub, J., Měsíček, J. and Javůrková, V., 1971. Folia Geobot. Phytotax.,
    Praha, 6: 179-214.

Holub, J., Měsíček, J. and Javůrková, V., 1972. Folia Geobot. Phytotax.,
    Praha, 7: 167-202.

Honsell, E., 1959. Ann. Bot. (Roma), 26: 177-188.

Hooper, M.D., 1961. In: Löve, A. and Löve, D., Opera Bot., 5.

Horánszky, A., Jankó, B. and Vida, G., 1972. Symp. Biol. Hung., 12: 177-182.

Horn, K., 1948. Blyttia, 6: 1-13.

Hou-Liu, P.Y., 1963. Acta Bot. Neerland., 12: 76-83.

Hovin, A.W. and Hill, H.D., 1966. Amer. Jour. Bot., 53: 702-708.

Howard, H.W., 1946. Nature (London), 158: 666.

Howard, H.W., 1947. Nature (London), 159: 66.

Hrubý, K., 1932. Preslia, 11: 40-44.

Hrubý, K., 1934. Bull. Intern. Acad. Czeck. Sci. Cl. Sci, Mat., Med.
    Prague, 35: 124-132.

Hubbard, C.D., 1954. Grasses. Penguin Books, Harmondsworth.

Hultén, E., 1945. Bot. Not., 1945: 127-148.

Hylander, N., 1957. Bull. Hard. Bot. Bruxelles, 27: 591-604.

Iljin, M.M. and Trukhaleva, N.A., 1960. Doklady Akad. SSSR, 132: 217-219.

Ingram, R., 1967. Watsonia, 6: 283-289.

Jacobsen, P., 1954. Hereditas, 40: 252-254.

Jalas, J., 1948. Hereditas, 34: 414-434.

Jalas, J., 1950. Ann. Bot. Soc. Vanamo, 24: 1-362.

Jalas, J. and Kaleva, K., 1966. Ann. Bot. Fenn., 3: 123-127.

Jalas, J. and Kaleva, K., 1967. Ann. Bot. Fenn., 4: 74-80.

Jalas, J. and Pohjo, T., 1965. Ann. Bot. Fenn., 2: 169-170.

Jalas, J. and Rönkkö, M.T., 1960. Arch. Soc. Zoo. Bot. Fenn. 'Vanamo',
    14 (2): 112-116.

384

Janaki-Ammal, E.K. and Saunders, B., 1952. Kew Bull., 1952: 539-542.

Jaretsky, R., 1928a. Jahrb. Wiss. Bot., 68: 1-45.

Jaretsky, R., 1928b. Jahrb. Wiss. Bot., 69: 357-490.

Jasiewicz, A. and Mizianty, M., 1975. Fragm. Fl. Geobot., 21: 277-288.

Johnson, M.A.T., 1982. Ann. Mus. Goulandris, 5: In Press.

Johnsson, H., 1941. Lunds Univ. Aarskr., Avd. 2, 37 (3): 1-43.

Johnsson, H., 1942. Hereditas, 28: 228-230.

Jones, K., 1955. In: Darlington, C.D. and Wylie, A.P., Chromosome Atlas
    of Flowering Plants.

Jones, K., 1958. New Phytol., 57: 191-210.

Jones, K., 1959. Welsh Pl. Breed. Stat. Rep. 1956-58: 68-71.

Jones, K. and Smith, J.B., 1966. Kew Bull., 20: 63-72.

Jonsell, B., Pálsson, J. and Portén, E.K., 1975. Svensk Bot. Tidskr., 69: 113-142.
    113-142.

Jørgensen, C.A., 1927. Jour. Genet., 18: 63-75.

Jørgensen, C.A., Sorensen, T. and Westergaard, M., 1958. Dansk Vid. Selsk.
    Biol. Skr., 9 (4): 1-172.

Juhl, H., 1952. Ber. Deutsch. Bot. Ges., 65: 330-337.

Kachidze, N., 1929. Planta, 7: 482-502.

Kaleva, K., 1969. Ann. Bot. Fennica, 6: 344-347.

Katznelson, J. and Zohary, D., 1967. Israel Jour. Bot., 16: 57-62.

Kay, Q.O.N., 1969. Watsonia, 7: 130-141.

Kerguélen, M., 1975. Lejeunia, 75: 9-16, 32-44, 145-182.

Kerguélen, M., 1976. Bull. Soc. Bot. Fr., 123: 317-324.

Khidir, M.O. and Knowles, P.F., 1970. Amer. Jour. Bot., 57: 123-129.

Khoshoo, T.N., 1959. Caryologia, 11: 297-318.

Kjellmark, S., 1934. Bot. Not. 1934: 136-149.

Kjellqvist, E., 1964. Bot. Not., 117: 389-396.

Kjellqvist, E. and Löve, Á., 1963. Bot. Not., 116: 241-248.

Kliphuis, E., 1962. Acta Bot. Neerl., 11: 90-92.

Kliphuis, E., 1962b. Proc. Kon. Akad. Wetenschap. Ser. C, 65 (3): 279-285.

Kliphuis, E., 1963. Med. Bot. Mus. Herb. Riksuniv. Utrecht, 195: 172-194.

Kliphuis, E., 1967. Acta Bot. Neerl., 15: 535-538.

Kliphuis, E., 1977. In: Löve, Á., Taxon, 26: 267-268.

Kliphuis, E. and Wieffering, J.H., 1972. Acta Bot. Neerl., 21: 598-604.

Knaben, G., 1948. Blyttia, 6: 17-32.

Knaben, G., 1950. Blyttia, 8: 129-155.

Knaben, G., 1954. Nytt Mag. Bot., 3: 117-138.

Knaben, G., 1958. Blyttia, 16: 61-80.

Knaben, G., 1959. Opera Bot., 2 (3): 1-74.

Knaben, G. and Engelskjön, T., 1967. Acta Boreali, 21: 1-57.

Knaben, G. and Engelskjon, T., 1968. Acta Univ. Berg. No. 4: 0-71.

Kollmann, F., 1973. Israel Jour. Bot., 22: 92-115.

Kollmann, F. and Feinbrun, N., 1968. Notes Roy. Bot. Gard. Edinb., 28 (2): 1ꟼꞡ⹀₁₀₀,

Kovanda, M., 1966. Preslia (Praha), 38: 48-52.

Kovanda, M., 1970a. In: Löve, Á., Taxon, 19: 106.

Kovanda, M., 1970b. Roz. Ceskosl. Akad. Ved., 80: 1-95.

Kožuharov, S. and Kuzmanov, B., 1962. Ann. Univ. Sofia, 57: 103-107.

Kožuharov, S. and Kuzmanov, B., 1964. Compt. Rend. Akad. Bulgar. Sc., 17: 491-494.

Kožuharov, S. and Kuzmanov, B., 1964b. Ann. Univ. Sofia, 57: 103-107.

Kožuharov, S. and Kuzmanov, B., 1965a. Mitt. Bot. Inst. Sofia, 15: 255-258.

Kožuharov, S. and Kuzmanov, B., 1965b. Caryologia, 18 (2): 349-351.

Kožuharov, S. and Kuzmanov, B., 1967. Compt. Rend. Acad. Bulg. Sci., 20: 469-472.

Kožuharov, S. and Kuzmanov, B., 1968. In: Löve, Á., Taxon, 17: 199-204.

Kožuharov, S. and Kuzmanov, B., 1969. Gen. Pl. Breeding, 2 (3): 181-185.

Kožuharov, S. and Kuzmanov, B., 1970. Mitt. Bot. Inst. Sofia, 20: 99-107.

Kožuharov, S. and Nicolova, T., 1975. In: Problems of Balkan Flora and
    Vegetation. Proc. 1st Internat. Symp. Balkan Flora and Veg. Varna,
    June, 1973- pp. 69-77. Sofia, Bulgarian Acad. Sciences.

Kožuharov, S. and Petrova, A.V., 1973. Compt. Rend. Acad. Bulg. Sci., 26 (6):
    829-832.

Kožuharov, S. and Petrova, A.V., 1974. In: Löve, Á., Taxon, 23: 377.

Kožuharov, S. and Petrova, A.V., 1975. In: Löve, Á., Taxon, 24: 369.

Kožuharov, S., Petrova, A.V. and Stoeva, M.P., 1974. Comp. Rend. Acad.
    Bulg. Sci., 27: 383-386.

Kožuharov, S., Popova, M. and Kuzmanov, B., 1968. Genetics Pl. Breeding,
    1 (3): 251-255.

Krendl, F., 1967. Öesterr. Bot. Zeits., 114: 508-549.

Krendl, F., 1974. In: Ehrendorfer, F. and Krendl, F., Bot. Jour. Linn.
    Soc., 68: 268-272.

Krendl, F., 1976. Ann. Natürhist. Mus. Wien, 80: 67-86.

Kress, A., 1963. Ber. Bay. Bot. Gesell., 36: 33-39.

Kress, A., 1963. Phyton, 10: 225-236.

Kubién, E., 1968. In: Skalińska, M., Pogan, E. and Jankun, A., Acta Biol.
    Cracov., Bot., 11: 199-224.

Kuhn, E., 1928. Jahrb. Wiss. Bot., 68: 382-430.

Küpfer, P., 1968. Bull. Soc. Neuchat. Sci. Nat., 91: 87-104.

Küpfer, P., 1969. In: Löve, A., Taxon, 18: 436-467.

Küpfer, P., 1969. Bull. Soc. Neuchat. Sci. Nat., 92: 31-48.

Küpfer, P., 1971. Acta Colloque Flore Veg. Chaines Alp. Jurass. Paris,
    167-185.

Küpfer, P. and Favarger, C., 1967. Compt. Rend. Acad. Sci. Paris, 264:
    2463-2465.

Kuzmanov, B., 1973. In: Löve, A., Taxon, 22: 651.

Kuzmanov, B. and Ančev, M.E., 1973. In: Löve, A., Taxon, 22: 461.

Kuzmanov, B., Georgieva, S., Nikolova, V. and Penceva, I., 1981. In: Löve,
    Á., Taxon, 30: 701-702.

Kuzmanov, B. and Jurukova, P., 1977. In: Löve, A., Taxon, 26: 557-565.

Kuzmanov, B. and Kožuharov, S., 1967. Compt. Rend. Acad. Bulg. Sci., 20 (5): 469-472.

Kuzmanov, B. and Kožuharov, S., 1970. In: Löve, Á., Taxon, 19: 265-266.

Kuzmanov, B., Popova, M. and Kožuharov, S., 1968. Genet. Pl. Breed. (Sofia), 2: 325-330.

Laane, M.M., 1966. Blyttia, 24: 270-276.

Laane, M.M., 1971. Blyttia, 29 (4): 229-234.

Labadie, J.P., 1976. In: Löve, Á., Taxon, 25: 636-639.

Ladizinski, G., 1971. Israel Jour. Bot., 20: 24.

Lambert, A.-M., 1969. Inf. Ann. Caryosyst. Cytogén., 3: 3.

Lambert, A.-M. and Giesi, J., 1967. Inf. Ann. Caryosyst. Cytogén., 1: 27-35.

Landolt, E., 1954. Ber. Schweiz. Bot. Ges., 64: 9-83.

Landolt, E., 1956. Ber. Schweiz. Bot. Ges., 66: 92-117.

Langlet, O.F.J., 1927. Svensk. Bot. Tidskr., 21: 1-17.

Langlet, O.F.J., 1932. Svensk. Bot. Tidskr., 26: 381-400.

Langlet, O.F.J., 1936. Svensk. Bot. Tidskr., 30: 288-294.

Langlet, O.F.J. and Söderberg, E., 1927. Acta Horti. Berg., 9: 85-104.

Larsen, K., 1954. Bot. Tidsskr., 50: 163-174.

Larsen, K., 1955a. Bot. Tidsskr., 52: 8-17.

Larsen, K., 1955b. Bot. Not., 108: 263-275.

Larsen, K., 1956a. Bot. Not., 109: 293-307.

Larsen, K., 1956b. Bot. Tidsskr., 53: 49-56.

Larsen, K., 1957a. Bot. Tidsskr., 53: 291-294.

Larsen, K., 1957b. Bot. Not., 110: 265-270.

Larsen, K., 1958a. Bot. Tidsskr., 54: 44-56.

Larsen, K., 1958b. Bot. Not., 111: 301-305.

Larsen, K., 1960. Bot. Tidsskr., 55: 313-315.

Larsen, K., 1963. Dansk. Bot. Ark., 20: 211-275.

Larsen, K., 1965. In: Löve, Á. and Solbrig, O.T., Taxon, 14: 91.

Larsen, K. and Laegaard, S., 1971. Bot. Tidsskr., 66: 249-268.

Larter, L.N.H., 1952. Jour. Genet., 26: 255-283.

Lazare, J.-J., 1976. In: Löve, Á., Taxon, 25: 483.

Lesins, K. and Lesins, J., 1961. Canad. Jour. Genet. Cytol., 3: 7-9.

Leute, G.-H., 1974. In: Löve, Á., Taxon, 23: 811-812.

Levan, A., 1940. Hereditas, 26: 317-320.

Lévêque, M. and Gorenflot,,R., 1969. Bull. Soc. Bot. Nord., France, 22:
    27-58.

Liljefors, A., 1953. Act. Hort. Berg., 16: 277-329.

Lindqvist, K., 1960. Hereditas, 46: 75-151.

Litardière, R. de, 1943. Boissiera, 7: 155-165.

Litardière, R. de, 1945. Rev. Bot. Appl. Agric. Trop., 25: 16-18.

Litardière, R. de, 1947. Candollea, 11: 175-227.

Litardière, R. de, 1949. Mém. Soc. Hist. Nat. Afr. Nord., 2: 199-208.

Litardière, R. de, 1950. Bol. Soc. Brot., Sér. 2, 24: 79-87.

Lohammon, G., 1942. In: Löve, Á. and Löve, D., Bot. Not., 1942: 19-59.

Lopane, F., 1951. Caryologia, 4: 44-46.

Lorenzo-Andreu, A., 1951. An. Aula Dei., 2: 195-203.

Lorenzo-Andreu, A. and Garcia-Sanz, P., 1950. An. Aula Dei., 2: 12-20.

Löve, Á., 1940. Bot. Not., 1940: 157-169.

Löve, Á., 1941. Bot. Not., 1941: 155-172.

Löve, Á., 1942. Hereditas, 28: 289-296.

Löve, Á., 1954a. Vegetatio, 5: 212-224.

Löve, Á., 1954b. VIII Congr. Int. Bot. Rapp. & Comm., 9-10: 59-66.

Löve, Á., 1961. In: Löve, Á. and Löve, D., Opera Bot., 5: 1-581.

Löve, Á., 1967. In: Löve, Á., Taxon, 16: 445-461.

Löve, Á. and Kjellqvist, E., 1964. Portug. Acta Biol., Ser. A., 8: 69-80.

Löve, Á. and Kjellqvist, E., 1973. Lagascalia, 3 (2): 147-182.

Löve, Á. and Löve, D., 1942. Kungl. Fysiografiska Sallsk. Lund Forkadl., 12
    (6): 1-19.

Löve, Á. and Löve, D., 1944. Ark. Bot., 31A, 12: 1-22.

Löve, Á. and Löve, D., 1948. Rep. Dep. Agric. Univ. Inst. Appl. Sci.
     (Iceland), Ser. B., 3: 9-131.

Löve, Á. and Löve, D., 1951. Svenska Bot. Tidskr., 45: 368-399.

Löve, Á. and Löve, D., 1956. Acta Hort. Gotob., 20 (4): 65-290.

Löve, Á. and Löve, D., 1958. Nucleus, 1: 1-10.

Löve, Á. and Löve, D., 1958b. Bot. Not., 111: 376-388.

Löve, Á. and Löve, D., 1961a. Bot. Not., 114: 33-47.

Löve, Á. and Löve, D., 1961b. Opera Bot. 5: 1-581.

Löve, Á. and Löve, D., 1961c. Amer. Fern. Jour., 51: 127-128.

Löve, Á. and Löve, D., 1961d. Bot. Not., 114: 48-56.

Löve, Á., Löve, D. and Raymond, M., 1957. Canad. Jour. Bot., 35: 715-761.

Löve, D., 1942. Svensk Bot. Tidskr., 36: 262-270.

Lovis, J.D., 1963. Brit. Fern. Gaz., 9: 110-113.

Lovka, M. and Sušnik, F., 1973. In: Löve, Á Taxon, 22: 000.

Lovka, M., Sušnik, F., Löve, Á. and Löve, D., 1971. In: Löve, Á., Taxon, 20:
     788-791.

Lovka, M., Sušnik, F., Löve, Á. and Löve, D., 1972. In: Löve, Á., Taxon,
     21: 333-346.

Lövkvist, B., 1956. Symb. Bot. Upsal., 14 (2): 1-131.

Lövkvist, B., 1957. Bot. Not., 110: 237-250.

Lövkvist, B., 1962. Bot. Not., 115: 261-287.

Lövkvist, B., 1963. In: Weimarck, G., Skånes Flora.

Lundqvist, A., 1965. Hereditas, 54: 70-87.

Lungeanu, I., 1967. Acta Bot. Horti Bucurest, 1966: 37-59.

Lungeanu, I., 1971. In: Löve, Á., Taxon, 20: 792.

Lungeanu, I., 1972. In: Löve, Á., Taxon, 21: 679-684.

Lungeanu, I., 1973. In: Löve, Á., Taxon, 22: 651.

Lungeanu, I., 1975. In: Löve, Á., Taxon, 24: 503-504.

Majóvský, J. et al., 1970. Acta Fac. Rer. Nat. Univ. Com. Bot., 18: 45-60.

Majóvský, J. et al., 1974. Acta Fac. Rer. Nat. Univ. Com. Bot., 22: 1-20.

Majóvský, J. et al., 1976. Acta Fac. Rer. Nat. Univ. Com. Bot., 25: 1-18.

Majóvský, J. et al., 1978. Acta Fac. Rer. Nat. Univ. Com. Bot., 26: 1-42.

Malecka, J., 1962. Acta Biol. Cracov., Ser. Bot., 5: 117-136.

Malecka, J., 1967. Acta Biol. Cracov., Ser. Bot., 10: 195-206.

Malecka, J., 1969. Acta Biol. Cracov., Ser. Bot., 12: 57-70.

Malecka, J., 1970. Acta Biol. Cracov., Ser. Bot., 13: 155-168.

Malheiros, N., 1942. Agron. Lusit., 4: 231-236.

Malheiros, N., Castro, D. and Camara, A., 1947. Agron. Lusit., 9: 51-74.

Mansurova, V.V., 1948. Doklady Akad. SSSR, N.S. 61: 119-123.

Manton, I., 1932. Ann. Bot., 46: 509-556.

Manton, I., 1949. Watsonia, 1: 36.

Manton, I., 1950. Problems of Cytology and Evolution in the Pteridophyta,
    Cambridge.

Manton, I., 1955. In: J.E. Lousley (Ed.) Species Studies in the British
    Flora, Bot. Soc. Brit. Isles, London, pp. 90-103.

Manton, I. and Reichstein, T., 1961. Ber. Schweiz. Bot. Ges., 71: 370-383.

Marchant, C.J., 1968. Bot. Jour. Linn. Soc. (Lond.), 60: 411-417.

Marchant, C.J., 1972. Kew Bull., 26: 395-404.

Marchetti, D.B., 1962. Arch. Bot. Biografico, 38: 257-261.

Marchi, P., 1971. Inf. Bot. Ital., 3: 82-94.

Marchi, P., Capineri, R. and D'Amato, G., 1974. Inf. Bot. Ital., 6: 303-312.

Marklund, G., 1931. In: Holmberg, O.R., Skandinaviens Flora 1 (6, 1), 1-160.
    Stockholm.

Markova, M., 1970. Mitt. Bot. Inst. Sofia, 20: 81-92.

Markova, M. and Ančev, M.E., 1973. Compt. Rend. Acad. Bulg. Sci., 26:
    827-828.

Markova, M. and Ivanova, P., 1970. Mitt. Bot. Inst. Sofia, 20: 93-98.

Markova, M. and Ivanova, P., 1971. Mitt. Bot. Inst. Sofia, 21: 123-131.

Markova, M., Radenkova, J. and Ivanova, P., 1972. In: Löve, Á., Taxon, 21: 333-346.

Marks, G.E., 1956. New Phytol., 55: 120-129.

Marsden-Jones, E.M. and Turrill, W.B., 1957. The Bladder Campions. London, Ray Society.

Marsden-Jones, E.M. and Weiss, F.E., 1960. Proc. Linn. Soc. London, 171: 27-29.

Martinoli, G., 1943. Nuovo Giorn. Bot. Ital., 50: 1-23.

Martinoli, G., 1949. Caryologia, 1: 122-130.

Martinoli, G., 1950. Caryologia, 3: 72-78.

Martinoli, G., 1951. Caryologia, 4: 86-97.

Martinoli, G., 1953. Caryologia, 5: 253-281.

Martinoli, G., 1954. VIII Cong. Int. Bot. Compt. Rend. Séances of Rapp. Comm., 9-10: 78-79.

Mattick, G., 1950. In: Tischler, G., Die Chromosomenzahlen der Gefässpflanzen Mitteleuropas, Junk, S-Gravenhage.

Maude, P.F., 1939. New Phytol., 38: 1-31.

Maude, P.F., 1940. New Phytol., 39: 17-32.

McNaughton, I.H., 1960. Scottish Pl. Breed. Stat. Rep., 1960: 76-84.

Maugini, E., 1952. Caryologia, 5: 168.

Maugini, E., 1952b. Caryologia, 5: 101-112.

Maugini, E., 1953. Caryologia, 5: 249-252.

Meili-Frei, E., 1966. Ber. Schweiz. Bot. Ges., 75: 219-292.

Merxmüller, H., 1968. Mitt. Bot. Staatssamm., München, 7: 275.

Merxmüller, H., 1970. Taxon, 19: 140-145.

Merxmüller, H., 1974. Phyton (Austria), 16: 137-158.

Merxmüller, H., 1975. Anal. Inst. Bot. Cavanilles, 32 (2): 189-196.

Merxmüller, H. and Damboldt, J., 1962. Ber. Deutsch. Bot. Ges., 75 (7): 233-236.

Merxmüller, H. and Grau, J., 1963. Ber. Deutsch Bot. Ges., 76: 23-29.

Merxmüller, H. and Grau, J., 1969. Ber. Deutsch. Bot. Ges., 76: 23-29.

Měsíček, J. and Hrouda, L., 1974. Folia Geobot. Phytotax., 9: 359.

Mesquita, J.F., 1964. Bol. Soc. Brot., Sér. 2, 38: 119-136.

Messia, T.P. and Rejon, M.R., 1976. In: Löve, Á., Taxon, 25: 164.

Meusel, H. and Ohle, H., 1966. Österr. Bot. Zeitsch., 113: 191-210.

Meyer, D.E., 1957. Ber. Deutsch. Bot. Ges., 70: 57-66.

Meyer, D.E., 1958. Ber. Deutsch. Bot. Ges., 71: 11-20.

Meyer, D.E., 1959. Willdenowia, 2: 214-218.

Meyer, D.E., 1960. Amer. Fern Jour. 50: 138-145.

Meyer, D.E., 1961. Ber. Deutsch. Bot. Ges., 3: 393.

Meyer, D.E., 1962. Ber. Deutsch. Bot. Ges., 75: 24.

Meyer, D.E., 1967. Ber. Deutsch. Bot. Ges., 80: 28.

Miceli, P. and Garbari, F., 1976. Inf. Bot. Ital., 8: 207-216.

Miceli, P. and Garbari, F., 1979. Atti Soc. Tosc. Sci. Nat., Mem., Ser. B.,
86: 37-51.

Michaelis, P., 1925. Ber. Deutsch. Bot. Ges., 43 (2): 61-67.

Mičieta, K., 1981. Acta Fac. Rer. Nat. Univ. Comen., Bot., 28: 95-103.

Milan, L., 1975. Biol. Vestn. (Ljubljana), 23: 25-40.

Mills, J.N. and Stace, C.A., 1974. Watsonia, 10: 167-168.

Mitra, J., 1956. Bot. Gaz., 117: 265-293.

Mirkovic, D., 1966. Acta Bot. Croat., 25: 137-152.

Mizianty, M. and Frey, L., 1973. Fragm. Flor. Geobot., 19 (3): 265-270.

Montgomery, F.H. and Yang, S.J., 1960. Can. Jour. Bot., 38: 381-386.

Monti, G. and Garbari, F., 1974. Nuovo Giorn. Bot. Ital., 108: 19-26.

Moore, D.M., 1958. Jour. Ecol., 46: 527-535.

Moore, D.M., 1973. In: Webb, D.A. and Halliday, G., Watsonia, 9: 333-344.

Moore, R.J., 1967. In: Löve, A. and Solbrig, O.T., Taxon, 16: 62-66.

Mori, M., 1957. Caryologia 9: 365-368.

Morisset, P., 1964. Proc. Bot. Soc. Brit. Is., 5: 378-379.

Morrison, J.W., 1959. Can. Jour. Bot., 37: 527-538.

Morton, J.K., 1956. Watsonia, 3: 244-252.

Morton, J.K., 1974. Watsonia, 10: 169.

Mosquin, T., 1967. Evolution, 21: 713-719.

Muntzing, A., 1929. Hereditas, 13: 185-341.

Muntzing, A., 1931b. Hereditas, 15: 166-178.

Muntzing, A., 1935. Hereditas, 20: 103-136.

Muntzing, A., 1958. Hereditas, 44: 280-329.

Muntzing, A., 1966. Bull. Bot. Soc. Bengal, 20: 1-15.

Muntzing, A. and Muntzing, G., 1941. Bot. Not., 1941: 237-278.

Muntzing, A. and Nygren, A., 1955. Hereditas, 41: 405-422.

Muntzing, A. and Prakken, R., 1940. Hereditas, 26: 463-500.

Murín, A., 1964. Caryologia, 17: 575-578.

Murín, A. and Májovský, J., 1976. In: Löve, Á., Taxon, 25: 487-489.

Murín, A. and Májovský, J., 1978. In: Löve, Á., Taxon, 27: 376-378.

Murín, A. and Váchová, M., 1970. In: Májovský, J. et al., Acta Fac. Rer. Nat.
    Univ. Comen., 16. 1-20.

Murray, B.G., 1975. In: I.B.K. Richardson, Bot. Jour. Linn. Soc., 71:
    211-234.

Nannfeldt, J.A., 1940. Symb. Bot. Upsal., 4 (4): 1-86.

Nannfeldt, J.A., 1942. In: Löve, Á. and Löve, D., Bot. Not., 1942: 19-59.

Natho, G., 1959. Feddes Rep., 61: 211-273.

Natividade, J.V., 1937. Bol. Soc. Brot., Sér. 2: 12: 21-85.

Negodi, G., 1937. Nuovo Giorn. Bot. Ital., 44: 667-672.

Netroufal, F., 1927. Österr. Bot. Zeitschr., 76: 101-115.

Neves, J. de Barros, 1944. Diss. Univ. Coimbra: 1-200.

Neves, J. de Barros, 1950. Agron. Lusit., 12: 601-610.

Nevling, L.I., 1962. Rhodora, 64: 277-282.

Nielson, N., 1924. Hereditas, 5: 378-382.

Nilsson, Ö. and Lassen, P., 1971. Bot. Not., 124 (2): 270-276.

Nordborg, G., 1958. Bot. Not., 111: 240-248.

Nordborg, G., 1966. Opera Bot., 11 (2): 1-103.

Nordborg, G., 1967. Opera Bot., 16: 1-166.

Nordenskjöld, H., 1937. Hereditas, 23: 304-316.

Nordenskjöld, H., 1951. Hereditas, 37: 325-355.

Noronha-Wagner, M. de, 1949. Genet. Iber., 1: 59-67.

Novotna, I., 1962. Preslia, 34: 249-254.

Nygren, A., 1946. Hereditas, 32: 131-262.

Nygren, A., 1948. In: Löve, Á. and Löve, D., Rep. Dep. Agric. Univ. Inst.
    Appl. Sci. (Iceland), Ser. B., 3.

Nygren, A., 1949. Hereditas, 35: 215-220.

Nygren, A., 1951. Hereditas, 37: 372-381.

Nygren, A., 1955. Ann. Roy. Agric. Cell. Sweden, 21: 179-191.

Nygren, A., 1956. Kungl. Lantbrukshogsk. Annal., 22: 179-191.

Nygren, A., 1957. Kungl. Lantbrukshogsk. Ann., 23: 488-495.

Nygren, A., 1962. Svensk. Vetenskap. Arsbok, 6: 1-29.

Ochlewska, M., 1966. Acta Biol. Cracov., Ser. Bot., 8: 135-145.

Ockendon, D.J., 1971. New Phytol., 70: 599-605.

Oehkers, F., 1926. Jahrb. Wiss. Bot., 65: 401-446.

Olsson, U., 1967. Bot. Not., 120: 255-267.

Ouweneel, W.J., 1968. Med. Bot. Mus. Herb. Riksuniv. Utrecht, 279: 184-188.

Packer, J.G., 1961. In: Löve, Á. and Löve, D., Opera Bot. 5.

Padmore, P.A., 1957. Watsonia, 4: 19-27.

Palmgren, O., 1939. Bot. Not., 139: 246-248.

Palmgren, O., 1942. IN: Löve, Á. and Löve, D., Kungl. Fysiogr. Sällsk. Lund.
    Forhändl., 12.

Palmgren, O., 1943. Bot. Not., 1963: 348-352.

Papeš, D., 1972. Acta Bot. Croat., 31: 71-79.

Papeš, D., 1975. Bot. Soc. Brit. Is. Conf. Rep., 15: 90-100.

Parreaux, M.-J., 1971. Ann. Litt. Univ. Besançon, 1971, 113-126.

Parreaux, M.-J., 1972. Annal. Sci. Univ. Besançon, Bot., 3 Sér., 13: 69-155.

Pauwels, L., 1959. Bull. Soc. Roy. Bot. Belg. 91: 291-297.

Pavone, P., Terrasi, C.M. and Zizza, A., 1981. In: Löve, Á., Taxon, 30: 695-696.

Pereira, A. de L., 1942. Bol. Soc. Brot., Sér. 2, 16: 5-40.

Perrenoud, R. and Favarger, C., 1971. Bull. Soc. Neuchât. Sci. Nat., 94: 21-27.

Persson, K., 1974. Opera Bot., 35: 1-188.

Peterson, D., 1933. Bot. Not., 1933: 500-504.

Peterson, D., 1935. Bot. Not., 1935: 409-410.

Peto, F.H., 1933. Jour. Genet., 28: 113-156.

Pettet, A., 1964. Watsonia 6: 39 50

Phillips, H.M., 1938. Chronica Bot., 4: 385-386.

Phitos, D., 1963. Mitt. Bot. Staatssamm. München, 5: 121-124.

Phitos, D., 1964. Ber. Deutsch. Bot. Ges., 77: 49-54.

Phitos, D., 1965. Österr. Bot. Zeits., 112: 449-498.

Phitos, D., 1966. Ber. Deutsch. Bot. Ges., 79: 246-249.

Phitos, D., 1970. Ber. Beutsch. Bot. Ges., 83: 69-73.

Phitos, D. and Damboldt, J., 1969. Ber. Deutsch. Bot. Ges., 82: 600.

Phitos, D. and Damboldt, J., 1971. Ann. Naturhist. Mus Wien, 75: 157-162.

Phitos, D. and Kamari, G., 1974. Bot. Not., 127: 302-308.

Pigott, C.D., 1954. New Phytol., 53: 470-495.

Piotrowicz, M., 1958. Acta Biol. Cracov., Ser. Bot., 1: 159-169.

Podlech, D., 1962. Ber. Deutsch. Bot. Ges., 75: 237-244.

Podlech, D., 1964. Madroño, 17: 266.

Podlech, D., 1965. Feddes Rep., 71: 50-187.

Podlech, D., 1970. Mitt. Bot. Staatssamm. München, 8: 211-217.

396

Podlech, D., 1975. XII Intern. Bot. Congr., Abstr.: 10.

Podlech, D. and Damboldt, J., 1964. Ber. Deutsch. Bot. Ges., 76 (9): 360-369.

Pogan, E., 1968. In: Löve, Á., Taxon, 17: 421.

Pogan, E., 1972. Acta Biol. Cracov, Bot., 15: 69-76.

Pogliani, M., 1964. Nuovo Giorn. Bot. Ital., 71: 558.

Polatschek, A., 1966. Österr. Bot. Zeits., 113: 1-46, 101-147.

Pólya, L., 1948. Arch. Biol. Hung. II, 18: 145-148.

Pólya, L., 1949. Acta Geobot. Hung., 6: 124-137.

Pólya, L., 1950. Ann. Biol. Univ. Debrecen, 1: 46-56.

Popova, M.T., 1972. In: Löve, Á., Taxon, 21: 164-165.

Popova, M. and Česchmedjiev, I.V., 1966. Naucni Trud. (Viss. Selsk. Inst. Vassil. Kolarov Plovdiv.), 15: 43-48.

Popova, M.T. and Česchmedjiev, I.V., 1975. In: Löve, Á., Taxon, 24: 143.

Popova, M.T. and Česchmedjiev, I.V., 1978. In: Löve, Á., Taxon, 27: 384-385.

Poucques, M.L. de, 1948. Bull. Soc. Sci. Nancy, N.S., 7: 33-39.

Priadcencu, A., Miclea, C. and Moisescu, L., 1967. Rev. Roum. Biol., Bot., 12: 421-425.

Prime, C.T., 1954. Jour. Ecol., 42: 241-248.

Pritchard, T.O., 1957. Jour. Ecol., 45: 966.

Proctor, M.C.F., 1955. Watsonia, 3: 154-159.

Przywara, L., 1970. Acta Biol. Cracov, 13: 133-142.

Puff, C., 1976. Canad. Jour. Bot., 54: 1911-1925.

Queirós, M., 1973. Bol. Soc. Brot., Sér. 2, 47: 77-103, 299-314.

Quézel, P., 1967. Taxon, 16: 240.

Quézel, P. and Contandriopoulos, J., 1965. Candollea, 20: 51-90.

Quézel, P. and Contandriopoulos, J., 1966. Bull. Soc. Bot. Fr., 113 (5-6): 351-353.

Rahn, K., 1957. Bot. Tidsskr., 53: 369-378.

Rahn, K., 1966. In: Löve, Á., Taxon, 15: 122-127.

Raitanen, P.R., 1967. Ann. Bot. Fenn. 4: 471-485.

Rao, Y.S., 1951. Current Sci., 20: 72.

Ratter, J.R., 1964. Notes Roy. Bot. Gard. Edinb., 25: 293-302.

Ratter, J.A., 1967. Notes Roy. Bot. Gard. Edinb., 27: 300-333.

Ratter, J.A., 1976. Notes Roy. Bot. Gard. Edinb., 34: 411-428.

Raven, P.H., 1963. Quart. Rev. Biol., 38: 109-177.

Raven, P.H. and Moore, D.M., 1964. Watsonia, 6: 36-38.

Raynaud, C., 1971. In: Löve, Á., Taxon, 20: 795.

Reese, G., 1952. Ber. Deutsch. Bot. Ges., 64: 241-256.

Reese, G., 1953. Ber. Deutsch. Bot. Ges., 66: 66-73.

Reese, G., 1954. VIII Congr. Int. Bot. Rapp. & Comm., 9 and 10: 83-85.

Reese, G., 1961. In: Löve, Á. and Love, D., Opera Bot., 5.

Reese, G., 1963. Sch. Naturw. Ver. Schlesw. Holst., 34: 44-70.

Rejón, M.R., 1974. In: Löve, Á., Taxon, 23: 805-806.

Rejón, M.R., 1976. In: Löve, Á., Taxon, 25: 341-342.

Renner, O., 1951. Ber. Deutsch. Bot. Ges., 63 (5): 129-138.

Renzoni, G.C., 1963. Nuovo Giorn. Bot. Ital., 70: 493-504.

Renzoni, G.C. and Garbari, F., 1970. Nuovo Giorn. Bot. Ital., 104: 61-73.

Renzoni, G.C. and Garbari, F., 1971. Atti Soc. Tosc. Sci. Nat. Mem., Ser. B,
      78: 99-118.

Ricci, I., 1961. Ann. Bot. (Roma), 27: 36-39.

Ricci, I. and Colasante, M.A., 1975. Ann. Bot. (Roma), 32: 218.

Richards, A.J., 1969. In: Löve, Á., Taxon, 18: 560-562.

Richards, A.J., 1972. In: Löve, Á., Taxon, 21: 165-166.

Ritter, J., 1973. Compt. Rend. Acad. (Paris), Sec. Biol., 167: 240.

Roberts, R.H., 1966. Watsonia, 6: 260-267.

Rodrigues, J.E. de, 1953. Diss., Univ. Coimbra.

Rodrigues, J.E. de, 1954. Bol. Soc. Brot. Sér. 2, 28: 117-129.

Rohner, P., 1954. Mitt. Naturf. Gess. Bern, 11: 43-107.

Rohweder, H., 1934. Bot. Jahrb., 66: 249-368.

Rohweder, H., 1939. Beih. Bot. Centralbl., 59: 1-58.

Rönning, O.I., 1962. Kungl. Norske Vid. Selsk. Skr., 1961 (4): 1-50.

Rosenberg, O., 1926. Arkiv. Bot., 20B (3): 1-5.

Rosenthal, C., 1936. Jahrb. Wiss. Bot., 83: 809-844.

Rosser, E.M., 1955. Watsonia, 3: 228-232.

Rottgardt, K., 1956. Beitr. Biol. Pflanzen, 32: 225-278.

Rousi, A., 1956. Ann. Bot. Soc. Vanamo, 29 (2): 1-64.

Rousi, A., 1961. Hereditas, 47: 81-110.

Rousi, A., 1965a. Ann. Bot. Fenn., 2: 1-18.

Rousi, A., 1965b. Ann. Bot. Fenn., 2: 47-112.

Rousi, A., 1967. Zuchter, 36 (8): 352-359.

Rousi, A., 1969. Hereditas, 62: 193-213.

Ruggeri, C., 1961. Caryologia, 14: 111-120.

Rune, O. and Ronning, O.I., 1956. Svensk. Bot. Tidskr., 50: 115-128.

Runemark, H., 1962. Bot. Not., 115: 8.

Runemark, H., 1967a. Bot. Not., 120: 84-94.

Runemark, H., 1967b. Bot. Not., 120: 9-16.

Runemark, H., 1967c. Bot. Not., 120: 161-176.

Runemark, H., 1970. Willdenowia, 6: 137-138.

Runemark, H. and Heneen, W.K., 1968. Bot. Not., 121: 51-79.

Rutishauser, A., 1936. Mitt. Naturf. Ges. Schaffhausen, 13: 25-47.

Rutland, J.P., 1941. New Phytol., 40: 210-214.

Ryberg, M., 1960. Acta Horti Berg., 19: 121-248.

Rychlewski, J., 1968. In: Skalińska, M., et al., Acta Biol. Cracov., 11:
    199-224.

Santos, A.C. dos, 1945. Bol. Soc. Brot., 19: 519-522.

Sañudo, A. and Rejón, M.R., 1975. Anal. Inst. Bot. Cav., 32 (2): 633-648.

Sauer, W., 1971. Carinthia II, 80: 79-87.

Sauer, W. and Chmelitschek, H., 1976. Mitt. Bot. Staatssamm. München, 12: 513-608.

Saunte, L.H., 1955. Hereditas, 41: 499-515.

Saura, F., 1943. Rev. Fac. Agron. Vet. (Bs. Aires), 10: 344-353.

Scannerini, S., 1971. Inf. Bot. Ital., 3: 66-67.

Schafer, B. and La Cour, L., 1934. Ann. Bot., 48: 693-713.

Scheerer, H., 1939. Planta, 29: 636-642.

Scheerer, H., 1940. Planta, 30: 716-725.

Schmidt, A., 1961. Österr. Bot. Zeits., 108: 20-88.

Schmidt, A., 1962. Ber. Deutsch. Bot. Ges., 75: 78-84.

Schmidt, A., 1963. Österr. Bot. Zeits., 110: 285-293.

Schmidt, A., 1964. Flora, 154: 158-162.

Schmidt, A., 1964b. Ber. Deutsch. Bot. Ges., 77: 256-261.

Schmidt, A., 1965. Ber. Deutsch. Bot. Ges., 77 (1): 94-99.

Schneider, U., 1910. Flora, 106: 1-41.

Schneider, U., 1964. Feddes Rep., 69 (3): 180-195.

Schäfer, G., 1954. Planta, 43: 537-565.

Scholte, G., 1977. In: Löve, Á., Taxon, 26: 258-259.

Schönfelder, P., 1968. Oesterr. Bot. Zeits., 115: 363-371.

Schotsman, H.D., 1961a. Bol. Soc. Brot., Sér. 2, 35: 95-127.

Schotsman, H.D., 1961b. Bull. Soc. Bot. Suisse, 71: 5-17.

Schotsman, H.D., 1961c. Bull. Soc. Neuchât, Sci. Nat., 84: 89-101.

Schotsman, H.D., 1967. In: Jovet, P., Flora de France, 1, Paris.

Schotsman, H.D., 1969. Bull. Centre Etude Rech. Sci. Biarritz, 7: 869-872.

Schulle, H., 1933. Flora, 127: 140-184.

Schulz-Schaeffer, J. and Jura, P., 1967. Zeits. Pflanzenzucht, 57: 146-166.

Schürhoff, P.N., 1926. Die Zytologie der Pflanzen. Stuttgart, Verl. F. Enke, 1-792.

Schwarz, O., 1964. Drudea, 1964: 5-16.

Schwarz, O. and Bässlar, M., 1964. Bot. Jahrb., 111: 193-207.

Schwarzenbach, F., 1922. Flora, 115: 393-514.

Scrugli, A., 1974. Inf. Bot. Ital., 6: 37-43.

Scrugli, A., De Martis, B. and Mulas, B., 1976. Inf. Bot. Ital., 8: 82-91.

Sealy, J.R. and Webb, D.A., 1950. Jour. Ecol., 38: 223-236.

Seitz, F.W., 1951. Zeits. Forstgenet. Forstpflanzenzücht., 1: 22-32.

Seitz, F.W., 1954. VIII Cong. Internat. Bot., Rapp. Comm., 13: 28-29.

Seitz, W., 1969. Ber. Deutsch. Bot. Ges., 82: 651-655.

Shimoya, C., 1952. Rec. Trav. Lab. Bot. Geol. Zool Fac. Sci. Univ.

   Montpellier, Ser. Bot., 5: 87-89.

Shimoya, C. and Ferlan, L., 1952. Broteria, 21 (4): 171-176.

Sibilio, E., 1961. Delpinoa, N. Ser., 3: 225-238.

Sikka, S.M., 1940. Jour. Genet., 40: 441-510.

Simola, L.K., 1964. Ann. Acad. Sci. Fenn. Ser. A., IV Bot., 78: 1-19.

Simonet, M., 1934. Ann. Sci. Nat., Bot., Ser. 10, 16: 229-338.

Simonet, M., 1955. Ann. Sci. Nat. Bot., Ser. 11, 16: 503-528.

Sinskaya, E.H. and Maleyeva, Z.P., 1959. Bot. Zhurn., 44: 1103-1113.

Sintes, S. and Cauderon, Y., 1965. Compt. Rend. Acad. Sci. Paris, 260:

   4249-4251.

Skalińska, M., 1950. Acta Soc. Bot. Polon., 20 (1): 45-68.

Skalińska, M., 1950b. Bull. Int. Acad. Polon. Sci. Lett. B, Sci. Nat., 1:

   149-175.

Skalińska, M., 1952. Bull. Int. Acad. Polon. Sci. Lett. 3BI: 119-136.

Skalińska, M., 1958. Acta Biol. Cracov., Ser. Bot., 1: 45-54.

Skalińska, M., 1959. Rep. IX Int. Bot. Congr., 2: 365.

Skalińska, M., 1967. Acta Biol. Cracov., Bot., 10: 127-141.

Skalińska, M., 1970. Acta Biol. Cracov., Bot., 13: 111-117.

Skalińska, M. and Kubién, E., 1972. Acta Biol. Cracov., Bot., 15: 39-50.

Skalińska, M., Malecka, J. and Izmailow, R., 1974. Acta Biol. Cracov., 17:

   133-163.

Skalińska, M., Piotrowicz, M., Czapik, R. et al., 1959. Acta Soc. Bot.

Polon., 28: 487-529.

Skalińska, M., Piotrowicz, M., Sokolowska-Kulczycka, A. et al., 1961. Acta
     Soc. Bot. Polon., 30: 463-489.

Skalińska, M. and Pogan, A., 1973. Acta Biol. Cracov., Bot., 16: 145-201.

Skalińska, M. et al., 1964. Acta Soc. Bot. Polon., 33: 45-76.

Skalińska, M., Pogan, E. and Pogan, A.L., 1966. Acta Biol. Cracov., Bot., 9:
     31-58.

Skalińska, M., Pogan, E., Jankun, A. et al., 1968. Acta Biol. Cracov., Bot.,
     11 (2): 199-224.

Skalińska, M., Jankun, A., Wcislo, H. et al., 1971. Acta Biol. Cracov., Bot.,
     14: 55-102.

Skovsted, A., 1934. Dansk Bot. Ark., 9 (5): 1-52.

Smith, G.L., 1963. New Phytol., 62: 283-300.

Smith, P.M., 1967. Feddes Rep., 77: 61;64.

Snogerup, S., 1962. Bot. Not., 115: 357-375.

Snogerup, S , 1963. Bot. Not., 116: 142-156.

Snogerup, S., 1971. In: Nilsson, Ö. and Snogerup, S., Bot. Not., 124: 1-8,
     179-186, 311-316, 435-441.

Snogerup, S., 1972. In: Nilsson, Ö. and Snogerup, S., Bot. Not., 125: 1-8,
     131-138, 203-211.

Sokolovskaya, A.P., 1955. Bot. Zhurn., 40 (6): 850-853.

Sokolovskaya, A.P., 1958. Bot. Zhurn., 43: 1146-1155.

Sokolovskaya, A.P. and Strelkova, O.S., 1941. Doklady Akad. SSSR, N.S., 32:
     145-147.

Sokolovskaya, A.P. and Strelkova, O.S., 1960. Bot. Zhurn., 45: 369-381.

Solbrig, O.T., Khyos, D.M., Powell, M. and Raven, P.H., 1972. Amer. Jour.
     Bot., 59: 869-878.

Söllner, R., 1954. Ber. Schweiz. Bot. Ges., 64: 221-354.

Šopova, M., 1966. God. Zborn. Biol. (Skopje), 16 (4): 107-144.

Šopova, M., 1972. God. Zborn. Biol. (Skopje), 24: 73-81.

Sørensen, T. and Christiansen, H., 1964. Bot. Tidsskr., 59: 311-314.

Sörensen, T. and Gudjonsson, G., 1946. K. Dansk. Vid. Selsk. Biol. Skr., 4 (2): 1-48.

Sorsa, M., 1963. Ann. Acad. Sci. Fennicae, Ser. A, IV, Biol., 68: 1-14.

Sorsa, M., 1966. Hereditas, 56: 395-396.

Sorsa, V., 1958. Hereditas, 44: 541-546.

Sorsa, V., 1962. Ann. Acad. Sci. Fenn. Ser. A, IV, Biol., 58: 1-14.

Speta, F., 1976. Linz. Biol. Beitr., 8: 239-322.

Staudt, G., 1962. Canad. Jour. Bot., 40: 869-886.

Stearn, W.T., 1978. Ann. Mus. Goulandris, 4: 83-198.

Stearn, W.T. and Özhatay, N., 1977. Ann. Mus. Goulandris, 3: 45-50.

Stebbins, G.L., Jenkins, J.A. and Walters, M.S., 1953. Univ. Calif. Publs. Bot. 26: 401-430.

Steiner, E., 1955. Bull. Torrey Bot. Club, 82: 292-297.

Stern, Г.C., 1946. A study of the genus Paeonia. London.

Stern, F.C., 1956. Snowdrops and Snowflakes. London: Ballantyne & Co.

Stoeva, M.P., 1977. In: Löve, Á., Taxon, 26: 560.

Strandhede, S.-O., 1961. Bot. Not., 114: 417-434.

Strandhede, S.-O., 1963. Bot. Not., 116: 215-221.

Strandhede, S.-O., 1965. Opera Bot., 9 (2): 1-86.

Strandhede, S.-O. and Dahlgren, R., 1968. Bot. Not., 121: 305-311.

Strasburger, E., 1902. Jahrb. Wiss. Bot., 37: 477-526.

Strasburger, E., 1904. Jahrb. Wiss. Bot., 41: 88-164.

Strid, A., 1965. Bot. Not., 118: 104-122.

Strid, A., 1971. Bot. Not., 124: 490-496.

Stuart, D.C., 1970. Notes Roy. Bot. Gard. Edinb., 30: 189-196.

Styles, B.T., 1962. Watsonia, 5: 177-214.

Sugiura, T., 1938. Proc. Imp. Acad. Tokyo, 14: 391-392.

Sugiura, T., 1944. Cytologia, 13: 352-359.

Suomalainen, E., 1947. Ann. Acad. Sci. Fenn., Ser. A, 13: 1-65.

Sušnik, F., 1962. Biol. Vest., 10: 7-9.

Sušnik, F., 1967. Biol. Vest., 15: 63-66.

Sušnik, F., Druskovic, B., Löve, Á. and Löve, D., 1972. In: Löve, Á.,
    Taxon, 21: 333-346.

Sušnik, F. and Lovka, M., 1973. In: Löve, Á., Taxon, 22: 462-463.

Täckholm, G., 1922. Acta Horti Berg., 7: 97-381.

Talavera, S., 1976. Lagascalia, 6: 275-296.

Talavera, S., 1978. Lagascalia, 7: 133-142.

Tamayo, A.I. and Grande, M.D., 1967. Bol. Real Soc. Espanola Hist. Nat.
    (Biol.), 65: 447-453.

Tarnavschi, I.T., 1935. Bull.Fac. Ştiint. Çernauti, 9: 47-122.

Tarnavschi, I.T., 1938. Bull. Fac. Ştiint. Çernauti, 12: 68-106.

Tarnavschi, I.T., 1948. Bull. Jard. Mus. Bot. Univ. Chij. 28 Suppl.:
    1-130.

Tarnavschi, I.T. and Lungeanu, I., 1970. In: Löve, Á., Taxon, 19: 609-610.

Teppner H., 1971. Österr. Bot. Zeit., 119: 196-233.

Teppner, H., Ehrendorfer, F. and Puff, C., 1976. Taxon, 25: 95-97.

Thomas, P.T., 1940. Jour. Genet., 40: 141-155.

Tischler, G., 1934. Bot. Jahrb., 67: 1-36.

Titova, N.N., 1935. Soviet Bot., 2: 61-67.

Titz, W., 1964. Österr. Bot. Zeits., 111: 618-620.

Titz, W., 1966. Österr. Bot. Zeits., 113: 187-190.

Tombal, P., 1969. Inf. Ann. Caryosyst. Cytogénét., 3: 29-32.

Tournay, J.A., 1968. Bull. Jard. Bot. Bruxelles, 38: 295-381.

Tracey, R., 1977. Plant Syst. Evol., 128: 287-292.

Trela-Sawicka, Z., 1968. Acta Biol. Cracov. Bot., 11: 59-69.

Tschermak-Woess, E., 1947. Chromosoma, 3: 66-87.

Tschermak-Woess, E., 1967. Caryologia, 20: 135-152.

Turesson, B., 1972. Bot. Not., 125: 223-240.

Turesson, G., 1957. Bot. Not., 110: 413-422.

Turesson, G., 1958. Bot. Not., 111: 159-164.

Turesson, G. and Turesson, B., 1960. Hereditas, 46: 717-737.

Tutin, T.G., 1957. Watsonia, 4: 1-10.

Tutin, T.G., 1961. In: Löve, Á. and Löve, D., Opera Bot., 5.

Tzvelov, N.N. and Zhukova, P.G., 1974. Bot. Zhurn., 59: 265-269.

Uhl, C.H., 1948. Amer. Jour. Bot., 35: 695-706.

Uhl, C.H., 1961. In: Löve, Á. and Löve, D., Opera Bot., 5.

Uhríková, A., 1970. Acta Fac. Rer. Nac. Univ. Com. Bot., 18: 61-67.

Uhríková, A. and Feráková, V., 1978. In: Löve, Á., Taxon, 27: 379.

Uhríková, A. and Májovský, J., 1977. In: Löve, Á., Taxon, 26: 263.

Uhríková, A. and Schwarzová, T., 1978. In: Löve, Á., Taxon, 27: 379-380.

Ujhelyi, J., 1959. Feddes Rep., 62: 59-70.

Ujhelyi, J., 1960. Ann. Hist. Nat. Mus. Nat. Hung., Bot., 52: 185-200.

Ujhelyi, J., 1960b. Bot. Közlemenyek (Budapest), 48: 278-282.

Ujhelyi, J., 1961. Ann. Hist. Nat. Mus. Nat. Hung., Bot., 53: 208-224.

Ujhelyi, J., 1962. Ann. Hist. Nat. Mus. Nat. Hung., Bot., 54: 199-220.

Ujhelyi, J., 1970. Ann. Hist. Nat. Mus. Nat. Hung., Bot., 62: 175-195.

Ujhelyi, J., 1972. In: Vida, G. (Ed.) Symp. Biol. Hung., 12: 163-176.

Upadhya, M.D. and Swaminathan, M.S., 1963. Chromosoma, 14: 589-600.

Urbańska-Woytkiewicz, K., 1967. Acta Boreal., A, Scientia, 22: 1-14.

Urbańska-Worytkiewicz, K., 1968. Bull. Soc. Bot. Suisse, 78: 192-201.

Urbańska-Worytkiewicz, K., 1968b. In: Löve, Á., Taxon, 17: 419-422.

Urbańska-Worytkiewicz, K., 1970. Ber. Geobot. Inst. ETH, St. Rubel, 40: 79-166.

Urbańska-Worytkiewicz, K., 1975. Aquatic Bot., 1: 377-394.

Vaarama, A., 1941. Ann. Soc. Zool. Bot. Fenn. Vanamo, 16 (2): 1-28.

Vaarama, A., 1947. Arch. Soc. Vanamo, 2: 55-59.

Vaarama, A., 1948. In: Löve, Á. and Löve, D., Rep. Univ. Inst. Appl. Sci.,

Dep. Agric., Iceland, Ser. B., 3.

Vaarama, A., 1949. Hereditas, 35: 251-252.

Vaarama, A. and Hiirsalmi, H., 1967. Hereditas, 58: 333-358.

Vaarama, A. and Leikas, R., 1970. In: Löve, Á., Taxon, 19 (2): 264-269.

Váchová, M. and Feráková, V., 1977. In: Löve, Á., Taxon, 26: 264.

Váchová, M. and Feráková, V., 1978. In: Love, Á., Taxon, 27: 382-383.

Vignoli, L., 1945. Nuovo Giorn. Bot. Ital., 52: 1-10.

Valdés, B., 1970. Bol. Roy. Soc. Esp. Hist. Not. (Biol.) 67: 243-256.

Valdés, B., 1970b. Bol. Roy. Soc. Esp. Hist. Not. (Biol.) 68: 193-197.

Valdés, B., 1973. Lagascalia, 3: 99-141.

Valentine, D.H., 1978. In: Valentine, D.H. and Lamond, J., Not. Roy. Bot.
    Gard. Edinb., 36: 39-42.

Van Loon, J.C. and Oudemans, J.J.M.H., 1976. Acta Bot. Neerl., 25 (5): 329-336.

Van Loon, J.C., Gadella, T.W.J. and Kliphuis, E., 1971. Acta Bot. Neerl.,
    20: 157-166.

Van Witsch, H , 1932. Österr. Bot. Zeits., 81: 108-141.

Vaucher, C., 1966. Bull. Soc. Neuchât. Nat., 89: 75-85.

Verma, S.C., 1958. Acta Bot. Neerl. 7: 629-634.

Vermeulen, P., 1938. Chron. Bot., 4: 107-108.

Vermeulen, P., 1947. Studies on Dactylorchids. Utrecht.F. Schotarius and
    Jens, pp. 1-180.

Viegli, L., Cela Renzoni, G., Corsi, G. and Garbari, F., 1972. Inf. Bot.
    Ital., 4 (3): 229-236.

Vignoli, L., 1933. Lav. R. Istit. Bot. Palermo, 4: 5-19.

Villard, M., 1970. Bull. Soc. Bot. Suisse, 80: 96-188.

Vladesco, A., 1940. Bull. Sect. Sci. Acad. Romaine, 23: 258-268.

Walter, R., 1973. Acta Biol. Cracov, 16: 227-233.

Walters, S.M., 1949. Jour. Ecol., 37: 192-206.

Walters, S.M., 1967. Acta Fac. Rer. Nat. Univ. Commen. Bot., 14: 7-11.

Walters, S.M. and Bozman, V.G., 1967. Proc. Bot. Soc. Brit. Is., 7: 83.

Walther, E., 1949. Mitt. Thuring. Bot. Ges. Beih., 1: 1-108.

Warburg, E.F., 1938. New Phytol., 37: 189-209.

Warburg, E.F., 1939. In: Maude, P.F., New Phytol., 38: 1-31.

Warburg, E.F., 1952. In: Clapham, A.R., Tutin, T.G. and Warburg, E.F.,

    Flora of the British Isles. Cambridge University Press.

Wcislo, H., 1970. Acta Biol. Cracov., 13: 79-88.

Weimarck, G., 1971. Bot. Not., 124: 129-175.

Westergaard, M., 1938. Dansk Bot. Ark., 9 (5): 1-11.

Wettsteim-Westerheim, W. von, 1933. Züchter, 5: 280-281.

Whyte, R.O., 1930. Jour. Genet., 23: 93-121.

Wieffering, G.H., 1969. In: Löve, Á., Taxon, 18: 442.

Wilee, J., 1961. In: Löve, Á. and Löve, D., Opera Bot., 5.

Wilkinson, J., 1941. Ann. Bot. N.S., 5: 150-165.

Wilkinson, J., 1944. Ann. Bot., N.S., 8: 269-284.

Wilson, J.Y., 1956. Nature (London), 178: 195-196.

Wilson, J.Y., 1959. Cytologia, 23: 435-446.

Winge, Ö., 1917. Compt. Rend. Trav. Lab. Carlsberg, 11: 1-46.

Winge, O., 1940. Compt. Rend. Trav. Lab. Carlsberg, Ser. Physiol., 23: 41-74.

Witsch, H., 1932. Öesterr. Bot. Zeits., 81: 108-141.

Wolkinger, F., 1966. Ber. Deutsch. Bot. Ges., 79: 343-352.

Wolkinger, F., 1967. Phyton (Horn), 12: 91-95.

Wright, F.R.E., 1936. Jour. Bot., 74 (Suppl. 1): 1-8.

Wulff, H.D., 1936. Planta, 26: 275-290.

Wulff, H.D., 1937a. Jahrb. Wiss. Bot., 84: 812-840.

Wulff, H.D., 1937b. Ber. Deutsch. Bot. Ges., 55: 262-269.

Wulff, H.D., 1938. Ber. Deutsch. Bot. Ges., 56: 247-254.

Wulff, H.D., 1939a. Ber. Deutsch. Bot. Ges., 57: 84-91.

Wulff, H.D., 1939b. Ber. Deutsch. Bot. Ges., 57: 424-431.

Yeo, P.F., 1954. Watsonia, 3 (2): 101-108.

Yeo, P.F., 1956. Watsonia, 3 (5): 253-269.

Yeo, P.F., 1966. Watsonia, 6: 216-245.

Yeo, P.F., 1970. Candollea, 25 (1): 21-24.

Young, D.P., 1953. Bot. Not., 1953: 233-270.

Zakharyeva, O.I. and Makushenko, L.M., 1969. Bot. Zhurn., 54: 1213-1227.

Zeltner, L., 1961. Bull. Soc. Bot. Suisse, 71: 18-24.

Zeltner, L., 1962. Bull. Soc. Neuchât. Sci. Nat., 85: 83-95.

Zeltner, L., 1963. Bull. Soc. Neuchât. Sci. Nat., 86: 93-100.

Zeltner, L., 1966. Bull. Neuchât. Sci. Nat., 89: 61-73.

Zeltner, L., 1967. Bull. Soc. Neuchât. Sci. Nat., 90: 241-246.

Zeltner, L., 1970. Bull. Soc. Neuchât. Sci. Nat., 93: 1-164.

Žertová, A., 1962. Nov. Bot. Horti. Bot. Univ. Carol. Prag., 1962: 51.

Žertová, A., 1963. Biologia (Bratislava), 18: 697-700,

Žešiger, E., 1901. Bot. Schweiz. Bot. Ges., 71: 113-117.

Zhukova, P.G., 1965. Bot. Zhurn., 50: 1320-1322.

Zhukova, P.G. and Tikhonova, A.P., 1971. Bot. Zhurn., 56: 868-875.

Zickler, D., 1967. Inf. Ann. Caryosyst. Cytogénét., 1: 7.

Zickler, D., 1968. Inf. Ann. Caryosyst. Cytogénét., 2: 5-8.

Zickler, D. and Lambert, A.-M., 1967. Inf. Ann. Caryosyst. Cytogén., 1: 3-6.

Zucconi, L., 1956. Caryologia, 8: 379-388.

410

413

415

417